AF361831

Image and Signal Processing

Image and Signal Processing

Guest Editors

Gwanggil Jeon
Imran Ahmed

Basel • Beijing • Wuhan • Barcelona • Belgrade • Novi Sad • Cluj • Manchester

Guest Editors

Gwanggil Jeon
Department of Embedded
Systems Engineering
Incheon National University
Incheon
Korea, South

Imran Ahmed
School of Computing and
Information Sciences
Anglia Ruskin University
Cambridge
United Kingdom

Editorial Office
MDPI AG
Grosspeteranlage 5
4052 Basel, Switzerland

This is a reprint of the Special Issue, published open access by the journal *Technologies* (ISSN 2227-7080), freely accessible at: www.mdpi.com/journal/technologies/special_issues/Image_Signal.

For citation purposes, cite each article independently as indicated on the article page online and using the guide below:

Lastname, A.A.; Lastname, B.B. Article Title. *Journal Name* **Year**, *Volume Number*, Page Range.

ISBN 978-3-7258-2776-3 (Hbk)
ISBN 978-3-7258-2775-6 (PDF)
https://doi.org/10.3390/books978-3-7258-2775-6

Contents

Preface

In recent years, the rapid integration of artificial intelligence into image and signal processing has led to transformative advancements across a wide range of applications. Since the pivotal moment of deep learning's breakthrough in the 2012 ImageNet Challenge, this technology has become a driving force in modern life, reshaping industries and enhancing daily experiences. The tremendous progress in deep learning has been fueled by sophisticated artificial intelligence techniques, including neural networks, evolutionary algorithms, and systems rooted in fuzzy and rough logic. However, to achieve reliable, real-world applications, these deep learning models often require extensive data for effective training, posing a challenge that researchers are addressing with innovations in lightweight and explainable models.

This reprint presents a curated collection of 14 papers originally published in this Special Issue, "Deep Learning Applications in Image and Signal Processing", highlighting novel and cutting-edge research across diverse facets of deep learning. The topics range from foundational advances in deep learning algorithms to practical applications in data mining, computer vision, natural language processing, clustering, image filtering and enhancement, and video segmentation. Additional chapters explore specialized applications such as motion detection, pattern recognition, and content-based image retrieval, reflecting the broad impact of deep learning in contemporary technology.

Furthermore, this reprint delves into the application of deep learning in fields that include robotics, industrial automation, autonomous systems, and gaming—sectors that stand to benefit immensely from these advancements. Each chapter showcases innovative methods and insights, providing valuable perspectives on the future directions of deep learning in image and signal processing.

We extend our gratitude to the authors and reviewers whose contributions made this Special Issue, and now this book, possible. Their dedication and expertise have produced a comprehensive resource that we hope will inspire continued innovation and exploration in the field. As deep learning evolves, we anticipate that these insights will support the ongoing efforts to develop intelligent, efficient, and robust technologies for the future.

Gwanggil Jeon and Imran Ahmed
Guest Editors

Editorial

Guest Editorial on Image and Signal Processing

Gwanggil Jeon [1,*] and **Imran Ahmed** [2]

1   Department of Embedded Systems Engineering, Incheon National University, 119 Academy-ro, Yeonsu-gu, Incheon 22012, Republic of Korea
2   School of Computing and Information Science, Anglia Ruskin University, Cambridge CB1 1PT, UK; imran.ahmed@aru.ac.uk
*   Correspondence: gjeon@inu.ac.kr

Citation: Jeon, G.; Ahmed, I. Guest Editorial on Image and Signal Processing. *Technologies* **2024**, *12*, 176. https://doi.org/10.3390/technologies12100176

Received: 19 September 2024
Accepted: 24 September 2024
Published: 26 September 2024

1. Introduction

In recent years, we have witnessed significant advancements in technologies that integrate artificial intelligence into image and signal processing, along with their wide-ranging applications. Since its breakthrough in the 2012 ImageNet challenge, deep learning has profoundly influenced modern life more than any other technology. This progress is largely driven by deep learning techniques powered by artificial intelligence, including fuzzy and rough systems, neural networks, and evolutionary algorithms. Deep learning typically requires vast amounts of data to be properly trained for real-world applications. To address this challenge, researchers are increasingly focusing on developing lightweight and explainable deep learning models.

This Special Issue invited authors to submit novel and cutting-edge research on deep learning applications in image and signal processing. Topics of interest include, but were not limited to, new deep learning algorithms, deep learning for data mining, computer vision, forecasting, natural language processing, clustering, image filtering, restoration and enhancement, image and video segmentation, tracking, feature extraction and analysis, motion detection and estimation, pattern recognition, and content-based image retrieval. Additionally, we welcomed contributions on the application of deep learning in domains such as robotics, industrial automation, autonomous systems, and gaming. A total of 14 papers were published in this Special Issue.

2. Overview of Contributions

Tephra fallout during explosive eruptions poses significant risks to air traffic, infrastructure, and human health. The contribution by Dozzo et al., "Exploiting PlanetScope Imagery for Volcanic Deposits Mapping", introduces a new technique to map tephra-covered areas using PlanetScope satellite imagery [item 1 in Appendix A]. By analyzing pre- and post-eruption reflectance values in visible (RGB) and near-infrared (NIR) bands, the authors developed a "Tephra Fallout Index (TFI)" to identify affected regions. Using the Google Earth Engine, they established TFI thresholds to quantify tephra surface coverage for different eruptions. The method is applied to the 2021 Mt. Etna eruptions, which impacted the volcano's eastern flank multiple times in rapid succession, complicating field surveys. Comparisons with available field data show strong alignment with satellite-derived results. This technique offers valuable potential for real-time volcanic hazard assessment and broader applications in mapping other hazardous events, providing a fast and effective tool for disaster management and risk reduction.

The contribution by Hitimana et al., "An Intelligent System-Based Coffee Plant Leaf Disease Recognition Using Deep Learning Techniques on Rwandan Arabica Dataset", focuses on developing an efficient method for detecting and identifying coffee leaf diseases in Rwanda, a country where coffee is a critical agricultural commodity [item 2 in Appendix A]. Farmers currently rely on manual disease detection, which is prone to errors.

With advancements in deep learning, automated detection offers a promising solution to improve crop yields. A dataset of 37,939 coffee leaf images was collected, targeting diseases such as coffee rust, mines, and red spider mites. Five deep learning models—InceptionV3, ResNet50, Xception, VGG16, and DenseNet—were trained, validated, and tested with an 80/10/10 split over 10 epochs. The DenseNet model achieved the best performance with 99.57% accuracy. The proposed method proved efficient, outperforming existing approaches and offering potential for future portable applications in coffee leaf disease detection.

The contribution by Yao et al., "A Foreign Object Detection Method for Belt Conveyors Based on an Improved YOLOX Model", focuses on developing a foreign object detection method for belt conveyors used in coal transportation, a critical component in intelligent mining [item 3 in Appendix A]. In complex production environments, non-coal foreign objects frequently come into contact with belts, risking safety issues like scratches, deviation, and breakage. To address this, the authors establish a foreign object image dataset and enhance it using an IAT image enhancement module and CBAM attention mechanism. A rotating decoupling head and MO-YOLOX network structure are introduced to predict the angle of foreign objects with large aspect ratios. Experiments conducted in a mining lab show that the proposed method achieves 93.87% accuracy, 93.69% recall, and 93.68% mAP50, with an average inference time of 25 ms. These results highlight the system's efficiency in foreign object detection, contributing to safer coal transportation operations.

The contribution by Yoo et al., "A Novel Approach to Quantitative Characterization and Visualization of Color Fading", focuses on quantifying and analyzing color fading and darkening over time, triggered by light exposure [item 4 in Appendix A]. The researchers used the newly developed PicMan software to compare and map pixel-by-pixel color differences in digital images. Japanese wood-block prints, both with and without color fading, were selected for analysis. The study demonstrated that pixel-by-pixel, line-by-line, and area-by-area comparisons effectively quantified color changes, presenting results in RGB, HSV, and CIE L * a * b * values. The results were displayed in numerical, graphical, and image formats, each with distinct advantages for communication and analysis. Additionally, color change simulations for past and future moments were demonstrated using interpolation and extrapolation methods. This work offers valuable insights for practical applications in art conservation, museum displays, and cultural heritage preservation, assisting decision-making in storage, restoration, and public display planning.

The contribution by Wiseman, "Adapting the H.264 Standard to the Internet of Vehicles", proposes a two-step method to reduce data transmission on Internet of Vehicle networks [item 5 in Appendix A]. The first step reduces image color resolution from full color to just eight colors, which, while noticeable, is sufficient for typical vehicle applications. The second step modifies the quantization tables used by H.264 compression to better suit eight-color images. The first step alone reduces image size by over 30%, and combining both steps results in a size reduction of more than 40%. Together, these steps significantly decrease the amount of data transferred on vehicular networks, optimizing network efficiency without compromising the functionality of common vehicle applications.

The contribution by Svendsen and Kadry, "Comparative Analysis of Image Classification Models for Norwegian Sign Language Recognition", addresses the communication challenges faced by the Deaf population by exploring image classification models for sign language recognition [item 6 in Appendix A]. Focusing on Norwegian Sign Language (NSL), a relatively under-researched area, the authors created a new dataset with 24,300 images of 27 NSL alphabet signs. A comparative analysis of machine learning models, including a Support Vector Machine (SVM), K-Nearest Neighbor (KNN), and a Convolutional Neural Network (CNN), was conducted to identify the most effective approach for sign language recognition. The SVM and CNN outperformed other models, achieving 99.9% accuracy with high computational efficiency. This research contributes significantly to NSL recognition, offering a foundation for future studies and the development of assistive communication systems for the Deaf community.

The contribution by Duarte-Correa et al., "Identifying Growth Patterns in Arid-Zone Onion Crops (Allium Cepa) Using Digital Image Processing", focuses on improving onion crop performance by addressing challenges encountered during its phenological cycle [item 7 in Appendix A]. Utilizing unmanned aerial vehicles (UAVs) and digital image processing, the research monitored key factors such as humidity, weed growth, vegetation deficits, and reduced harvest performance. An algorithm was developed to identify patterns that most significantly affected crop growth. Despite an expected local yield of 40.166 tons/ha, only 25.00 tons/ha was achieved due to blight caused by constant humidity and limited sunlight. This resulted in poor leaf health, underdeveloped bulbs, and 50% of the crop being medium-sized. Additionally, approximately 20% of the total production was lost due to blight and unfavorable weather conditions. The study underscores the importance of technical solutions to enhance sustainable farming and crop management.

The contribution by Yoo et al., "Image-Based Quantification of Color and Its Machine Vision and Offline Applications", explores image-based colorimetry, leveraging the accessibility of smartphones with image sensors and increasing computational capabilities [item 8 in Appendix A]. Its low cost, portability, and compatibility with data processing make it suitable for a range of interdisciplinary applications, including art, fashion, food science, medicine, agriculture, geology, and more. The work focuses on the image-based quantification of color using specially developed software, demonstrating color extraction from a single pixel to customized regions of interest (ROIs) in digital images. Various color models like RGB, HSV, CIELAB, and Munsell are used to quantify colors from images of dyed T-shirts, tongues, and assays. The study also demonstrates histograms and statistical analyses of colors, proposing this method as a reliable and objective tool for color-based diagnostics and decision-making across diverse fields. The validity is verified through multiple examples in practical applications.

In the contribution by Shi et al., "Mobilenetv2_CA Lightweight Object Detection Network in Autonomous Driving", a lightweight object detection algorithm, based on MobileNetv2_CA, was proposed to address issues of high complexity, excessive parameters, and missed small targets in autonomous driving [item 9 in Appendix A]. Mosaic image enhancement is applied in pre-processing to improve feature extraction for small and complex targets. The Coordinate Attention (CA) mechanism is embedded into the MobileNetv2 backbone, alongside PANet and Yolo detection heads, enabling multi-scale feature fusion. This results in a lightweight object detection network. Tests show that the network achieved an 81.43% average detection accuracy on the VOC2007 + 2012 dataset, and 85.07% accuracy with a 31.84 FPS speed on the KITTI dataset. The network's parameter count is just 39.5 M, making it suitable for autonomous driving applications.

The contribution by Lemenkova and Debeir, "GDAL and PROJ Libraries Integrated with GRASS GIS for Terrain Modelling of the Georeferenced Raster Image", addresses limitations in traditional Geographic Information System (GIS) approaches by applying scripting methods and libraries such as Geospatial Data Abstraction Library (GDAL), PROJ, and GRASS GIS for geospatial data processing and morphometric analysis [item 10 in Appendix A]. The workflow involves converting Earth Global Relief Model (ETOPO1) data using GDAL, transforming it into various cartographic projections with PROJ, and analyzing topographic data with GRASS GIS modules. The study reveals patterns in topographic data, including elevation and depth distributions, and demonstrates the effectiveness of scripting techniques for topographic modeling and raster data processing. The integration of GDAL, PROJ, and GRASS GIS provides a more efficient and flexible approach to cartographic analysis, enhancing the informativeness and spatial data processing capabilities of traditional GIS software.

The contribution by Tychola et al., "Identifying Historic Buildings over Time through Image Matching", focuses on the identification of historic buildings over time using feature correspondence techniques [item 11 in Appendix A]. Photographs of landmarks in Drama, Greece, taken under varying conditions (e.g., lighting, weather, rotation, scale), were analyzed using traditional feature detection and description algorithms such as SIFT, ORB, and BRISK. These algorithms help identify homologous points in images to study changes in buildings over time. The research evaluates the performance of these algorithms in terms of accuracy, efficiency, and robustness, with SIFT and BRISK being the most accurate, while ORB and BRISK are the most efficient. The study highlights the role of computer vision in preserving historical architecture through accurate and efficient image matching techniques.

The contribution by Yoo et al., "Development of Static and Dynamic Colorimetric Analysis Techniques Using Image Sensors and Novel Image Processing Software for Chemical, Biological and Medical Applications", presents the development of colorimetric sensing techniques using image sensors and novel image processing software for chemical, biological, and medical applications [item 12 in Appendix A]. The system enables real-time monitoring and recording of colorimetric data from a point (s), line (s), and area (s) of interest in images and videos, supporting manual and automatic data collection. These techniques can be applied in process control, optimization, and machine learning. Video clips of chromatographic experiments with colored inks and blinking LED lights were analyzed, extracting colorimetric data as a function of time. The results were visualized through time-lapse images and RGB intensity graphs. The analysis was demonstrated using RGB, HSV, and CIE L*a*b* values for both static and dynamic colorimetric information, showcasing the effectiveness of the novel image processing software for color-based analysis.

The contribution by Kim and Koo, "Embedded System Performance Analysis for Implementing a Portable Drowsiness Detection System for Drivers", proposes an efficient platform for running Ghoddoosian's drowsiness detection algorithm, which utilizes temporal blinking patterns [item 13 in Appendix A]. Unlike previous implementations on powerful desktop computers, the study tested embedded systems suitable for vehicles. After comparing the Jetson Nano and Beelink (Mini PC), the Beelink was determined to be more efficient, with a processing time of 22.73 ms compared to the Jetson Nano's 94.27 ms. The Beelink's portability and power efficiency made it ideal for in-vehicle use. Additionally, a threshold optimization algorithm was developed to balance sensitivity and specificity in detecting drowsiness. This research advances drowsiness detection by identifying a real-time, practical platform for implementation in vehicles, bridging the gap between theoretical algorithms and practical application in road safety.

The contribution by Ramírez-Arias et al., "Evaluation of Machine Learning Algorithms for Classification of EEG Signals", focuses on improving the classification of motor movements using brain–computer interfaces (BCIs) by analyzing electroencephalographic (EEG) signals and training machine learning (ML) algorithms [item 14 in Appendix A]. EEG signals from 30 Physionet subjects, related to left hand, right hand, fist, foot, and relaxation movements, were processed using electrodes C3, C1, CZ, C2, and C4. Feature extraction techniques were applied, and nine ML algorithms were trained and tested. LabVIEW™ 2015 was used for signal processing, and MATLAB 2021a for algorithm training and evaluation. Among the algorithms, Medium-ANN achieved the highest performance with an AUC of 0.9998, Cohen's Kappa of 0.9552, and a loss of 0.0147. This approach is promising for applications like robotic prostheses, especially where resources are limited, such as embedded systems or edge computing devices.

3. Conclusions

This Special Issue presents 14 groundbreaking research findings on advanced image and signal processing. The insights shared herein are anticipated to foster further advancements and research in the domain of image and signal processing in the future.

Funding: This research received no external funding.

Acknowledgments: I thank the authors who published their research results in this Special Issue and the reviewers who reviewed their papers. I also thank the editors for their hard work and perseverance in making this Special Issue a success.

Appendix A

1. Dozzo, M.; Ganci, G.; Lucchi, F.; Scollo, S. Exploiting PlanetScope Imagery for Volcanic Deposits Mapping. Technologies 2024, 12, 25. https://doi.org/10.3390/technologies12020025.
2. Hitimana, E.; Sinayobye, O.J.; Ufitinema, J.C.; Mukamugema, J.; Rwibasira, P.; Murangira, T.; Masabo, E.; Chepkwony, L.C.; Kamikazi, M.C.A.; Uwera, J.A.U.; et al. An Intelligent System-Based Coffee Plant Leaf Disease Recognition Using Deep Learning Techniques on Rwandan Arabica Dataset. Technologies 2023, 11, 116. https://doi.org/10.3390/technologies11050116.
3. Yao, R.; Qi, P.; Hua, D.; Zhang, X.; Lu, H.; Liu, X. A Foreign Object Detection Method for Belt Conveyors Based on an Im-proved YOLOX Model. Technologies 2023, 11, 114. https://doi.org/10.3390/technologies11050114.
4. Yoo, W.S.; Kang, K.; Kim, J.G.; Yoo, Y. A Novel Approach to Quantitative Characterization and Visualization of Color Fad-ing. Technologies 2023, 11, 108. https://doi.org/10.3390/technologies11040108.
5. Wiseman, Y. Adapting the H.264 Standard to the Internet of Vehicles. Technologies 2023, 11, 103. https://doi.org/10.3390/technologies11040103.
6. Svendsen, B.; Kadry, S. Comparative Analysis of Image Classification Models for Norwegian Sign Language Recognition. Technologies 2023, 11, 99. https://doi.org/10.3390/technologies11040099.
7. Duarte-Correa, D.; Rodríguez-Reséndiz, J.; Díaz-Flórez, G.; Olvera-Olvera, C.A.; Álvarez-Alvarado, J.M. Identifying Growth Patterns in Arid-Zone Onion Crops (Allium Cepa) Using Digital Image Processing. Technologies 2023, 11, 67. https://doi.org/10.3390/technologies11030067.
8. Yoo, W.S.; Kang, K.; Kim, J.G.; Yoo, Y. Image-Based Quantification of Color and Its Machine Vision and Offline Applications. Technologies 2023, 11, 49. https://doi.org/10.3390/technologies11020049.
9. Shi, P.; Li, L.; Qi, H.; Yang, A. Mobilenetv2_CA Lightweight Object Detection Network in Autonomous Driving. Technolo-gies 2023, 11, 47. https://doi.org/10.3390/technologies11020047.
10. Lemenkova, P.; Debeir, O. GDAL and PROJ Libraries Integrated with GRASS GIS for Terrain Modelling of the Georeferenced Raster Image. Technologies 2023, 11, 46. https://doi.org/10.3390/technologies11020046.
11. Tychola, K.A.; Chatzistamatis, S.; Vrochidou, E.; Tsekouras, G.E.; Papakostas, G.A. Identifying Historic Buildings over Time through Image Matching. Technologies 2023, 11, 32. https://doi.org/10.3390/technologies11010032.
12. Yoo, W.S.; Kim, J.G.; Kang, K.; Yoo, Y. Development of Static and Dynamic Colorimetric Analysis Techniques Using Image Sensors and Novel Image Processing Software for Chemical, Biological and Medical Applications. Technologies 2023, 11, 23. https://doi.org/10.3390/technologies11010023.
13. Kim, M.; Koo, J. Embedded System Performance Analysis for Implementing a Portable Drowsiness Detection System for Drivers. Technologies 2023, 11, 8. https://doi.org/10.3390/technologies11010008.
14. Ramírez-Arias, F.J.; García-Guerrero, E.E.; Tlelo-Cuautle, E.; Colores-Vargas, J.M.; García-Canseco, E.; López-Bonilla, O.R.; Galindo-Aldana, G.M.; Inzunza-González, E. Evaluation of Machine Learning Algorithms for Classification of EEG Signals. Technologies 2022, 10, 79. https://doi.org/10.3390/technologies10040079.

Communication

Exploiting PlanetScope Imagery for Volcanic Deposits Mapping

Maddalena Dozzo [1,2], Gaetana Ganci [1,*], Federico Lucchi [3] and Simona Scollo [1]

1 Istituto Nazionale Geofisica e Vulcanologia, Sezione di Catania-Osservatorio Etneo, Piazza Roma 2, 95125 Catania, Italy; maddalena.dozzo@ingv.it (M.D.); simona.scollo@ingv.it (S.S.)
2 Dipartimento di Scienze della Terra e del Mare (DiSTeM), Università degli Studi di Palermo, Via Archirafi 22, 90123 Palermo, Italy
3 Dipartimento di Scienze Biologiche, Geologiche e Ambientali, Alma Mater Studiorum, Università di Bologna, Piazza Porta San Donato 1, 40126 Bologna, Italy; federico.lucchi@unibo.it
* Correspondence: gaetana.ganci@ingv.it

Abstract: During explosive eruptions, tephra fallout represents one of the main volcanic hazards and can be extremely dangerous for air traffic, infrastructures, and human health. Here, we present a new technique aimed at identifying the area covered by tephra after an explosive event, based on processing PlanetScope imagery. We estimate the mean reflectance values of the visible (RGB) and near infrared (NIR) bands, analyzing pre- and post-eruptive data in specific areas and introducing a new index, which we call the 'Tephra Fallout Index (TFI)'. We use the Google Earth Engine computing platform and define a threshold for the TFI of different eruptive events to distinguish the areas affected by the tephra fallout and quantify the surface coverage density. We apply our technique to the eruptive events occurring in 2021 at Mt. Etna (Italy), which mainly involved the eastern flank of the volcano, sometimes two or three times within a day, making field surveys difficult. Whenever possible, we compare our results with field data and find an optimal match. This work could have important implications for the identification and quantification of short-term volcanic hazard assessments in near real-time during a volcanic eruption, but also for the mapping of other hazardous events worldwide.

Keywords: volcanic hazard; Mount Etna; tephra fallout; Google Earth Engine

Citation: Dozzo, M.; Ganci, G.; Lucchi, F.; Scollo, S. Exploiting PlanetScope Imagery for Volcanic Deposits Mapping. *Technologies* **2024**, *12*, 25. https://doi.org/10.3390/technologies12020025

Academic Editor: Valeri Mladenov

Received: 28 November 2023
Revised: 25 January 2024
Accepted: 2 February 2024
Published: 8 February 2024

1. Introduction

Volcanic eruptions close to inhabited areas represent a major natural hazard due to lava flows, tephra fallout, mudflows, toxic gasses, and other phenomena that can be triggered during volcanic activity. Explosive eruptions can prove extremely dangerous for air traffic, causing severe damage to aircraft jet engines, obscuring the windscreen and landing lights, and resulting in other damaging effects [1,2]. These eruptions also pose risks to agriculture, roof stability, and human health [3,4]. Indeed, fine ash is associated with long-term health effects, such as silicosis and chronic pulmonary diseases [5]. For the characterization of explosive eruptions and their hazards, field investigations of tephra deposits constitute the first and, today, an essential step. Tephra samples are collected in the field and then analyzed in the laboratory to evaluate the main features of the eruptive event (e.g., total mass, total grain-size distribution, etc.). The volume of volcanic products can be mapped with field measurements by multiplying the covered area, which can be determined with a good precision, and estimating the average thickness or mass on square meter. Furthermore, tephra could be cleaned up or carried away by the action of the winds, making a rapid analysis necessary. In addition, field activities could also be impractical in some contexts and prove risky for volcanologists in proximal areas. Indeed, volcanoes are often located in remote areas, and they are generally inaccessible during eruption and for extended periods after the eruptive event ends, and their products can be scattered or dispersed over regional or global scales. Moreover, depending on the eruption, tephra deposits can cover larger areas, some of them very difficult to sample in one or two days. The values of mass (kg/m^2) are in general calculated and used to draw an isomass map of the tephra deposit. Isomass maps and isopach maps (distribution of mass/area and tephra thickness) are necessary to estimate erupted mass or volume [6–8], whereas the distribution of largest clasts (depicted in isopleth maps) is typically used to estimate column height and wind speed [9]. These parameters are important for evaluating the areas that may be affected by volcanic ash and forecasting regions that should be avoided by aircraft (e.g., [10]). Dedicated empirical and analytical models are generally used to determine the plume height, erupted mass, initial grain-size distribution, mass eruption rate, and duration. Many numerical models are currently available to reproduce the dispersion and fallout processes, and they have been successfully applied to several recent eruptions [11–20]. However, it is well-known that the reduction of uncertainty in model outcomes cannot be done without integrating observational data coming from an advanced monitoring system [21,22]. Radar, cameras, and infrasound sensors constitute valuable tools for monitoring active or potentially active volcanoes and have recently been used for forecasting volcanic ash transport and dispersion during an eruption, facilitating real-time management of volcanic hazards [23–26]. In particular, these instruments have been used to directly estimate the mass eruption rate (MER) in near real-time and determine the eruptive column height [27–31]. However, these pioneering applications are either not operational or limited to a few of the world's best-monitored volcanoes (e.g., Etna volcano, Italy). In recent years, satellite detection techniques have been investigated. The use of remote sensing techniques aimed at analyzing the tephra deposit is essential to have a more rapid response during a volcanic crisis. It is necessary since emergency responders and government agencies need to make fast decisions that should be based on accurate forecasts of tephra dispersal and assessments of the expected impact. By differencing successive digital topographies, deposit thickness can be mapped, but this method is limited by the availability of topographic data of the surface underneath the volcanic deposit of interest, and such measurements are difficult, if not impossible, to be made retrospectively [32]. Moreover, this method can only be used if deposit thickness is more than 1 m, making it impractical for tephra fallout arising from a single episode of lava fountain because of the high measurement error, which is equal to 1 m [33].

The technological advancements and increasing availability of high-resolution satellite imagery offer new possibilities for mapping volcanic deposits related to a single eruptive event. Spatial resolution is one of the main limitations in deposit mapping, together with the revisit time of satellites, particularly in rapidly evolving situations. The constant improvements in satellite sensors over the years allowed us to achieve, with the launch of the PlanetScope microsatellite constellation, very high spatial resolution (~3 m) images, with a daily temporal cadence.

In this work, we present a new method to identify the fallout deposits related to individual eruptive episodes in near real-time, exploiting the low cost and high spatial resolution PlanetScope imagery. Near-real-time tephra mapping is of primary importance since it could help field volcanologists in faster sampling and provide a synoptic view of the areas affected by volcanic deposits where sampling is not possible. The programming codes were developed on the Google Earth Engine computing platform (GEE) considering the variations in measured reflectance with and without fallout in urban paved areas. To test the method, we chose Mt. Etna volcano as a case study.

During 2021–2023, Mt. Etna produced more than 60 lava fountains characterized by persistent tephra fallout, which often affected the same portion of the volcano, sometimes two or three times within a day. The extremely high frequency of these explosive events made field surveys difficult, since sampling soon after each eruptive event is fundamental. Moreover, at Etna, field surveys are carried out in certain areas based on simulations, which can differ from the real tephra deposit if the eruption has different eruption source parameters. As an example, the column height may change during the course of an eruptive event, depending on the eruption style, and this reflects on the area affected by the tephra fallout. Indeed, models that consider a constant column height for an entire eruptive event will provide greater impacted areas than those that include decreasing column heights [34].

2. Materials and Methods

The techniques currently used to identify fallout deposits can prove extremely time-consuming and are unsuitable for the study of individual eruptive episodes with near-real-time results. Technological advancements and the increasing availability of high-resolution satellite imagery have recently offered new possibilities for mapping volcanic deposits related to an individual eruptive event. PlanetScope satellite imagery (PS), operated by Planet Labs Inc., is acquired from a constellation of over 180 microsatellites ('CubeSat' or 'Doves'), roughly $10 \times 10 \times 30$ cm in size (i.e., three-unit or 3U CubeSats), orbiting at an altitude of ~400 km. The constellation consists of multiple launches of groups ('flocks') of single satellites from 2014 to 2022, allowing on-orbit capability to improve steadily, with technology improvements being implemented at a rapid pace. The first available image dates to July 2014, acquired by the first generation of PlanetScope satellites (known as 'Dove Classic' or 'PS 2'). The second generation of PlanetScope satellites ('Dove-R' or 'PS2.SD') has been in operation since November 2018, having sensor characteristics which allow excellent spectral resolution. The third and latest generation of PlanetScope satellites (known as 'SuperDove' or 'PSB.SD') is currently in orbit, providing a limited number of images with 8 separate spectral bands since 2022.

Figure 1 shows the three generations of PlanetScope imagery with related spectral bands and wavelengths, including the spectral response curves.

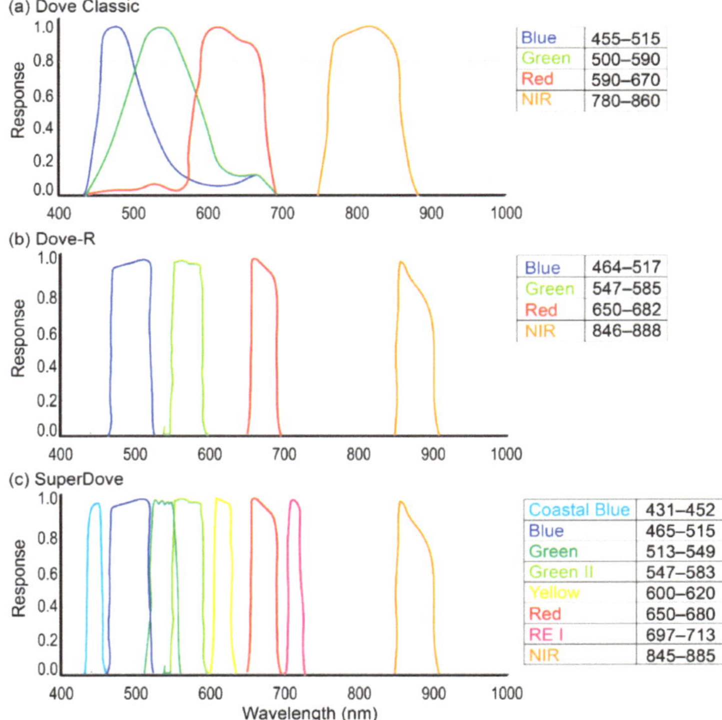

Figure 1. Spectral responses for PlanetScope: (**a**) Dove Classic 'PS2', (**b**) Dove-R 'PS2.SD', and (**c**) SuperDove 'PSB.SD' sensors (from [35]).

The main feature of PlanetScope imagery lies in its high spatial resolution (~3 m) and daily image capture rate for each point on the Earth's surface. These characteristics, together with the policy of free access to data for study or research purposes, make PlanetScope imagery ideal for monitoring volcanic activity, as well as for studying other dynamic geophysical phenomena [36]. Several PlanetScope data products with different processing levels are available; in this work we used the orthorectified 'Analytic Surface Reflectance' data products with four spectral bands (RGB NIR). In addition to orthorectification, radiometric corrections are applied to correct for any sensor artifacts and transform the data to scaled at-sensor radiance.

The methodology introduced to process PlanetScope imagery is illustrated in the flowchart of Figure 2.

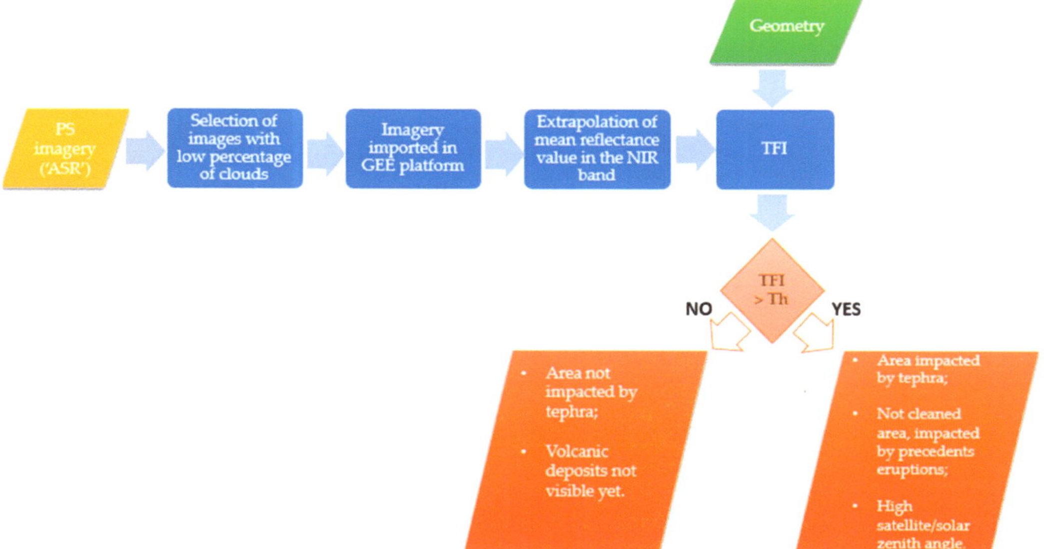

Figure 2. Flowchart with the main steps of the methodology: PlanetScope PS imagery (in ASR—analytic surface reflectance format) and selected geometries represent the data input; blue rectangles report the main processing steps (where GEE stands for Google Earth Engine and TFI includes the Tephra Fallout Index computation), while orange parallelograms are the possible outputs to highlight areas impacted by tephra fallout depending on the TFI values with respect to the selected threshold, Th.

Firstly, PlanetScope satellite imagery in the 'Analytic Surface Reflectance' format was downloaded as GeoTiff and imported on the cloud-based geospatial analytic Google Earth Engine (GEE) platform, selecting those images without volcanic plume (which in general cover a large area) and with a low percentage of weather clouds (less than 25%), since clouds could alter or mask, together with other factors (e.g., snow, solar zenith angle, satellite zenith angle), the reflectance value of the ground. Among those images, we performed our computations considering only selected geometries not affected by any clouds. GEE allows users to import their own imagery and manage and process enormous volumes of data, making its use ideal for this study. Geometries, thus provided as input (Figure 2), were chosen, possibly with a progressively greater distance from the source area and localized in different points of the volcano flanks in order to analyze a larger area. We picked out quite large, paved zones, without vegetation, to distinguish the volcanic deposit. Moreover, we chose similar areas, both in terms of surface composition and texture, to have an effective comparison in terms of emissivity. For each image, we also took into account the satellite zenith angle and the solar zenith angle, available from the image properties. The objective of the proposed method was to find any variations between the pre- and post-eruptive reflectance values in the selected geometries, exploiting the different ways that different materials exhibit (in this case tephra and cement) to reflect light.

We expected to find higher variations in the average value of the NIR band after the occurrence of explosive events characterized by intense tephra fallout since this band is particularly sensitive to materials and their texture [37].

In order to magnify the temporal variations potentially observed in the NIR band, we introduced a new index, which we called the Tephra Fallout Index (TFI) to be computed for selected paved areas. This index is the result of the average as a percentage between the NIR band mean value of the date before and after the eruptive event considering those days not affected by any fallout, defined as follows:

$$TFI = \frac{\left| R_{NIRe} - \left(R_{NIRpd} + R_{NIRfd} \right)/2 \right|}{\left(R_{NIRpd} + R_{NIRfd} \right)/2} \times 100 \tag{1}$$

where, once a geometry is defined, R_{NIRe} is the mean reflectance value in the NIR band of the date considered, i.e., the day of eruption; R_{NIRpd} is the mean reflectance value in the NIR band of the date prior to the eruptive event; and R_{NIRfd} is the mean reflectance value in the NIR band for the first acquisition in the days following the eruption. All these physical quantities are computed in the selected geometry. Eventually, the methodology aims to identify a threshold (Th) for the TFI in order to spatially define those areas affected by the presence of tephra fallout. The threshold value is a dynamic limit, which we defined by observing the minimum value in each investigated area, and so it can differ between one eruptive event and another. It is, therefore, necessary to analyze the values of the same day or the day following the eruption, depending on the time of passage of the satellite and based on the quality of the image, which must be free from snow and clouds in the examined area. It is worth noting that days too distant from the eruption need to be taken with caution since the volcanic material could have been cleaned up or carried away by the action of the winds. To obtain a spatially coherent threshold value for the TFI, it is essential to consider the same pre-eruption date and post-eruption date in the whole area investigated. Moreover, the TFI value is affected by the absolute radiometric uncertainty of the PlanetScope sensor, which should be quantified on a case-by-case basis.

3. Case Study: Mt. Etna 2021–2022 Paroxysmal Eruptions

Mount Etna has a persistent activity characterized by degassing and explosive phenomena at its four summit craters, Voragine (VOR), Bocca Nuova (BN), North-East Crater (NEC) and South-East Crater (SEC), as well as by frequent flank eruptions [38–40]. Since 1989, the volcanic activity has mostly taken place at SEC, which has produced several eruptions involving multiple eruptive episodes, defined as 'episodic' eruptions [41,42], characterized by long-lived episodic activity (e.g., [43]). Each episode lasts a few hours and is characterized by explosive activity ranging from powerful Strombolian activity to lava fountaining [44]. Usually, violent Strombolian activity produces both weak plumes that last from hours to months and eruptive columns up to 15 km a.s.l., along with associated tephra falls extending several tens of kilometers from the vent [45,46], often accompanied by short-lasting lava overflows from the crater rim [45]. Several paroxysms have occurred at the summit craters over the last decades. Between 2011 and 2013, more than 50 eruptions of this type were recorded from the SEC [47]. In 2015, the paroxysmal episodes at the Voragine crater generated an eruptive column up to 15 km a.s.l., defined as a sub-Plinian eruption. In this type of eruption, the contribution in terms of tephra can be very significant, producing considerable volumes [48–50]. In 2019, several Strombolian and effusive eruptions affected the South-East Crater, which then renewed activity in the spring of 2020, with Strombolian activity leading to four major episodes in December of the same year.

For this study, we consider the eruptions that occurred in 2021, when the SEC was affected by several paroxysmal episodes [47,51], often more times within a day. We also benefited from numerous ground-based data useful to validate our satellite-based results. Between February and April 2021, 17 paroxysms occurred, the last one being the longest of the entire series (540 min). During these episodes, the eruptive column reached up to 11 km a.s.l., leading to the dispersion of ashes and lapilli up to hundreds of km from the point of emission, as well as to the fall of bombs coarser than 15 cm at more than 7 km away and lava flows that affected the Valle del Bove and the southern flank of the volcano [52]. After a break of about one month, the paroxysmal activity resumed on the night between 18 and 19 May and continued until the end of 2021, with interspersed episodes of inactivity, concluding in February 2022 with two very violent explosive events on 10 and 21 February.

We analyzed about one hundred PlanetScope images from February to October 2021 (available in the Supplementary Material), considering the villages located on the north-east, east, south-east, and south flanks of the volcano (Figure 3), which were the most affected by tephra fallout in 2021. Areas adjacent to the summit craters of Etna and more distal areas were chosen to have an overall view of the distribution of the tephra deposit.

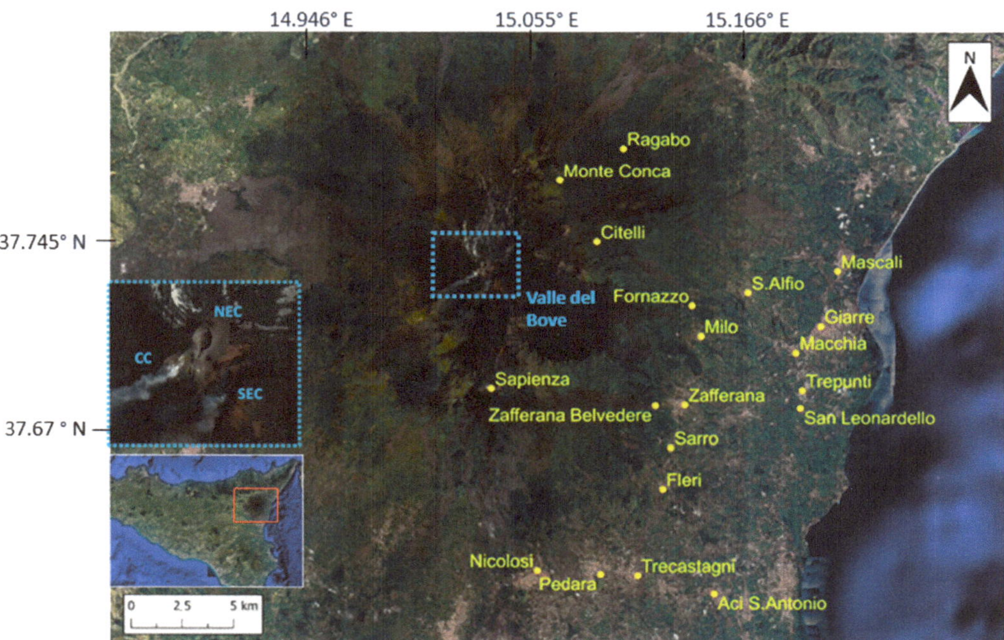

Figure 3. Google map of Mt. Etna, where the localization of the analyzed areas and villages located on the north-east, east, south-east, and south flanks of Mount Etna is reported. In the bottom left inset, the map of Sicily with the Etna volcano framed in red is reported, and in the top inset, there is a zoom on the Summit Craters.

We selected those images that were not influenced by the volcanic plume passage and had a low percentage of clouds, since clouds could alter or mask the reflectance value of the ground. We imported the images in the GEE platform. In the case of cloudy images, we only considered locations where the area of interest had no clouds. The same reasoning was applied to the areas covered by snow, focusing mainly on the areas located on the north-east flank of the volcano due to their altitude, which could alter the reflectance value too. The chosen PlanetScope images were uploaded to Google Drive and imported individually in the "Assets" section of Google Earth Engine, to which the platform is connected. Since the analyzed area includes about 400 km^2, multiple PlanetScope images were often processed for a single date. Thus, we defined a geometry for each paved area of interest in a specific village and calculated the mean reflectance value in the NIR band and the Tephra Fallout Index.

As we can see from Figure 4, showing pictures of Piazza del Duomo (Giarre), as an example of a paved area we analyzed, before and after the eruptive event of 7 March 2021, the reflectance of the square changes with respect to both the color and the texture of the fallout. The two pictures shown in Figure 4 are exemplificative of this study, since the square changes its appearance and its reflectance significantly when entirely covered by the tephra. Moreover, we chose specific areas based on the INGV-OE bulletins (www.ct.ingv.it; accessed on 22 November 2023), which also report field data for some villages. Other examples of selected areas are the parking in front of Rifugio Citelli, Rifugio Sapienza, and Monte Conca or the one in front of the Hotel Belvedere (Zafferana Etnea).

Figure 4. Piazza del Duomo (Giarre) before (**a**) and after (**b**) the paroxysmal event on 7 March 2021 (courtesy of T. Caggegi). On the left inset the satellite view of Mt. Etna is depicted, with Giarre highlighted in yellow.

4. Results

We found a general decrease in the reflectance values and high values of the TFI in correspondence with the eruptive episodes. Taking the paved area in Nicolosi as an example, after the eruptive episode of 20 June 2021, there is a variation in the TFI from ~0.05 (5%) (considering the day before and after the eruption) to 0.13 (13%) (in correspondence with the eruption) (Figure 5).

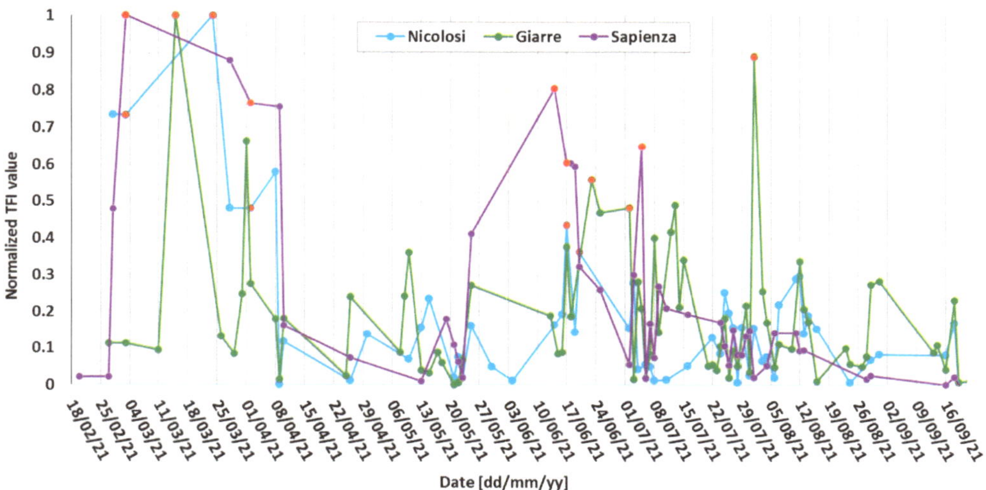

Figure 5. Normalized TFI values related to a square in Nicolosi (900 m^2 surface area, blue dotted line), Piazza del Duomo, Giarre (2600 m^2 surface area, green dotted line), and the parking area next to Rifugio Sapienza (14,250 m^2 surface area, purple dotted line). The days affected by eruptive episodes are highlighted with red dots: (i) Nicolosi: 28/02/2021, 3/03/2021, 24/03/2021, 2/04/2021, 17/06/2021, 20/06/2021; (ii) Giarre: 15/03/2021, 23/06/2021, 2/07/2021, 1/08/2021; (iii) Sapienza: 3/03/2021, 2/04/2021, 14/06/2021, 17/06/2021, 5/07/2021.

Figure 5 shows the TFI normalized between 0 and 1 for Nicolosi, Giarre, and Sapienza. We observed higher values in correspondence with the days of eruption or the following day (Figure 5), expecting a higher surface coverage density here as well.

In the Nicolosi TFI time series it can be observed how all the days affected by the eruptive activity (in red) exhibit normalized TFI values above ~0.3, while the others are below this value. Some exceptions to this rule are due to high solar zenith angles, the lack of cloud-free satellite data on the immediately preceding or succeding day with respect to the eruption, or proximity to the eruption date. Indeed, days too distant from the eruption need to be considered with caution, since the volcanic material could have been cleaned up or carried away by the action of the winds. As an example, on 28 March and 8 April, two high values of TFI occur in correspondence with high values of solar zenith angle (about 40 degrees), higher than the same values found in the other images (30 degrees).

As for the Giarre TFI time series, we observed that the days affected by the eruptive activity are above ~0.5, while the others are below this value. In addition, in this case, on 1 April, we found a high value of TFI in correspondence with a high solar zenith angle (about 40 degrees).

In the Rifugio Sapienza TFI time series we observed how all the days affected by the eruptive activity (in red) are above ~0.6, while the others are below this value. Again, on 28 March and 9 April, two high TFI values are characterized by a solar zenith angle of approximately 40 degrees. Moreover, from Figure 5, we can see that the eruption that occurred on 17 June also impacted the following date (18 June), characterized by a slightly lower value in comparison, which could represent the presence of volcanic tephra deposit not yet cleaned.

To estimate the accuracy of the TFI calculation, we verified the absolute radiometric uncertainty in the ASR images acquired on Mt. Etna during 2021. From the technical note on the quality assessment for PlanetScope (DOVE) provided by ESA (https://earth.esa.int/eogateway/search?text=planetscop; accessed on 22 November 2023), we found that the biggest uncertainty for the NIR band in the two-stripe 'Dove Classic' (δ) is of 6.5%, which leads to $2 \times \delta \approx 13\%$ of uncertainty in the TFI values.

After temporal coherence on each area analyzed was verified, we verified spatial coherence too, creating normalized TFI maps for each eruption and reporting all the analyzed areas. For a more targeted analysis, various data were cross-referenced to have a visual comparison. The data were obtained from the INGV-OE bulletins (www.ct.ingv.it; accessed on 22 November 2023) and from the SEVIRI satellite sensor available on the Eumetview platform (https://view.eumetsat.int/; accessed on 22 November 2023), which allows the visualization of the dispersion of volcanic clouds from 2020 to the present, providing updates every 15 min through the volcanic ash RGB product, where the ash is represented in red color. From the Eumetview platform, the main direction of the volcanic plume can be observed, and the areas that could have been involved are immediately noticed. The best view of the plume from the satellite was selected, considering the time of the eruption, as well as any changes in its direction. By cross-referencing the information from PlanetScope imagery with that from SEVIRI, we were able to extrapolate a threshold to distinguish the presence of volcanic tephra in a certain area, rather than in another for the same eruptive event.

Three lava fountains that occurred at Mt Etna are detailed below, reporting for each eruption the TFI value calculated on each paved area (reported in brackets). For each geometry, we made a punctual calculation of the index and compared the result with the other areas where the same pre- and post- eruptive dates are available. Indeed, the TFI value is not reported on the map for the villages covered by clouds or snow or cut off in one or both satellite PS image.

On the map inset, the corresponding RGB product retrieved from Eumetview and used to draw the volcanic plume profile is shown. The delimitation of the plume profile helps to immediately visualize the areas potentially involved, considering that there may not be a direct match between the volcanic cloud and the volcanic deposit on the ground.

4.1. 2 March 2021 Paroxysmal Eruption

During an eruptive event, even if short-lived, conditions may vary in terms of wind direction and speed, and this will inevitably affect the ash fallout. Figure 6 reports the TFI results for the eruption that occurred on 2 March 2021. On this date, at 10:45 UTC, weak Strombolian activity started at the SEC. This activity increased from 11:34 UTC with a consequent emission of tephra, and at 12:24 UTC, the Strombolian activity suddenly changed to a lava fountain, producing an eruptive column that reached a height of approximately 9000 m from the top of the volcano [23,53] and ended at 14:50 UTC. The volcanic cloud dispersed on the southern slope of the volcano causing ash and lapilli fallout in the villages of Nicolosi, Aci Sant'Antonio, Pedara, and finally in Catania. The lava fountain activity at the SEC ceased in the afternoon. The total volume of the erupted material on 2 March is equal to 1.03×10^6 m^3 [47]. The density map reporting the normalized TFI values related to the eruption is compared with the corresponding ash RGB products calculated from SEVIRI and available on Eumetview (Figure 6). In this example, two different frames are reported to highlight the change in the direction of the plume, thus increasing the area involved by the plume passage (dashed black line). To study this eruption, the image of 3 March 2021 was taken into consideration for the TFI calculation, as the eruption occurred in the afternoon.

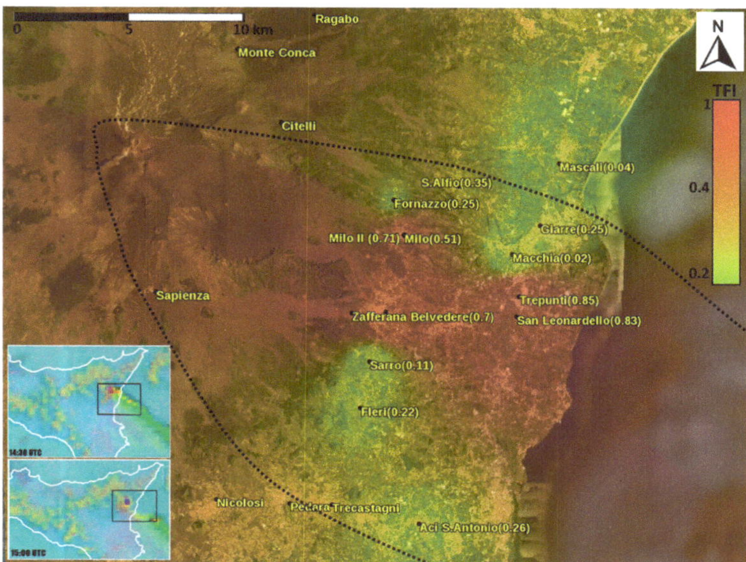

Figure 6. Density TFI map related to the 2 March 2021 eruption, integrated in a GIS. The profile of the volcanic plume is shown (dashed black line), and all the selected geometries are reported in yellow, with the name of the corresponding village and the respective normalized TFI value in brackets. On the left insets are reported the ash RGB products calculated on the same date at 14:30 UTC and at 15:00 UTC.

Reporting the data related to the eruption on 2 March 2021 (Figure 6), we found higher normalized TFI values (shown in red on the map) in the geometries located at the middle of the volcanic plume profile, with still appreciable values in the south. From the observation of the ash RGB products it is possible to notice that the ash (in red-orange) changes direction from 14:30 UTC to 15:00 UTC, involving these two areas and confirming the success of our method. Here, the visual comparison with data from SEVIRI is particularly important to verify the quality of our results, allowing us to better constrain the areas affected by the tephra fallout, since the plume does not follow a clear direction. In this example, it is not possible to define a threshold to distinguish the areas affected by the tephra fallout because of missing data in the northern part of Etna, which is not involved in the ash fallout.

4.2. 22 May 2021 Paroxysmal Eruption

A resumption of Strombolian activity was registered at the SEC on 19 May, after more than a month of inactivity. On 22 May 2021 at 08:14 UTC, sporadic intra-crater explosions were observed and, after about an hour, the explosive activity increased, with weak ash emissions, dispersing rapidly. The activity increased again in the early evening hours, accompanied by abundant emission of tephra. After having recorded a further increase in Strombolian activity, it evolved to a lava fountain at 20:27 UTC [47], and the emitted products dispersed in the ENE direction. This eruptive phase ended at 22:08 UTC with the formation of two lava overflows, fed by the crater, one dispersing in the SW direction and the other in the E direction. A few hours after the end of the lava fountain, starting from 01:51 UTC, a new resumption of explosive activity was observed at the SEC, with abundant emission of ash, which dispersed in the E direction. This phase continued until approximately 02:03 UTC, then moved on to discontinuous explosions with weak emissions of ash, which quickly dispersed. The total volume of material emitted by this eruption, with a maximum height of the eruptive column of approximately 8 km, is equal to 0.87×10^6 m^3 [47].

The density map reporting the normalized TFI values related to the eruption that occurred on 22 May 2021 is reported in Figure 7, along with the corresponding ash RGB product calculated from the SEVIRI sensor on the same date by Eumetview. To study this eruption, the image of 23 May 2021 was taken into consideration, as the eruption occurred overnight.

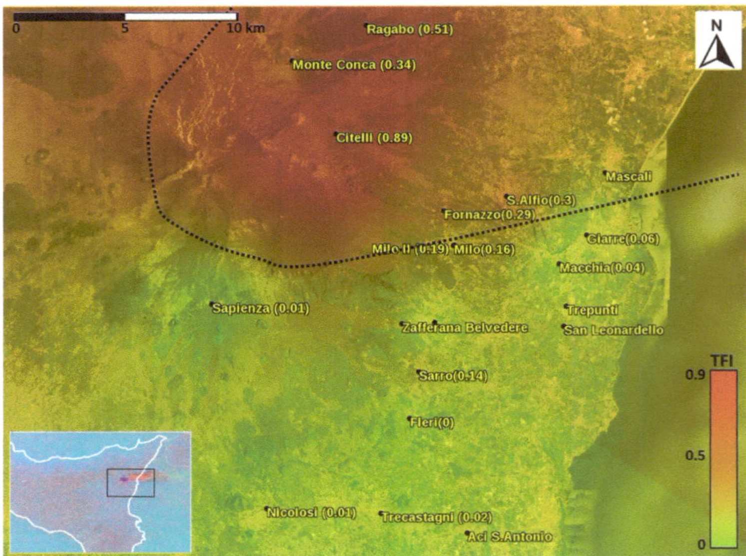

Figure 7. Density TFI map related to 22 May 2021 eruption integrated in a GIS. The profile of the volcanic plume is shown (dashed black line), and all the selected geometries are reported in yellow, with the name of the corresponding village and the respective normalized TFI value in brackets. On the left inset, the ash RGB product calculated at 21:45 UTC on the same date is reported.

From Figure 7, it can be observed that the villages affected by the passage of the ash cloud have a higher normalized TFI value than the other areas, ranging from 0.29 (Fornazzo) to 0.89 (Citelli), whereas the other paved areas show values from 0.19 (Milo) to 0 (Fleri). In the north-east area, we therefore find the highest values, confirming the direction of the dispersion of the products emitted, as reported in the bulletin on the INGV-OE website related to that day. Among the localities affected by the passage of the volcanic cloud, Fornazzo is the one which presents a lower percentage, constituting the threshold limit value for this eruption among the areas on which the calculation could be performed.

4.3. 1 July 2021 Paroxysmal Eruptions

During the week of the 1 July, three episodes of lava fountaining occurred at the SEC, with the formation of eruptive columns, which reached heights varying from a minimum of 5 km to a maximum of 10 km a.s.l. (8 km in the case of the 1 July). The activity started as Strombolian at 22:40 UTC to evolve to lava fountain activity at 22:50 UTC, finishing at 00:27 UTC [47]. All lava fountains were accompanied by lava overflows southwestward. The tephra fallout on this date affected the eastern/south-eastern sector of the volcano. The total volume of material emitted by the eruption is equal to 0.91×10^6 m^3 [47]. To study this eruption, the image of 2 July 2021 was taken into consideration for the calculation, as the eruption occurred overnight (Figure 8).

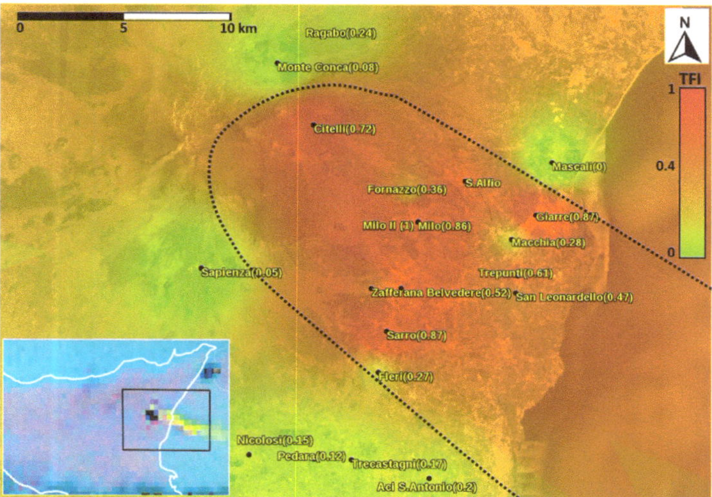

Figure 8. Density TFI map related to the 1 July 2021 eruption, integrated in a GIS. The profile of the volcanic plume is shown (dashed black line) and all the selected geometries are reported in yellow, with the name of the corresponding village and the respective normalized TFI value in brackets. On the left insets, the ash RGB product calculated at 23:45 UTC on the same date is reported.

Figure 8 reports the density TFI map after the eruption occurred on the 1 July 2021, with the corresponding ash RGB product calculated from a satellite image of Etna dated 1 July 2021 by Eumetview. From Figure 8, it is possible to observe that the villages affected by the passage of the ash cloud have a higher normalized TFI value compared to the other areas, with values up to 1. On the other hand, the areas which do not seem to have been affected by the tephra fallout show lower values, with a minimum value of 0. We also found a quite high value in the Ragabo area, which is probably due to the accumulation of tephra coming from different episodes, since this area is not usually cleaned. In the example above, we can define a threshold based on the lowest value present within the delimited area affected by the passage of the volcanic plume. It can be represented by the value of 0.27, corresponding to Fleri.

5. Discussion and Conclusions

Satellite remote sensing techniques constitute a valuable tool for volcano monitoring, and their use, aimed at analyzing the tephra deposits, is fundamental for achieving a more rapid response during a volcanic crisis. This is particularly important, considering the fact that field surveys can be impractical in certain contexts, representing a source of risk for operators.

In this work we identify, for the first time, the volcanic deposit related to individual eruptive events in near real-time by processing PlanetScope satellite imagery, which offers high spatial resolution and a daily revisit time for each point on the Earth's surface.

Since tephra deposits alter the spectral signature of the reflectance detected by satellite, for eruptions having sufficient data in terms of image quality, PlanetScope imagery allows for the evaluation of the areas in which the tephra fallout took place. Hence, the introduction of a new index, which we call the Tephra Fallout Index (TFI), maps the areas involved by the tephra fallout and quantifies the related surface coverage density.

We applied our technique to Mt. Etna (Italy), which, during 2021, was affected by recurring eruptive events. Whenever possible, we compared our results with field data available on the bulletins on the INGV-OE website and found an optimal agreement. Some areas were specifically chosen for their presence on the bulletins to have easier feedback on the areas where the tephra fallout occurred. Higher TFI values were observed in temporal correspondence with the same day or the following day of the eruption (depending on the time of satellite's passage with respect to the eruption time) and in spatial correspondence with the area below the volcanic plume observed by the SEVIRI sensor. The normalized TFI maps provide a surprising spatial coherence that is in agreement with the satellite observations of the volcanic plume. Moreover, after comparing the TFI values with field measurements, we could observe that they provide interesting information about the surface coverage density of the fallout deposit.

We also identified a threshold to distinguish areas affected by tephra deposits. The threshold value is a dynamic limit, given by the observation of the minimum value in each investigated area, and thus, it differs between one eruptive event and another. Moreover, the TFI is affected by a certain percentage of uncertainty (in our case 13%), due to the absolute radiometric uncertainty on the NIR band.

It is worth considering that while some eruptions produce volcanic deposits visible only after a few days, due to the accumulation of multiple deposits, for others, the tephra deposit may not have been cleaned up and remains observable for a while. Thus, the fact of not knowing whether an area was cleaned or not after an eruption constitutes a limit for this analysis. This problem is greater in cases where it is not possible to study the eruption immediately after its occurrence due to the timing of the satellite passage compared to the eruption itself or in case where no dates are available for calculation (if several eruptive events occurred in the previous or following days).

Since the introduced method is based on satellite optical data analysis, it is affected by cloud covering, which can alter, together with other factors, such as the snow, the solar zenith angle, and the satellite zenith angle, the reflectance values measured at the satellite sensors. For this reason, we discarded the areas affected by cloud or snow cover, with a consequent lack of data in some areas, and we identified as anomalous the data with high satellite or solar zenith angle. Anyway, the fast growth of satellites in orbit and their integration with other satellite data, such as synthetic aperture radar (SAR) satellite instruments, could solve this problem since they are not adversely impacted by cloud, rain (partially) or ash cover, or time of day [54]. Our methodology could indeed integrate multispectral imagery with data coming from high spatial resolution X-band SAR imagery, using sources such as the ICEYE constellation with 27 active satellites or Capella Space with 9 active satellites.

Furthermore, a future aspiration is to obtain a shorter revisit time in the passage of satellites and, therefore, a greater number of data available. In this regard, several studies have highlighted a positive trend in the last ten years regarding the launch of satellite sensors and their use for volcano monitoring [55]. As for PlanetScope microsatellites, their number and quality have been increasing at a rapid pace. This growth, starting from 2022 with the advent of the third generation of PlanetScope satellites, has led to the acquisition of multispectral images with eight separate spectral bands which could prove to be the key for future studies.

Future improvements will include punctual ground-based validation, including the use of disdrometers (e.g., [56–58]) to investigate the relationship between the thickness of the deposit and the TFI values. Indeed, TFI supplies information about the ash coverage density, expecting higher TFI values in the areas covered mostly by tephra, and lower values where the coverage is rarefied, but not with regards to the deposit thickness.

Considering the limitations described above, this work could have important implications not only for the identification and the quantification of short-term volcanic hazard assessments in near real-time during a volcanic eruption but also for the mapping of other hazardous events worldwide.

Supplementary Materials: The following supporting information can be downloaded at: https://www.mdpi.com/article/10.3390/technologies12020025/s1, The list of the Planetscope images analyzed in this study is provided as the Geometries.xlsx file, with a sheet for each selected geometry.

Author Contributions: Conceptualization, G.G.; methodology, M.D. and G.G.; validation, S.S. and F.L.; formal analysis, M.D., G.G. and S.S.; investigation, M.D., G.G., S.S. and F.L.; resources, G.G.; data curation, S.S., M.D. and F.L.; writing—original draft preparation, M.D., G.G., S.S. and F.L.; writing—review and editing, M.D. and G.G.; visualization, M.D. and G.G.; supervision, G.G., S.S. and F.L.; funding acquisition, G.G. All authors have read and agreed to the published version of the manuscript.

Funding: This work was supported by the INGV project Pianeta Dinamico (CUP D53J19000170001) funded by MUR ("Fondo finalizzato al rilancio degli investimenti delle amministrazioni centrali dello Stato e allo sviluppo del Paese", legge 145/2018), Volcanological Theme SAFARI (An Artificial Intelligence-based StrAtegy For volcAno hazaRd monItoring from space).

Institutional Review Board Statement: Not applicable.

Informed Consent Statement: Not applicable.

Data Availability Statement: The PlanetScope imagery used in this work was provided by Planet Labs through their Education and Research Program. The datasets generated during this study are available from the corresponding author on reasonable request.

Acknowledgments: The authors gratefully thank the Planet and the Education and Research program (https://www.planet.com/markets/education-and-research/; accessed on 22 November 2023) for providing the PlanetScope images to this study and acknowledge EUMETSAT (www.eumetsat.int; accessed on 22 November 2023) for providing ash RGB products. We would like to thank the INGV-MUR "Working Earth-VT_SAFARI-2023-25" project. The English style was corrected by Stephan Conway.

Conflicts of Interest: The authors declare no conflicts of interest.

References

1. Casadevall, T.J. Volcanic ash and aviation safety: Proceedings of the first international symposium on volcanic ash and aviation safety. *US Geol. Surv. Bull.* **1994**, *2047*, 1–450.
2. Miller, T.P.; Casadevall, T.J. Volcanic ash hazards to aviation. In *Encyclopedia of Volcanoes*; Sigurdsson, H., Houghton, B.F., McNutt, S.R., Rymer, H., Stix, J., Eds.; Academic Press: San Diego, CA, USA, 2000; pp. 915–930.
3. Andronico, D.; Del Carlo, P. PM_{10} measurements in urban settlements after lava fountain episodes at Mt. Etna, Italy: Pilot test to assess volcanic ash hazard to human health. *Nat. Hazards Earth Syst. Sci.* **2016**, *16*, 29–40. [CrossRef]
4. Horwell, C.J.; Sargent, P.; Andronico, D.; Castro, L.; Tomatis, M.; Hillman, S.E.; Michnowicz, S.A.; Fubini, B. The iron-catalysed surface reactivity and health-pertinent physical characteristics of explosive volcanic ash from Mt. Etna, Italy. *J. Appl. Volcanol.* **2017**, *6*, 12. [CrossRef]
5. Horwell, C.J.; Baxter, P.J. The respiratory health hazards of volcanic ash: A review for volcanic risk mitigation. *Bull. Volcanol.* **2006**, *69*, 1–24. [CrossRef]
6. Pyle, D.M. The thickness, volume and grainsize of tephra fall deposits. *Bull. Volcanol.* **1989**, *51*, 1–15. [CrossRef]
7. Fierstein, J.; Nathenson, M. Another look at the calculation of fallout tephra volumes. *Bull. Volcanol.* **1992**, *54*, 156–167. [CrossRef]
8. Bonadonna, C.; Houghton, B.F. Total grain-size distribution and volume of tephra-fall deposits. *Bull. Volcanol.* **2005**, *67*, 441–456. [CrossRef]
9. Carey, S.; Sparks, R.S.J. Quantitative models of the fallout and dispersal of tephra from volcanic eruption columns. *Bull. Volcanol.* **1986**, *48*, 109–125. [CrossRef]
10. Tupper, A.; Textor, M.; Herzog, A.; Graf, F.; Richards, S. Tall clouds from small eruptions: The sensitivity of eruption height and fine ash content to tropospheric instability. *Nat. Hazards* **2009**, *51*, 375–401. [CrossRef]
11. Suzuki, T. A theoretical model for dispersion of tephra. In *Arc Volcanism, Physics and Tectonics*; Shimozuru, D., Yokoyama, I., Eds.; Terra Scientific: Tokyo, Japan, 1983; pp. 95–113.
12. Armienti, P.; Macedonio, G.; Pareschi, M.T. A numerical model for simulation of tephra transport and deposition: Applications to 18 May 1980, Mount St. Helens eruption. *J. Geophys. Res.* **1988**, *93*, 6463–6476. [CrossRef]
13. Bursik, M.; Sparks, R.S.J.; Gilbert, J.S.; Carey, S. Sedimentation of tephra by volcanic plumes: I. Theory and its comparison with a study of the Fogo A plinian deposit, Sao Miguel (Azores). *Bull. Volcanol.* **1992**, *54*, 329–344. [CrossRef]
14. Heffter, J.L.; Stunder, B.J.B. Volcanic ash forecast transport and dispersion (VAFTAD) model. *Weather Forecast.* **1993**, *8*, 533–541. [CrossRef]

15. Searcy, C.; Dean, K.; Stringer, W. PUFF: A high resolution volcanic ash tracking model. *J. Volcanol. Geotherm. Res.* **1998**, *80*, 1–16. [CrossRef]

16. Hurst, A.W.; Turner, R. Performance of the program ASHFALL for forecasting ashfall during the 1995 and 1996 eruptions of Ruapehu volcano. *N. Z. J. Geol. Geophys.* **1999**, *42*, 615–622. [CrossRef]

17. Koyaguchi, T.; Ohno, M. Reconstruction of Eruption Column Dynamics on the Basis of Grain Size of Tephra Fall Deposits: 1. Methods. *J. Geophys. Res.* **2001**, *106*, 6499–6512. [CrossRef]

18. Bonadonna, C.; Macedonio, G.; Sparks, R.S.J. Numerical modelling of tephra fallout associated with dome collapses and Vulcanian explosions: Application to hazard assessment on Montserrat. *Geol. Soc. Lond. Mem.* **2002**, *21*, 517–537. [CrossRef]

19. Bonadonna, C.; Philips, J.C. Sedimentation from strong volcanic plumes. *J. Geophys. Res.* **2003**, *108*, 2340–2368. [CrossRef]

20. Bonadonna, C.; Philips, J.C.; Houghton, B.F. Modeling tephra sedimentation from a Rapehau weak plume eruption. *J. Geophys. Res.* **2005**, *110*, B08209. [CrossRef]

21. Bonadonna, C.; Folch, A.; Loughlin, S.; Puempel, H. Future developments in modelling and monitoring of volcanic ash clouds: Outcomes from the first IAVCEI-WMO workshop on Ash Dispersal Forecast and Civil Aviation. *Bull. Volcanol.* **2012**, *74*, 1–10. [CrossRef]

22. Folch, A. A review of tephra transport and dispersal models: Evolution, current status, and future perspectives. *J. Volcanol. Geotherm. Res.* **2012**, *235–236*, 96–115. [CrossRef]

23. Scollo, S.; Prestifilippo, M.; Bonadonna, C.; Cioni, R.; Corradini, S.; Degruyter, W.; Rossi, E.; Silvestri, M.; Biale, E.; Carparelli, G.; et al. Near-Real-Time Tephra Fallout Assessment at Mt. Etna, Italy. *Remote Sens.* **2019**, *11*, 2987. [CrossRef]

24. Beckett, F.; Witham, C.; Leadbetter, S.; Crocker, R.; Webster, H.; Hort, M.C.; Jones, A.; Devenish, B.; Thomson, D. Atmospheric Dispersion Modelling at the London VAAC: A Review of Developments since the 2010 Eyjafjallajökull Volcano Ash Cloud. *Atmosphere* **2020**, *11*, 352. [CrossRef]

25. Dioguardi, F.; Beckett, F.; Dürig, T.; Stevenson, J.A. The impact of eruption source parameter uncertainties on ash dispersion forecasts during explosive volcanic eruptions. *J. Geophys. Res. Atmos.* **2020**, *125*, e2020JD032717. [CrossRef]

26. Mastin, L.G.; Pavolonis, M.; Engwell, S.; Clarkson, R.; Witham, C.; Brock, G.; Lisk, I.; Guffanti, M.; Tupper, A.; Schneider, D.; et al. Progress in protecting air travel from volcanic ash clouds. *Bull. Volcanol.* **2022**, *84*, 9. [CrossRef]

27. Bear-Crozier, A.; Pouget, S.; Bursik, M.; Jansons, E.; Denman, J.; Tupper, A.; Rustowicz, R. Automated detection and measurement of volcanic cloud growth: Towards a robust estimate of mass flux, mass loading and eruption duration. *Nat. Hazards* **2020**, *101*, 1–38. [CrossRef]

28. Freret-Lorgeril, V.; Bonadonna, C.; Corradini, S.; Guerrieri, L.; Lemus, J.; Donnadieu, F.; Scollo, S.; Gurioli, L.; Rossi, E. Tephra characterization and multi-disciplinary determination of Eruptive Source Parameters of a weak paroxysm at Mount Etna (Italy). *J. Volcanol. Geotherm. Res.* **2021**, *421*, 10743. [CrossRef]

29. Mereu, L.; Scollo, S.; Bonadonna, C.; Donnadieu, F.; Freret Lorgeril, V.; Marzano, F.S. Ground-based remote sensing and uncertainty analysis of the mass eruption rate associated with the 3–5 December 2015 paroxysm of Mt. Etna. In Proceedings of the EGU General Assembly 2022, Vienna, Austria, 23–27 May 2022. EGU22-9221. [CrossRef]

30. Mereu, L.; Scollo, S.; Garcia, A.; Sandri, L.; Bonadonna, C.; Marzano, F.S. A New Radar-Based Statistical Model to Quantify Mass Eruption Rate of Volcanic Plumes. *Geophys. Lett.* **2023**, *50*, e2022GL100596. [CrossRef]

31. Calvari, S.; Bonaccorso, A.; Ganci, G. Anatomy of a Paroxysmal Lava Fountain at Etna Volcano: The Case of the 12 March 2021, Episode. *Remote Sens.* **2021**, *13*, 3052. [CrossRef]

32. Slatcher, N.; James, M.R.; Calvari, S.; Ganci, G.; Browning, J. Quantifying Effusion Rates at Active Volcanoes through Integrated Time-Lapse Laser Scanning and Photography. *Remote Sens.* **2015**, *7*, 14967–14987. [CrossRef]

33. Ganci, G.; Cappello, A.; Bilotta, G.; Herault, A.; Zago, V.; Del Negro, C. Mapping Volcanic Deposits of the 2011–2015 Etna Eruptive Events Using Satellite Remote Sensing. *Front. Earth Sci.* **2015**, *6*, 83. [CrossRef]

34. Scollo, S.; Boselli, A.; Coltelli, M.; Leto, G.; Pisani, G.; Prestifilippo, M.; Spinelli, N.; Wang, X. Volcanic ash concentration during the 12 August 2011 Etna eruption. *Geophys. Res. Lett.* **2015**, *42*, 2634–2641. [CrossRef]

35. Frazier, A.; Hemingway, B.L. A Technical Review of Planet Smallsat Data: Practical Considerations for Processing and Using PlanetScope Imagery. *Remote Sens.* **2021**, *13*, 1390. [CrossRef]

36. Aldeghi, A.; Carn, S.; Rudiger, E.-W.; Groppelli, G. Volcano Monitoring from Space Using High-Cadence Planet CubeSat Images Applied to Fuego Volcano, Guatemala. *Remote Sens.* **2019**, *11*, 2151. [CrossRef]

37. Salamati, N.; Fredembach, C.; Süsstrunk, S. Material Classification Using Color and NIR Images. In Proceedings of the IS&T/SID 17th Color Imaging Conference (CIC) 2009, Albuquerque, NM, USA, 11–13 November 2009.

38. Zuccarello, F.; Bilotta, G.; Ganci, G.; Proietti, C.; Cappello, A. Assessing impending hazards from summit eruptions: The new probabilistic map for lava flow inundation at Mt. Etna. *Sci. Rep.* **2023**, *13*, 19543. [CrossRef] [PubMed]

39. Del Negro, C.; Cappello, A.; Bilotta, G.; Ganci, G.; Hérault, A.; Zago, V. Living at the edge of an active volcano: Risk from lava flows on Mt. Etna. *GSA Bull.* **2019**, *132*, 1615–1625. [CrossRef]

40. Palano, M.; Viccaro, M.; Zuccarello, F.; Gresta, S. Magma transport and storage at Mt. Etna (Italy): A review of geodetic and petrological data for the 2002–03, 2004 and 2006 eruptions. *J. Volcanol. Geotherm. Res.* **2017**, *347*, 149–164. [CrossRef]

41. Andronico, D.; Corsaro, R.A. Lava Fountains during the Episodic Eruption of South-East Crater (Mt. Etna), 2000: Insights into Magma-Gas Dynamics within the Shallow Volcano Plumbing System. *Bull. Volcanol.* **2011**, *73*, 1165–1178. [CrossRef]

42. Andronico, D.; Cannata, A.; Di Grazia, G.; Ferrari, F. The 1986–2021 paroxysmal episodes at the summit craters of Mt. Etna: Insights into volcano dynamics and hazard. *Earth-Sci. Rev.* **2021**, *220*, 103686.
43. Parfitt, E.A.; Wilson, L. The 1983-86 Pu'u 'O'o Eruption of Kilauea Volcano, Hawaii: A Study of dike Geometry and Eruption Mechanisms for a Long-Lived Eruption. *J. Volcanol. Geotherm. Res.* **1994**, *59*, 179–205. [CrossRef]
44. Allard, P.; Burton, M.; Muré, F. Spectroscopic evidence for a lava fountain driven by previously accumulated magmatic gas. *Nature* **2005**, *433*, 407–410. [CrossRef]
45. Alparone, S.; Cannata, A.; Gresta, S. Time variation of spectral and wavefield features of volcanic tremor at Mt. Etna (January June 1999). *J. Volcanol. Geotherm. Res.* **2007**, *161*, 318–332. [CrossRef]
46. Scollo, S.; Prestifilippo, M.; Pecora, E.; Corradini, S.; Merucci, L.; Spata, G.; Coltelli, M. Eruption column height estimation of the 2011–2013 Etna lava fountains. *Ann. Geophys.* **2014**, *57*, 1–5. [CrossRef]
47. Calvari, S.; Nunnari, G. Comparison between Automated and Manual Detection of Lava Fountains from Fixed Monitoring Thermal Cameras at Etna Volcano, Italy. *Remote Sens.* **2022**, *14*, 2392. [CrossRef]
48. Wadge, G.; Walker, G.P.L.; Guest, J.E. The output of the Etna volcano. *Nature* **1975**, *255*, 385–387. [CrossRef]
49. Guest, J.E. Styles of eruption and flow morphology on Mt. Etna. *Mem. Soc. Geol. Ital.* **1982**, *23*, 49–73.
50. Romano, R.; Sturiale, C. The historical eruptions of Mt Etna (Volcanological data). *Mem. Soc. Geol. It.* **1982**, *23*, 75–97.
51. Ganci, G.; Bilotta, G.; Zuccarello, F.; Calvari, S.; Cappello, A. A Multi-Sensor Satellite Approach to Characterize the Volcanic Deposits Emitted during Etna's Lava Fountaining: The 2020–2022 Study Case. *Remote Sens.* **2023**, *15*, 916. [CrossRef]
52. Marchese, F.; Filizzola, C.; Lacava, T.; Falconieri, A.; Faruolo, M.; Genzano, N.; Mazzeo, G.; Pietrapertosa, C.; Pergola, N.; Tramutoli, V.; et al. Mt. Etna Paroxysms of February–April 2021 Monitored and Quantified through a Multi-Platform Satellite Observing System. *Remote Sens.* **2021**, *13*, 3074. [CrossRef]
53. Corradini, S.; Guerrieri, L.; Lombardo, V.; Merucci, L.; Musacchio, M.; Prestifilippo, M.; Scollo, S.; Silvestri, M.; Spata, G.; Stelitano, D. Proximal monitoring of the 2011-2015 Etna lava fountains using MSG-SEVIRI data. *Geosciences* **2018**, *8*, 140. [CrossRef]
54. Plank, S. Rapid Damage Assessment by Means of Multi-Temporal SAR—A Comprehensive Review and Outlook to Sentinel-1. *Remote Sens.* **2014**, *6*, 4870–4906. [CrossRef]
55. Ramsey, M.S.; Harris, A.J.L. Volcanology 2020: How will thermal remote sensing of volcanic surface activity evolve over the next decade? *J. Volcanol. Geotherm. Res.* **2013**, *249*, 217–233. [CrossRef]
56. Scollo, S.; Coltelli, M.; Prodi, F.; Folegani, M.; Natali, S. Terminal settling velocity measurements of volcanic ash during the 2002–2003 Etna eruption by an X-band microwave rain gauge disdrometer. *Geophys. Res. Lett.* **2005**, *32*, L10302. [CrossRef]
57. Bonadonna, C.; Genco, R.; Gouhier, M.; Pistolesi, M.; Cioni, R.; Alfano, F.; Hoskuldsson, A.; Ripepe, M. Tephra sedimentation during the 2010 Eyjafjallajökull eruption (Iceland) from deposit, radar, and satellite observations. *J. Geophys. Res.* **2011**, *116*, B12202. [CrossRef]
58. Kozono, T.; Iguchi, M.; Miwa, T.; Miki, D. Characteristics of tephra fall from eruptions at Sakurajima volcano, revealed by optical disdrometer measurements. *Bull. Volcanol.* **2019**, *81*, 41. [CrossRef]

Article

An Intelligent System-Based Coffee Plant Leaf Disease Recognition Using Deep Learning Techniques on Rwandan Arabica Dataset

Eric Hitimana [1,*], Omar Janvier Sinayobye [1], J. Chrisostome Ufitinema [2], Jane Mukamugema [2], Peter Rwibasira [2], Theoneste Murangira [3], Emmanuel Masabo [1], Lucy Cherono Chepkwony [4], Marie Cynthia Abijuru Kamikazi [1], Jeanne Aline Ukundiwabo Uwera [1], Simon Martin Mvuyekure [5], Gaurav Bajpai [6] and Jackson Ngabonziza [7]

[1] Department of Computer and Software Engineering, University of Rwanda, Kigali P.O. Box 3900, Rwanda; sijaom2@gmail.com (O.J.S.); masabem@gmail.com (E.M.); abijurucyn@gmail.com (M.C.A.K.); alineuwera00@gmail.com (J.A.U.U.)
[2] Department of Biology, University of Rwanda, Kigali P.O. Box 3900, Rwanda; jufitine@gmail.com (J.C.U.); jmugema@gmail.com (J.M.); rwibasira@gmail.com (P.R.)
[3] Department of Computer Science, University of Rwanda, Kigali P.O. Box 2285, Rwanda; tmurangira@gmail.com
[4] African Center of Excellence in Data Science, University of Rwanda, Kigali P.O. Box 4285, Rwanda; luciejs@gmail.com
[5] Rwanda Agriculture Board, Kicukiro District, Rubilizi, Kigali P.O. Box 5016, Rwanda; msmartin202@gmail.com
[6] Directorate of Grants and Partnership, Kampala International University, Ggaba Road, Kansanga, Kampala P.O. Box 20000, Uganda; gb.bajpai@gmail.com
[7] Bank of Kigali Plc, Kigali P.O. Box 175, Rwanda; ngabojck@gmail.com
* Correspondence: e.hitimana@ur.ac.rw

Citation: Hitimana, E.; Sinayobye, O.J.; Ufitinema, J.C.; Mukamugema, J.; Rwibasira, P.; Murangira, T.; Masabo, E.; Chepkwony, L.C.; Kamikazi, M.C.A.; Uwera, J.A.U.; et al. An Intelligent System-Based Coffee Plant Leaf Disease Recognition Using Deep Learning Techniques on Rwandan Arabica Dataset. *Technologies* **2023**, *11*, 116. https://doi.org/10.3390/technologies11050116

Academic Editor: Pedro Antonio Gutiérrez

Received: 26 July 2023
Revised: 26 August 2023
Accepted: 27 August 2023
Published: 1 September 2023

Abstract: Rwandan coffee holds significant importance and immense value within the realm of agriculture, serving as a vital and valuable commodity. Additionally, coffee plays a pivotal role in generating foreign exchange for numerous developing nations. However, the coffee plant is vulnerable to pests and diseases weakening production. Farmers in cooperation with experts use manual methods to detect diseases resulting in human errors. With the rapid improvements in deep learning methods, it is possible to detect and recognize plan diseases to support crop yield improvement. Therefore, it is an essential task to develop an efficient method for intelligently detecting, identifying, and predicting coffee leaf diseases. This study aims to build the Rwandan coffee plant dataset, with the occurrence of coffee rust, miner, and red spider mites identified to be the most popular due to their geographical situations. From the collected coffee leaves dataset of 37,939 images, the preprocessing, along with modeling used five deep learning models such as InceptionV3, ResNet50, Xception, VGG16, and DenseNet. The training, validation, and testing ratio is 80%, 10%, and 10%, respectively, with a maximum of 10 epochs. The comparative analysis of the models' performances was investigated to select the best for future portable use. The experiment proved the DenseNet model to be the best with an accuracy of 99.57%. The efficiency of the suggested method is validated through an unbiased evaluation when compared to existing approaches with different metrics.

Keywords: coffee leaf diseases; arabica coffee; deep learning; VGG16; DenseNet

1. Introduction

In Rwanda, agriculture accounts for a third of the GDP (gross domestic product) and makes up most jobs (approximately 80%) [1]. Additionally, a significant source of export value, particularly from the production of tea and coffee, accounts for more than 20% of Rwanda's overall exports by value across all sectors: more than $100 million/year [2].

Coffee is a \$60 million industry in Rwanda that is primarily supplied by small-holder growers in the country's several agroecological zones. Along with the supply chain, the estimated 350,000 farmers whose livelihoods depend on growing coffee face jeopardy [3]. Therefore, the government has a top priority for the future development of this cash crop for export. Among the varieties of coffee plants in Rwanda, coffee arabica is the one shown promising resistance to climate change.

Small-scale farmers are primarily responsible for cultivating coffee, utilizing farming methods that involve fragmented land and numerous small plots spread across hilly areas. Typically, farmers own around two to six plots, depending on the number of coffee trees in each plot. Due to the scattered nature of these plots and the distance between them and the farmers' homes, the frequency of plant and land management activities is reduced. In addition, the mix-up of different crops with coffee along with separate small farms contributes to the spread of coffee leaf diseases. To rearrange land usage patterns, the Ministry of Agriculture and Animal Resources is executing a policy for land consolidation. Apart from the land management policies, the local farmers working unprofessionally are recommended to work cooperatively. This exercise helps them to get support from government agencies, such as training, and other inputs impacting the high quality of coffee production [3].

It has been reported that one of the crops at risk from climate change and the spread of disease/pest infections is coffee [4]. Furthermore, these circumstances arise from a variety of fungal species and other causes. The disease-causing agents, present on the leaves or other parts of the tree, are highly transmissible and can rapidly spread if not promptly addressed. According to the study, approximately 10% of the global plant economy is currently being impacted by the destructive consequences of plant infections and infestations [5].

Coffee farmers in Rwanda, like those in other regions, face continuous threats from various pests and diseases [6]. While some of these problems are minor and have a limited impact on crop yield and quality, others, such as coffee berry disease, coffee leaf rust, and coffee wilt disease (tracheomycosis), pose significant dangers. These serious diseases can not only affect individual farmers, but also have a major economic impact on countries or regions heavily reliant on coffee for foreign exchange earnings [7]. For instance, coffee wilt disease has been present in Africa since the 1920s, but since the 1990s, there have been widespread and recurring outbreaks. This results in substantial losses in countries such as Uganda, where over 14 million coffee trees have been destroyed, as well as in the Democratic Republic of Congo [8,9]. Once this disease takes hold on a farm, it becomes extremely challenging to control. Since coffee is a perennial crop, certain pests and diseases can survive and multiply throughout the growing season, continuously affecting the coffee plants, although their populations and impact may vary over time [10]. Other pests and diseases may only attack coffee during periods when conditions are favorable. Regardless, the damage caused by these pests and diseases can be significant, affecting both crop yield and quality [11].

Some pests and diseases, such as the white coffee stem borer, coffee wilt disease, parasitic nematodes, and root mealy bugs, could kill coffee plants outright. On the other hand, pests, such as the coffee berry borer, green scales, leaf rust, and brown eye spot, may not directly kill the plants but can severely hinder their growth by causing defoliation, ultimately impacting the quality of the coffee berries [12].

The process of diagnosing plant diseases is complex and entails tasks such as analyzing symptoms, recognizing patterns, and conducting various tests on leaves. These procedures require significant time, resources, and skills to complete [13]. In many instances, an incorrect diagnosis can result in plants developing immunity or reduced susceptibility to treatment. The intricacy of plant disease diagnosis has led to a decrease in both the quantity and quality of crop yields among farmers [14].

The drawn-out process frequently results in a widespread infection with significant losses [15]. Coffee is one of the most well-known drinks in the world and might go extinct without conservation, monitoring, and seed preservation measures, according to scientists. Global warming, deforestation, illness, and pests are all factors in the decline [16]. By implementing effective crop protection systems, early monitoring and accurate diagnosis of crop diseases can be achieved, which, in turn, can help prevent losses in production quality.

Recognizing various types of coffee plant diseases is of utmost significance and is deemed a critical concern. Timely detection of coffee plant diseases can lead to improved decision-making in agricultural production management. Infected coffee plants typically exhibit noticeable marks or spots on their stems, fruits, leaves, or flowers. Importantly, each infection and pest infestation leaf have distinct patterns that can be utilized for diagnosing abnormalities. The identification of plant diseases necessitates expertise and human resources. Moreover, the process of manually examining and identifying the type of plant infection is subjective and time-consuming. Additionally, there is a possibility that the disease identified by farmers or experts could be misleading at times [17]. As a result, the use of an inappropriate pesticide or treatment might occur during the evaluation of plant diseases, ultimately leading to a decline in crop quality and potentially causing environmental pollution.

The application of computer vision and artificial intelligence (AI) technologies has been expressed as instrumental tools in combating plant diseases [18–20]. There are multiple methods available to address the problem of detecting plant infections with the help of technologies, as the initial signs of infection manifest as various spots and patterns on leaves [21]. The introduction of machine learning and deep learning techniques has led to significant advancements in plant disease recognition, revolutionizing research in this field. These techniques have facilitated automatic classification and feature extraction, enabling the representation of original image characteristics. Moreover, the availability of datasets, GPU machines, and software supporting complex deep learning architectures with reduced complexity has made the transition from traditional methods to deep learning platforms feasible. CNNs have particularly gained widespread attention due to their remarkable capabilities in recognition and classification. CNNs excel in extracting intricate low-level features from images, making them a preferred choice for replacing traditional methods in automated plant disease recognition and yielding improved outcomes [22].

The research problem is based on the numerous efforts of government agencies and farmers in the use of manual methods to detect coffee diseases. In addition, a huge monetary effort is used to train farmers in coffee disease identification. However, the trained methods result in wrong findings [23]. To remedy the detected diseases, they may use the wrong pesticides, which do not treat the matter but affect environmental degradation.

This study aimed to develop and train five deep learning models on the collected dataset of coffee arabica leaves and determine the best model yielding the best results by leveraging pre-trained models and transferring knowledge approaches. The objective was to identify the most effective transfer learning technique for achieving accurate classification and optimal recognition accuracy in a multi-class coffee leaf disease context. The main contributions of this study are (1) to assess, collect, and classify the coffee leaves dataset in the Rwandan context; (2) to apply different data preprocessing techniques on the labeled data set; and (3) to determine the best transfer learning technique for achieving the most accurate classification and optimal recognition on multi-class plant diseases.

The remaining sections of the paper are structured as follows. Section 2 details the related works of this research. Section 3 outlines the materials and methods employed in this study. The findings and results are presented in Section 4. Section 5 delves into the discussion of the various experiments conducted. Finally, in Section 6, the research concludes by summarizing the key points and outlining potential future directions for research.

2. Related Works

Several methods have been suggested by researchers to achieve the precise detection and classification of plant infections. Some of these methods employ conventional image processing techniques that involve manual feature extraction and segmentation [24].

Among many methods, the use of K-means clustering for image leaf segmentation by extracting infected regions and later performing classification using a multi-class support vector machine is investigated [25]. The probabilistic neural network method was used to extract methodologies with statistical features on cucumber plant infection [26]. The preprocessing of images, from red, green, and blue (RGB) conversion to gray; HE; K-means clustering; and contour tracing is computed, and the results are used for classifications using support vector machine (SVM), K-NN, and convolutional neural networks (CNN). The experiment was carried out on tomato leaf infection detection [27] and grapes [28]. The automatic detection of leaf damage on coffee leaves has been conducted using image segmentation with Fuzzy C-means clustering applied to the V channel of the YUV color space image [29]. The automatic identification and classification of plant diseases and pests as well as the severity assessment, specifically focusing on coffee leaves in Brazil, is investigated. They targeted two specific issues: leaf rust caused by Hemileia vastatrix and leaf miner caused by Leucoptera coffee. Various image processing techniques were employed, including image segmentation using the K-means algorithm, the Otsu method, and the iterative threshold method, performed in the YCgCr color space. Texture and color attributes were calculated for feature extraction. For classification purposes, an artificial neural network trained with backpropagation and an extreme learning machine was utilized. The images utilized were captured using an ASUS Zenfone 2 smartphone (ZE551ML) with a resolution of 10 Megapixels (4096 × 2304 pixels). The database used in the study consisted of 690 images [30].

Moreover, the existing models heavily depend on manual feature engineering techniques, classification methods, and spot segmentation. However, with the advent of artificial intelligence in the field of computer vision, researchers have increasingly utilized machine learning [31] and deep learning [32] models to improve recognition accuracy significantly.

A CNN-based predictive model for classification and image processing in paddy plants is proposed [33]. Similarly, the utilization of a CNN for disease detection in paddy fields using convolutional neural networks with four to six layers to classify various plant species is elaborated on [34]. The application of CNN with a transfer learning approach to classify, recognize, and segment different plant diseases is tested [35]. Although CNNs have been extensively used with promising results, there is a lack of diversity in the datasets employed [36]. To achieve the best outcomes, training deep learning models with larger and more diverse datasets is crucial. While previous studies have demonstrated significant achievements, there is still room for improvement in terms of dataset diversity, particularly in capturing realistic images from actual agricultural fields with diverse backgrounds.

Deep-learning models based on CNNs have gained popularity in image-based research due to their effectiveness in learning intricate low-level features from images. However, training deep CNN layers can be computationally intensive and challenging. To address these issues, researchers have proposed transfer learning-based models [37–39]. These models leverage pre-trained networks, such as VGG-16, ResNet, DenseNet, and Inception [40], which have been well-established and widely used in the field. Transfer learning allows for the models to leverage the knowledge gained from pre-training on large datasets, enabling faster and more efficient training on specific image classification tasks.

The focus of the automatic and accurate estimation of disease severity to address concerns related to food security, disease management, and yield loss prediction was investigated on beans [41]. They applied deep learning techniques to analyze images of Apple black rot from the Plant Village dataset, which is caused by the fungus Botryosphaeria obtusa. The study compared the performance of different deep learning models, including VGG16, VGG19, Inception-v3, and ResNet50. The results demonstrated that the deep VGG16 model, trained with transfer learning, achieved the highest accuracy of 90.5% on the hold-out test set.

The classification of cotton leaves based on leaf hairiness (pubescence) used a four-part deep learning model named HairNet. HairNet demonstrated impressive performance, achieving 89% accuracy per image and 95% accuracy per leaf. Furthermore, the model successfully classified the cotton based on leaf hairiness, achieving an accuracy range of 86–99% [42]. A deep learning approach was developed to automate the classification of diseases in banana leaves. The researchers utilized the LeNet architecture, a CNN through a 3700 image dataset. The implementation of the approach utilized deeplearning4j, an open-source deep-learning library that supports GPUs. The experiment was applied to detect two well-known banana diseases, namely Sigatoka and Speckle [35].

The application of emerging technologies, such as image processing, machine learning, and deep learning in the agriculture sector, is transforming the industry, leading to increased productivity, sustainability, and profitability while reducing environmental impact. A lot of authors have investigated different algorithms for different or specific plant types to ensure common solutions; however, the solution is problem-specific. It has been observed that most of the modeling has been attempted on the Plant Village dataset [43] to check the performance of the models selected.

Table 1 showcases different methods used for plant leaf classification, along with the corresponding accuracy percentages achieved on different types of leaves. The "Proposed model" refers to DenseNet, which obtained an accuracy of 99.57% on coffee leaf classification.

Table 1. Comparison of our resulting model with existing deep learning models.

Ref. No and Year	Method	Accuracy (%)	Plant Name
[44]—2021	Proposed FCNN & SCNN Hybrid Principal	92.01	Crop Leaf
[45]—2021	Component Analysis	95.10	Plant Leaf
[46]—2020	Hybris PCA & Optimization Algorithm	90.20	Olive Leaf
[47]—2020	ResNet50	99.00	Okra Leaf
[48]—2020	Deep CNN	98.00	Coffee Leaf
[49]—2022	Deep Transfer EffientNet	98.70	Grape Leaf
Proposed model	DenseNet	99.57	Coffee Leaf

3. Materials and Methods

For proper plant disease management, early detection of diseases in coffee leaves is required to facilitate farmers. This section provides a complete description of the methodology used to collect coffee leaves and the methods used to experiment with the modeling techniques. Discussion of the process to collect leaves and several transfer-learning algorithms have been elaborated on to investigate the best model responding to the research scope. The architecture and training process of each model with the experimental setup on the used dataset is also discussed.

Rwanda has many high mountains and steep-sloped hills, with much of the farmland suffering from moderate to severe soil erosion, and the appearance of coffee diseases and pest are based on climate variability [50]. Among different types of coffee plants, such as arabica and robusta [51], this study focuses on the most popular variety known as arabica [52] that exists in Rwanda. We surveyed and visited 10 coffee washing stations located in different 5 districts, such as Ngoma, Rulindo, Gicumbi, Rutsiro, and Huye. The districts selected represent all 27 districts of Rwanda caring about climate variations [50]. In each district, we sampled 30 farmers giving 150 sample sizes. The visit was done to cooperate with agronomists who know the coffee pests and diseases to support coffee leaf labeling activities and to engage farmers to assess if they have the capacity to identify different coffee leaf diseases. The visit was attempted in the harvesting session, which is in March 2021, and in the summer session, which is in June and July 2021. The dataset images were collected from four distinct provinces located in the Eastern (sunny region with high low altitude with no hills), Northern (the cold region with high altitude), Southern (the cold region with modulated altitude), and Western (cold, highlands with high altitude). The quantitative and qualitative methodology was adopted to investigate the disease occurrence distribution in Rwanda as shown in Figure 1.

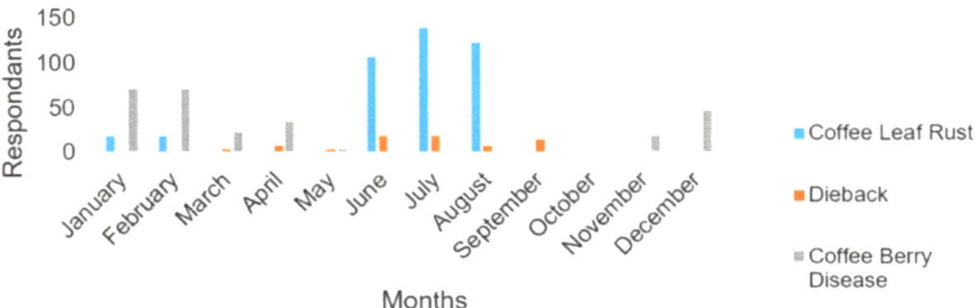

Figure 1. Knowledge of coffee diseases alongside disease occurrence in months.

According to our respondents, coffee leaf rust, known as "coffee leaf rust", is the most dangerous disease ravaging coffee in Rwanda. As shown in Figure 1, the disease occurs mostly in June, July, and August.

The process of data collection was followed by the experiment of coffee disease detection using deep learning techniques. Figure 2 details the architectural flow of the implementation.

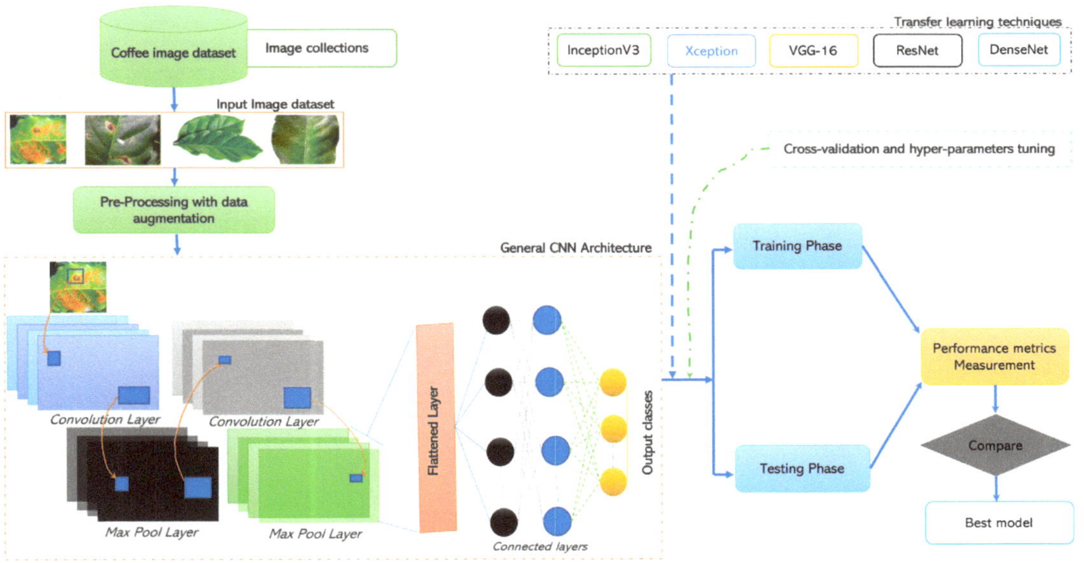

Figure 2. Proposed architectural implementation flow.

The suggested pipeline for detecting coffee leaf diseases begins with preparing the dataset and concludes with making predictions using different models and comparison analysis. To accomplish this, the Python 3.10 programming language, TensorFlow 2.9.1, numpy version 1.19.2, and matplotlib version 3.5.2 libraries were employed for dataset preparation and development environment setup. Those tools have proven to be useful for data preprocessing and modeling purposes [53,54]. The experiment used CNN deep learning models, such as InceptionV3, Resnet50, VGG16, Xception, and DenseNet models. The experiment used infrastructure with an HP Z240 workstation equipped with two Intel(R) Xeon(R) Gold 6226R and Tesla V100s 32GB memory NVIDIA GPU of 64 cores in total, which significantly accelerated the training process of deep neural networks. In the subsequent sections, each stage of the proposed coffee plant leaf disease detection pipeline will be thoroughly discussed.

3.1. Dataset

The researchers collected 37,939 images dataset in RGB format. The coffee images had at least four classes in the dataset, namely the class rust, red spider mite, miner, and healthy. The dataset's classes were made up of these directories, each of which corresponded to a certain disease.

Figure 3 shows the details of the sample dataset classes used in the experiment. Due to the severity of the matter, in a specific class, you may find different images with similar infections at different stages. This is because, at a certain stage, the model can be able to track and classify the real name or approximate name of the diseases.

Figure 3. Few sampled coffee leaf image datasets. (**a**) Rust infection; (**b**) Red spider mites' infection in different stages; (**c**) Health leaf; (**d**) Miner infections.

Before supplying the images from the dataset to the CNN architectures, we preprocessed them to make sure the input parameters matched the requirements of the CNN model. Each input image was downsized to 224×224 dimensions after preprocessing. To guarantee that all the data were described under the same distribution, normalization (i.e., image/255.0) was then applied, which improved training convergence and stability [55].

3.2. Used Deep Learning Models

In the following section, this study details all the different models and tools used. The modeling of the coffee leaf images was conducted using different deep-learning techniques, such as InceptionV3, Resnet50, VGG16, Xception, and DenseNet as shown in Figure 2.

3.2.1. InceptionV3

InceptionV3, developed by Google Research, belongs to the Inception model series, and serves as a deep convolutional neural network structure. Its primary purpose is to facilitate image recognition and classification assignments [56–58].

Its architecture is known for its deep structure and the use of Inception modules. These modules consist of parallel convolutional layers with different filter sizes, allowing the network to capture features at multiple scales. By incorporating these parallel branches, the model can effectively handle both local and global features in an image [59]. One of the key innovations in InceptionV3 is the use of 1×1 convolutions, which serve as bottleneck layers. These 1×1 convolutions help reduce the number of input channels and computational complexity, making the network more efficient.

The Inception V3 model consists of a total of 42 layers, surpassing the layer count of its predecessors, Inception V1 and V2. Nonetheless, the efficiency of this model is remarkable [60]. It can be fine-tuned on specific datasets or used as a feature extractor in transfer learning scenarios, where the pre-trained weights are utilized to extract meaningful features from images and train a smaller classifier on top of them. With its powerful deep learning architecture that excels in image recognition and classification tasks, this model was selected in this study to investigate its performance.

3.2.2. ResNet50

This research experiment suggested the use of ResNet-50 as Residual Network-50 introduced by Microsoft Research [61]. It is a variant of the ResNet family of models, which are renowned for their ability to train very deep neural networks by mitigating the vanishing gradient problem. It is known for its residual connection enabling the network to learn residual mappings instead of directly learning the desired underlying mapping. The residual connections facilitate passing information from earlier layers directly to later layers, helping to alleviate the degradation problem caused by increasing network depth.

The ResNet-50 architecture consists of 50 layers, including convolutional layers, pooling layers, fully connected layers, and shortcut connections. It follows a modular structure, where residual blocks with varying numbers of convolutional layers are stacked together [62]. Each residual block includes a set of convolutional layers, followed by batch normalization and activation functions, with the addition of the original input to the block. This ensures that the gradient flows through the skip connections and facilitates the learning of residual mappings.

The model was applied to plant disease detection [63,64] by extracting contextual dependencies within images, focusing on essential features of disease identification. The method was chosen to take advantage of its learning of residual mappings and feed the model with the coffee image classes and their features. The pre-training enables the model to learn generic visual features that can be transferred to different image-related tasks.

3.2.3. VGG16

The Visual Geometry Group 16 (VGG16) is a convolutional neural network architecture developed by the Visual Geometry Group at the University of Oxford. It is known for its simplicity and effectiveness in the image classification tasks model [65].

The VGG16 architecture consists of 16 layers, including 13 convolutional layers and 3 fully connected layers. It follows a sequential structure, where convolutional layers are stacked together with max pooling layers to progressively extract features from input images. The convolutional layers use small 3×3 filters, which help capture local patterns and details in the images [66]. The architecture maintains a consistent configuration throughout the network, with the number of filters increasing as the spatial dimensions decrease. This uniformity simplifies the implementation and enables the straightforward transfer of learned weights to different tasks [67].

The pre-training model of VGG16 enables the model to learn general visual representations, fine-tuned or used as feature extractors for specific tasks. Its deep structure and small receptive field have been considered in this research context to capture hierarchical features in coffee leaf images and avail all possible found classes.

3.2.4. Xception

Detailed as Extreme Inception, a deep convolutional neural network architecture introduced by François Chollet, the creator of Keras [68]. The model is based on the Inception architecture but incorporates key modifications to improve its performance and efficiency. Its architecture aims to enhance the depth-wise separable convolutions introduced in Inception modules. In depth-wise separable convolutions, the spatial convolution and channel-wise convolution are decoupled, reducing the number of parameters and computational complexity.

The architecture of Xception introduces the notation of an extreme version of Inception, where the traditional convolutional layer is replaced by a depth-wise separable convolution.

The extreme version of the Inception module enables it to capture spatial and channel-wise information more effectively. Xception has been pre-trained on large-scale image classification datasets, such as ImageNet, and has demonstrated impressive performance in various computer vision tasks [69].

It is used as a feature extractor or fine-tuned on specific datasets, enabling it to generalize well to various image-related tasks.

3.2.5. DenseNet

Dense Convolutional Network is a deep convolutional neural network architecture known for its dense connectivity pattern and efficient parameter sharing [70]. This sharing facilitates feature reuse and gradient flow throughout the network. It uses the concept of dense blocks, where each layer is connected to every other layer in a feed-forward manner. DenseNet takes this concept further by concatenating feature maps from all previous layers. This dense connectivity pattern enables direct connections between layers at different depths, facilitating the flow of information and gradients through the network [71].

The DenseNet architecture consists of dense blocks followed by transition layers. A dense block is a series of convolutional layers, where each layer's input is concatenated with the feature maps of all preceding layers. Transition layers are used to down-sample feature maps and reduce spatial dimensions. This architecture enables the model to capture both local and global features effectively.

The operational mechanism of a dense block as shown in Figure 4, supports the subsequent layers by applying batch normalization (BN), ReLu activation, convolution, and pooling to modify the outcome. It has achieved state-of-the-art results on various image classification benchmarks. In the coffee leaves context, the DenseNet model has been used to classify the leaf based on the list of trained dataset classes.

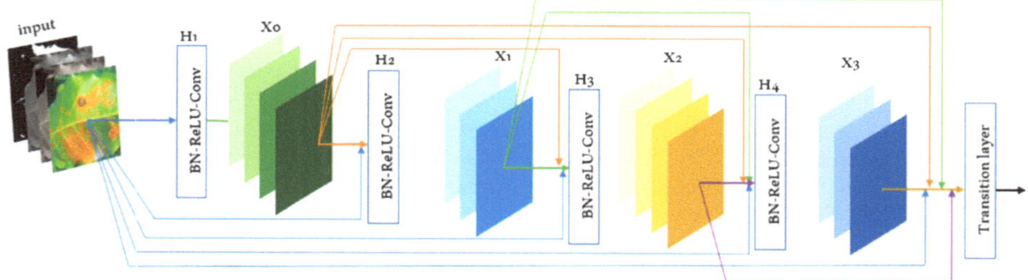

Figure 4. DenseNet Architecture.

3.3. Performance Measurement

The experimental setup has been conducted using the methodology, methods, and infrastructures discussed in the above sections. To measure the performance of the transfer learning techniques, different metrics were considered. The performance accuracy matrix, precision-recall metric, and receiver operating characteristic (ROC), with the area under the curve (AUC), are being used to evaluate segmentation performance. The performance of the classifier is measured using evaluation metrics to select the best-performing ones for further use.

3.3.1. Precision-Recall Curve

The confusion matrix is a useful tool for assessing performance by comparing actual and predicted values. It provides insights into sensitivity, which represents the true positive rate and indicates the ability to correctly identify healthy and diseased leaves. Precision–recall curves are used in binary classification to study the output of a classifier.

To extend the precision–recall curve and average precision to multi-class or multi-label classification, it was necessary to binarize the output. One curve could be drawn per label, but one could also draw a precision–recall curve by considering each element of the label indicator matrix as a binary prediction (micro-averaging).

$$\text{Precision} = \frac{\text{True positives}}{\text{True Positives} + \text{False Positives}} \tag{1}$$

$$\text{Recall} = \frac{\text{True positives}}{\text{True Positives} + \text{False Negatives}} \tag{2}$$

The performance evaluation of plant disease classification involved analyzing the output, which could be binary or multiclass. Specificity, accuracy was referred to as the positive predicted value and defined in Equation (1). Recall, also known as the probability of detection, was calculated by dividing the number of correctly classified positive outcomes by the total number of positive outcomes (Equation (2)).

3.3.2. Receiver Operating Characteristic (ROC) Curve

The curve is mainly used to understand deterministic indicators of categorization sorting and computational modeling issues. ROC curves feature true positive rate (TPR) on the Y axis and false positive rate (FPR) on the X axis. The meaning is that the top left corner of the plot is the "ideal" point—a FPR of zero and a TPR of one. This is not very realistic, but it does mean that a larger area under the curve (AUC) is usually better. The "steepness" of ROC curves is important since it is ideal to maximize the TPR while minimizing the FPR. ROC curves are typically used in binary classification, where the TPR and FPR can be defined unambiguously.

Average precision (AP) summarizes such a plot as the weighted mean of precisions achieved at each threshold, with the increase in recall from the previous threshold used as the weight:

$$AP = \sum_{n=0}^{n} (R_n - R_{n-1}) P_n \tag{3}$$

where P_n and R_n are the precision and recall at the nth threshold. A pair (R_k, P_k) is referred to as an operating point. AP and the trapezoidal area under the operating points are calculated using the function sklearn.metrics.auc of Python package to summarize a precision–recall curve that led to different results.

3.3.3. Matthews Correlation Coefficient (MCC)

As an alternate approach that is not influenced by the problem of imbalanced datasets, the Matthews correlation coefficient is a technique involving a contingency matrix. This method calculates the Pearson product-moment correlation coefficient [72] between predicted and actual values. It is expressed in Equation (4) where TP is true positive.

$$MCC = \frac{TP \times TN - FP \times FN}{\sqrt{(TP + FP) \times (TP + FN) \times (TN + FP) \times (TN + FN)}} \tag{4}$$

(Worst value: -1; best value: $+1$)

MCC stands out as the sole binary classification measure that yields a substantial score solely when the binary predictor effectively predicts most of the positive and negative data instances accurately [73]. It assumes values within the range of -1 to $+1$. The extreme values of -1 and $+1$ signify completely incorrect classification and flawless classification, respectively. Meanwhile, MCC = 0 is the anticipated outcome for a classifier akin to tossing a coin.

3.3.4. F1 Scores

Among the parametric group of F-measures, which is named after the parameter value $\beta = 1$, the F1 score holds the distinction of being the most frequently employed metric. It is determined as the harmonic average of precision and recall (refer to the formulas (1) and (2)), and its shape is expressed in the Equation (5):

$$\text{F1 score} = \frac{2 \times \text{TP}}{2 \times \text{TP} + \text{FP} + \text{FN}} \tag{5}$$

(Worst value: -1; best value: $+1$)

The F1 score spans the interval $[0, 1]$, with the lowest value achieved when TP (true positives) equals 0, signifying the misclassification of all positive samples. Conversely, the highest value emerges when FN (false negatives) and FP (false positives) both equal 0, indicating flawless classification. There are two key distinctions that set apart the F1 score from MCC and accuracy: firstly, F1 remains unaffected by TN (true negatives), and secondly, it does not exhibit symmetry when classes are swapped.

4. Results

In this study, each experiment involved evaluating the training accuracy and testing accuracy. The losses incurred during the testing and training phases were computed for every model. The collected coffee leaves dataset was utilized to train the DCNN with transfer learning models. The selected pre-trained models are ResNet-50, Inception V3, VGG-16, Xception, and DenseNet.

4.1. Description of Dataset

To conduct our experimental analysis, the dataset was partitioned into three subsets: training samples, testing samples, and validation samples. Among the coffee plant leaf disease classes, a total of 37,939 images were available and trained with a ratio of 80:10:10. Out of these, 30,053 samples were used for training, 3793 for validation, and 4093 for testing. It is important to note that all these sets, including the training, testing, and validation sets, encompassed all four classes representing coffee plant leaf diseases used in this research context.

4.2. Preprocessing and Data Augmentation

The dataset consisted of four diseases of one type of crop species (coffee arabica). For our experimental purposes, we utilized color images from the collected dataset, as it was shown that they aligned well with the transfer learning models. To ensure compatibility with different pre-trained network models that require varying input sizes, the images were downscaled to a standardized format of 256×256 pixels. For VGG-16, DenseNet-121, Xception, and ResNet-50, the input size was set to $224 \times 224 \times 3$ (height, width, and channel depth) while for Inception V3, the input shape was $299 \times 299 \times 3$.

Although the dataset contained many images, approximately 37,939, depicting various coffee leaf diseases, these images accurately represented real-life images captured by farmers using different image acquisition techniques, such as high-definition cameras and smartphones, and downloaded from the internet. Due to the substantial size of the dataset, there was a risk of overfitting. To overcome the overfitting, regularization techniques were employed, including data augmentation after preprocessing.

In order to maintain the data augmentation capabilities, this study applied several transformations to the preprocessed images. Those transformations include clockwise and anticlockwise rotation, horizontal and vertical flipping, zoom intensity, and rescaling of the original images. This technique not only prevented overfitting and reduced model loss, but also enhanced the model's robustness, resulting in improved accuracy when tested with the real-life coffee plant images.

4.3. Network Architecture Model

The selection of pre-trained network models was based on their suitability for the task of plant disease classification. Detailed information about the architecture of each model can be found in Table 2. These models employ different filter sizes to extract specific features from the feature maps. The filters play a crucial role in the process of feature extraction. Each filter, when convolved with the input, extracts distinct features, and the specific features extracted from the feature maps depend on the values assigned to the filters. This research experiment utilized the original pre-trained network models, incorporating the specific combinations of convolution layers and filter sizes employed in each model.

Table 2. Pre-trained network architecture models' parameters.

Parameters	InceptionV3	Xception	ResNet50	DenseNet	VGG16
Total layers	314	135	178	430	22
Max pool layers	4	4	1	1	5
Dense layers	2	2	2	2	2
Drop-out layers	-	-	2	-	2
Flatten layers	-	-	1	-	1
Filter size	$1 \times 1, 3 \times 3, 5 \times 5$	3×3	3×3	$3 \times 3, 1 \times 1$	3×3
Stride	2×2	2×2	2×2	2×2	1
Trainable parameters	23,905,060	22,963,756	25,689,988	8,091,204	15,244,100

Table 2 provides various parameters for different network models, including InceptionV3, Xception, ResNet50, VGG16, and DenseNet. The parameters include the total number of layers, max pool layers, dense layers, dropout layers, flatten layers, filter size, stride, and trainable parameters. These parameters are essential in understanding the architecture and complexity of each model.

In our experiment, each model was standardized with a learning rate of 0.01, a dropout rate of 2, and had four output classes for classification.

The coffee leaves dataset was divided into training, testing, and validation samples. For training the Inception V3, VGG16, ResNet50, Xception, and DenseNet models, 80% of the coffee leaf samples were utilized. Each model underwent ten epochs, and it was observed that all models started to converge with high accuracy after four epochs. The recognition accuracy of the InceptionV3 model is illustrated in Figure 5a, reaching a training accuracy of 99.34%. Figure 5b depicts the log loss of the InceptionV3 model.

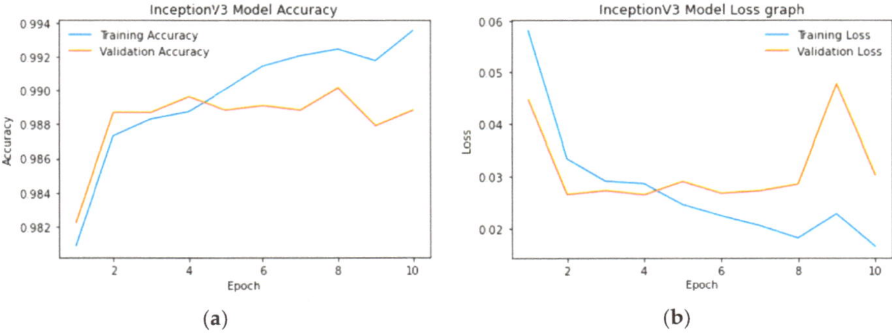

Figure 5. InceptionV3 model performance analysis using the collected dataset. (**a**) Model training and validation accuracy; (**b**) Model training and validation loss.

During this research experiment, the second model considered is the ResNet50 model from the same dataset. Following the standardization of hyperparameters, the model underwent training using 80% of the dataset. Subsequently, 10% of the samples were allocated for testing while the remaining 10% were utilized for validation and testing purposes. From Figure 6a, it can be observed that the model recognition accuracy is around 96% in the first three epochs, and therefore, its stability increased to get an accuracy of 98.70%. This performance is lower than the one represented by InceptionV3 shown in Figure 5. On the other hand, the training and validation losses of the ResNet50 model were around 0.056% and 0.057%, respectively.

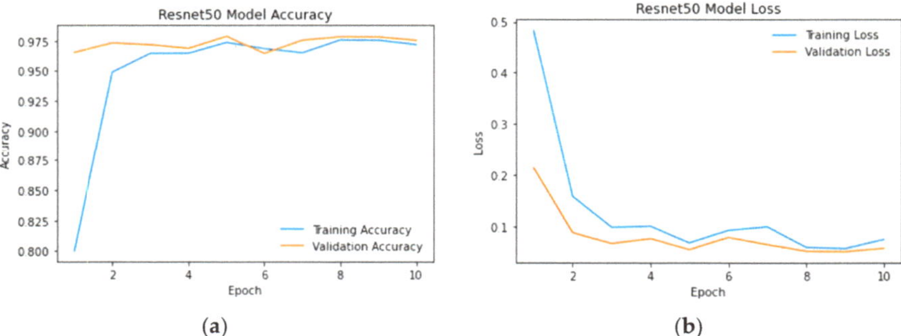

(**a**) (**b**)

Figure 6. ResNet50 model performance analysis using the collected dataset. (**a**) Model training and validation accuracy; (**b**) Model training and validation loss.

Figure 7 demonstrates the behavior of the Xception model on the used datasets after adjusting the hyperparameters. The training and validation accuracy reached 99.40% and 98.84%, respectively, with around four epochs showing less steadiness. Its training and validation losses are shown to be 0.014% and 0.033%, respectively. This execution surpasses that of what the ResNet50 demonstrated, as delineated in Figure 6.

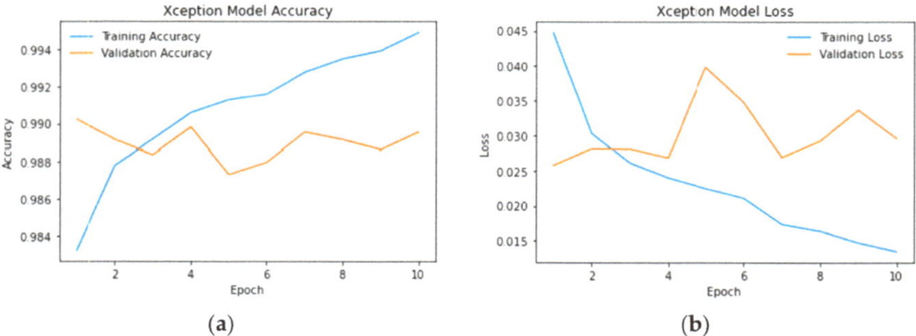

(**a**) (**b**)

Figure 7. Xception model performance analysis using the collected dataset. (**a**) Model training and validation accuracy; (**b**) Model training and validation loss.

The VGG16 model was used as the fourth model using the same dataset. After standardizing the hyperparameters, the model was trained with 80% of the dataset. Subsequently, 10% of the samples were allocated for testing while the remaining 10% were used for validation and testing purposes. By considering Figure 8a, it can be observed that the model achieved a recognition accuracy of approximately 98% in the initial four epochs, and it gradually increased to reach an accuracy of 98.81%. This performance is less than that of the Xception model, as depicted in Figure 6. Furthermore, the training and validation losses of the VGG16 model were approximately 0.0291% and 0.066%, respectively.

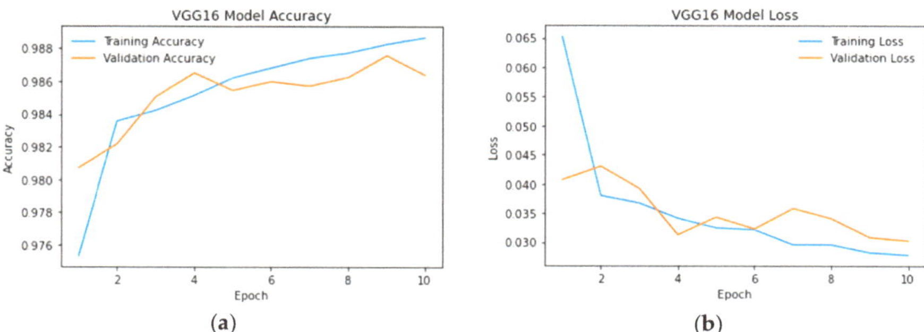

(a)   **(b)**

Figure 8. VGG16 model performance analysis using the collected dataset. (**a**) Model training and validation accuracy; (**b**) Model training and validation loss.

Figure 9 demonstrates the behavior of the DenseNet model on the used datasets after adjusting the hyperparameters. The training and validation accuracy reached 99.57% and 99.09%, respectively, with around four epochs showing less steadiness. Its training and validation losses are shown to be 0.0135% and 0.0225%, respectively. This execution surpasses that of all other demonstrated models.

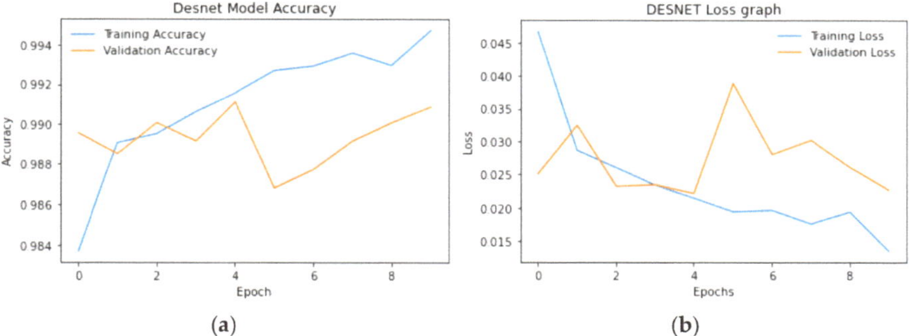

(a)   **(b)**

Figure 9. DenseNet model performance analysis using the collected dataset. (**a**) Model training and validation accuracy; (**b**) Model training and validation loss.

Figure 10 depicts the behaviors of all five used models on the collected dataset of coffee leaf diseases using Receiver Operating Characteristic (ROC) Curves. It is used to understand indicators of categorization sorting and computational modeling challenges. The curves feature true positive rate (TPR) on the Y axis and false positive rate (FPR) on the X-axis.

It illustrates how the true positive rate (the percentage of correctly classified lesion images) and false positive rate (the percentage of incorrectly classified non-lesion images) change as the classifier's threshold for distinguishing between lesions and non-lesions is adjusted while evaluating test set images.

Figure 11 illustrates the performance of the five employed models on the gathered coffee leaf diseases dataset using precision–recall curves. These curves help serve as a measure to assess the effectiveness of a classifier, especially in situations where there is a significant class imbalance. These curves depict the balance between precision, which gauges the relevance of results, and recall, which measures the comprehensiveness of the classifier's performance.

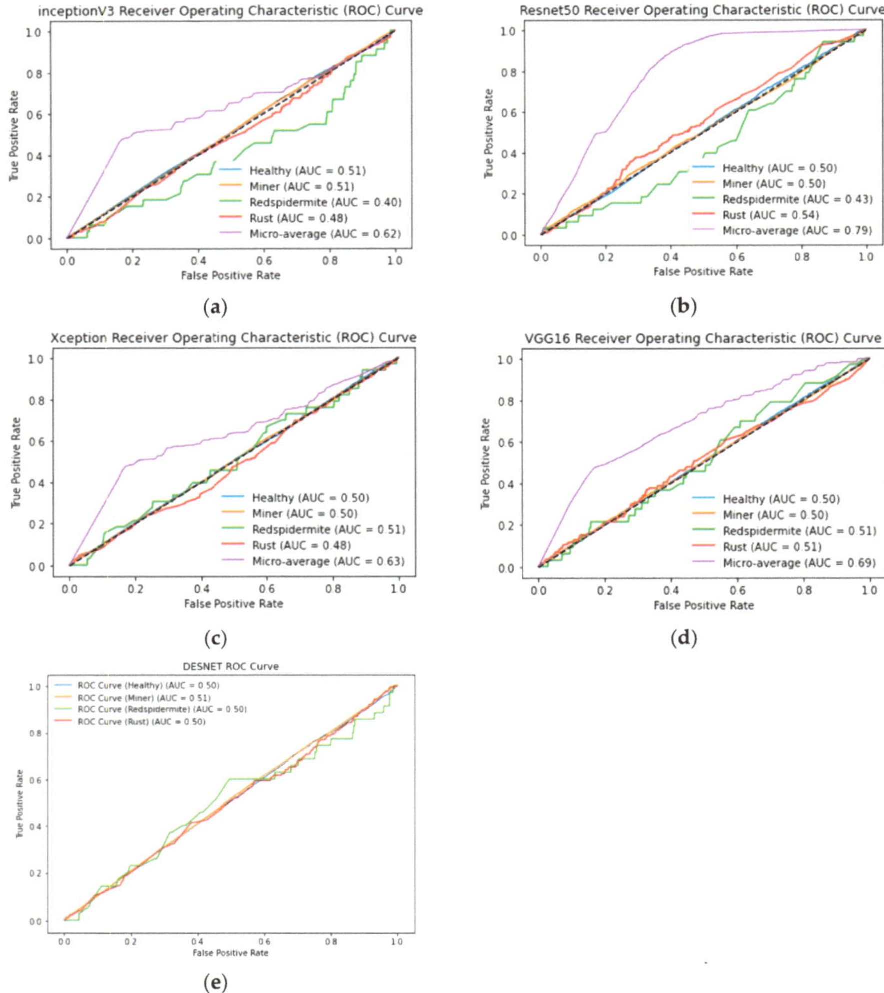

Figure 10. Receiver Operating Characteristic (ROC) Curves. (**a**) Details the behaviors of the InceptionV3 model; (**b**) Details the behaviors of the ResNet model; (**c**) Details the behaviors of the Xception model; (**d**) Details the behaviors of the VGG16 model; (**e**) Demonstrates the behaviors of the DenseNet model.

Figure 12 depicts the performance comparison of the five employed models on the gathered coffee leaf diseases dataset using F1 score and MCC metrics. The graph shows the efficiency of the DenseNet Model with an F1 score and MCC of 0.98 and 0.94, respectively. The second proven model is to be VGG16 with an F1 score and MCC of 0.9 and 0.89, respectively. The worst model on the used dataset is shown to be Xception with the F1 score and MCC of 0.48 and 0.4, respectively.

Table 3 provides a comparison of different network models based on their training and validation performance.

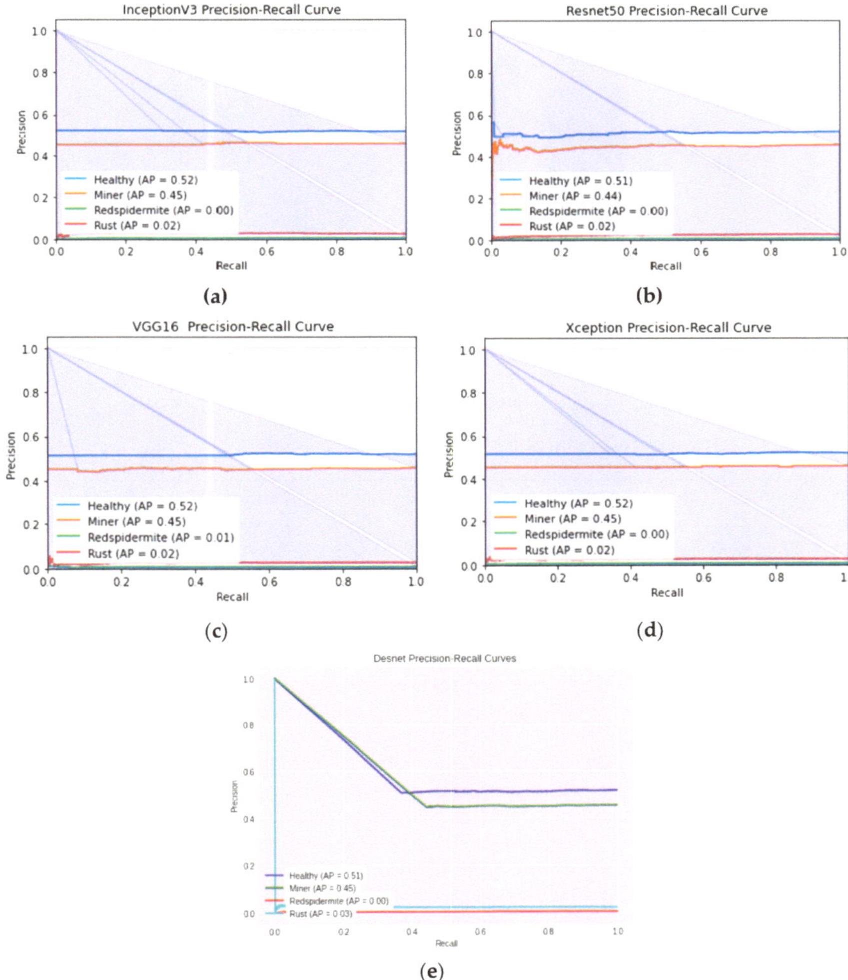

Figure 11. Precision–Recall curves of the tested models. (**a**) Details the behaviors of the InceptionV3 model; (**b**) Details the behaviors of the ResNet model; (**c**) Details the behaviors of the Xception model; (**d**) Details the behaviors of the VGG16 model; (**e**) Demonstrates the behaviors of the DenseNet model.

Table 3. Summary of network models comparison of performance analysis from the coffee leaf dataset.

Network Models	Training Accuracy (%)	Training Loss (%)	Validation Accuracy (%)	Validation Loss (%)
InceptionV3	99.34	0.0167	99.01	0.0306
ResNet50	98.70	0.0565	97.80	0.0577
Xception	99.40	0.0140	98.84	0.0337
VGG16	98.81	0.0291	97.53	0.0668
DenseNet	99.57	0.0135	99.09	0.0225

Regarding statistical examination, the ANOVA (Analysis of Variance) test has been executed, and the outcomes are exhibited in Table 4.

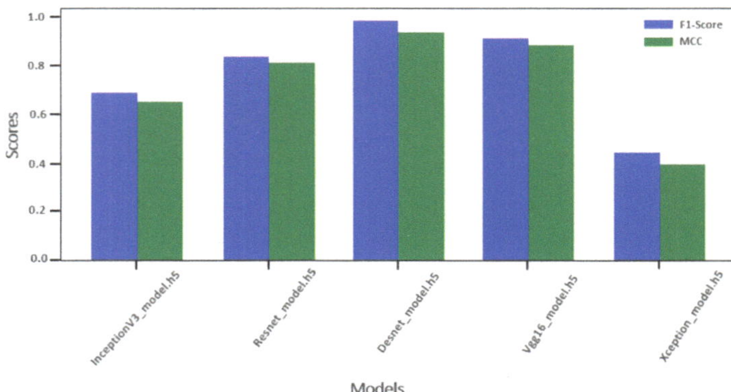

Figure 12. Models' comparison using F1-Score and MCC.

Table 4. The results of the analysis of variance test.

ANOVA Table	SS	DF	MS	F-Value	*p*-Value
Treatment (between columns)	0.029	4	0.007	233.3333	$p < 0.0001$
Residual (within columns)	0.002	75	0.0003		
Total	0.031	79			

The outcomes shown in Table 4 reveal a noteworthy distinction in the selected deep learning algorithm compared to the other methods. This is evident from the ANOVA results provided. The "Treatment" row (which corresponds to the differences between columns) exemplifies this with a substantial F-value of 233.333 and an extremely low *p*-value, less than 0.0001. It is noteworthy that the residual variance is merely 0.002, indicating limited variability among the diverse methods. This suggests that the variation observed in the outcome measure is primarily attributed to the effect of the chosen technique. The variance in the outcome measure was computed across all groups and amounted to 0.031 in the sum of squares. While the ANOVA outcomes point to the superior performance of the selection algorithm concerning the outcome measure compared to other methods, it is important to acknowledge that this is a preliminary observation.

The findings do not provide insights into the magnitude or direction of the effect, nor do they elucidate the specific differences between DenseNet and alternative methods. To ascertain if two samples are extracted from a common population, one can employ a non-parametric method known as the Wilcoxon signed-rank test.

The outcomes of this examination are exhibited in Table 5. Within this table, the assessment aimed to compare the efficacy of the presented models on the dataset.

Table 5. The results of the Wilcoxon signed-rank test.

	DTO + DT	PSO + DT	GWO + DT	GA + DT
Theoretical median	5.75×10^{-8}	5.75×10^{-8}	5.75×10^{-8}	5.75×10^{-8}
Actual median	3.57×10^{-5}	3.57×10^{-5}	3.57×10^{-5}	3.57×10^{-5}
Number of values	37,964	37,964	37,964	37,964
Wilcoxon Signed-Rank Test	0	0	0	0
Sum of signed ranks (W)	37,891	37,891	37,891	37,891
Sum of positive ranks	1,682,355	1,682,355	1,682,355	1,682,355
Sum of negative ranks	−1,644,464	−1,644,464	−1,644,464	−1,644,464
p-value (two-tailed)	0	0	0	0
Exact or estimate?	Exact	Exact	Exact	Exact
Significant (alpha = 0.05)?	Yes	Yes	Yes	Yes
How big is the discrepancy?	3.56×10^{-5}	5.06×10^{-8}	9.14×10^{-8}	4.83×10^{-8}

In our comprehensive assessment of the five deep learning models for image classification, we conducted an in-depth analysis to discern their unique capabilities on top of different optimization methods. The results, presented in Table 5, reveal subtle distinctions among these models. Notably, statistical tests, including the Wilcoxon Signed-Rank Test, indicate statistically significant differences in their median performance scores. However, it is crucial to emphasize that these differences, while statistically significant, are practically negligible. Each of the five models, namely InceptionV3, ResNet, DenseNet, VGG16, and Xception, consistently delivered competitive results, reflecting the maturity and robustness of contemporary deep learning architectures. Our study highlights nuanced performance differences while emphasizing the pivotal balance between statistical significance and practical utility, ultimately leading us to select DenseNet as the optimal choice for our image classification task. Nevertheless, it is essential to acknowledge the overall excellence demonstrated by each model, showcasing the prowess of contemporary deep-learning techniques.

5. Discussion

In the farming industry, especially for coffee plantations, caring about the importance of coffee consumption worldwide and the drawbacks of coffee diseases and pests affecting production, timely detection of diseases is crucial for achieving high yields. To support improving productivity, the incorporation of the latest technologies is needed for the early diagnosis of coffee diseases from leaves. The literature survey suggested that using deep learning models contributed efficiently to image classification while transfer learning-based models are effective in reducing training computation complexity by addressing the need for extensive datasets. Therefore, this study reveals the application of five pre-trained models in the Rwandan coffee leaf disease dataset to measure performances and provide advice for portable hand-held devices to facilitate farmers.

The performances of models, such as Inception V3, Xception, VGG-16, ResNet-50, and DenseNet, are evaluated with different metrics to identify the most suitable model for the accurate classification of coffee plant leaf diseases. The evaluation metrics, such as ROC, and precision–recall values, were measured.

Figure 9 illustrates a graphical representation of the pre-trained network models based on the evaluation metric, such as ROC. The VGG16 and DenseNet present good performance compared to other used models on all disease classes. The AUC for all discussed diseases in this context appeals to be in the range of 0.5 to 1. This indication means that the model can correctly classify coffee rust, minor, health, and red spider mites surveyed to be abundant in Rwanda. To tackle the problem of vanishing gradients induced by skip connections, we utilized regularization methods, such as batch normalization. The use of deeper models presented several difficulties, such as overfitting, covariant shifts, and longer training times. To surmount these obstacles, we conducted experiments to finely adjust the hyperparameters.

The assessment of model performance was measured using the AP metric as shown in Equation (3). In the performed experiment of the dataset used on the selected pre-trained models, Figure 10 shows the results of different models. The illustration demonstrated that DenseNet and VGG16 have better AP for the used classes than InceptionV3, Xception, and ResNet50. DenseNet demonstrates AP values of 51%, 40%, 0%, and 3% for health, miner, and red spider mite class, respectively. VGG16 demonstrates AP values of 52%, 45%, 1%, and 2% for health, miner, and red spider mite class, respectively. The VGG16 expressed to grab some detections on red spider mites compared to others. The observation is that lack of enough images in this class. The evaluation outcomes revealed that DenseNet and VGG16 performed better than InceptionV3, Xception, and ResNet50 models.

Table 1 presents different research references, the year of publication, the methods used, accuracy percentages, and the corresponding plant names for leaf classification. The "proposed model" labeled as DenseNet achieved the highest accuracy of 99.57% in classifying coffee leaves. Table 3 shows the comparison of different models and their score accuracies. The training accuracy and loss represent how well the models performed on the training data while the validation accuracy and loss show their performance on previously unseen validation data. Among the models, DenseNet achieved the highest training accuracy (99.57%) and validation accuracy (99.09%), indicating its excellent ability to learn and generalize from the data. On the other hand, ResNet50 had the lowest validation accuracy (97.80%) and the highest validation loss (0.0577), suggesting it might slightly struggle to generalize to new data compared to the other models. To emphasize the model evaluation criteria, we performed statistical tests with ANOVA and Wilcoxon, as shown in Tables 4 and 5, to check the variability of models on our dataset. It reaffirms our decision to choose the 'DenseNet' model based on a comprehensive evaluation of various factors, including not only ANOVA or Wilcoxon tests, but also median discrepancies and other metrics discussed.

6. Conclusions and Future Directions

In this study, we investigated the coffee farming industry in Rwanda, focusing on various identified coffee leaf diseases. Our research involved a successful analysis of different transfer learning models, specifically chosen to accurately classify five distinct classes of coffee plant leaf diseases. We standardized and evaluated cutting-edge deep learning models using transfer learning techniques, considering the classification accuracy, precision, recall, and AP score as the evaluation metrics. After analyzing several pre-trained architectures, including InceptionV3, Xception, and ResNet50, we found that DenseNet and VGG16 performed exceptionally well. Based on our findings, we proposed a model training pipeline that was followed throughout the experiment.

DenseNet model training was found to be more straightforward, primarily attributed to its smaller number of trainable parameters and lower computational complexity. This quality makes DenseNet particularly well-suited for coffee plant leaf disease identification, especially when incorporating new coffee leaf diseases that were not part of the initial training data, as it reduces the overall training complexity. The experimented model's quality has been tested using statistical tests, such as Wilcoxon and ANOVA. The proposed model demonstrated exceptional performance, achieving an impressive classification accuracy of 99.57%, along with high values for AUC and AP metrics.

In our future endeavors, we aim to tackle challenges associated with real-time data collection. We plan to develop a multi-object deep learning model capable of detecting coffee plant leaf diseases not just from individual leaves, but also from a bunch of leaves as well. Moreover, we are currently working on the implementation of a mobile application that will leverage the trained model obtained from this study. This application will provide valuable assistance to farmers and the agricultural sector by enabling the real-time identification of leaf diseases in Rwanda based on the samples taken.

Author Contributions: Conceptualization, E.H. and G.B.; methodology, E.H., E.M., G.B., S.M.M. and P.R.; software, E.H., G.B. and J.N.; validation, E.H., J.N. and G.B.; formal analysis, E.H., S.M.M. and O.J.S.; investigation, E.H., M.C.A.K., J.M., E.M., J.A.U.U., L.C.C. and T.M.; resources, E.H.; data curation, J.N.; writing—original draft preparation, E.H.; writing—review and editing, E.H., G.B., J.C.U. and M.C.A.K.; visualization, J.N.; supervision, G.B. and P.R.; project administration, E.H.; funding acquisition, O.J.S. and P.R. All authors have read and agreed to the published version of the manuscript.

Funding: This research was funded through a grant offered by the University of Rwanda in partnership with SIDA (Swedish International Development Agency) under the programme UR-Sweden program. The grant supported all research activities, such as data collection, purchase of equipment and materials, fieldwork, etc. The APC was also funded by the same grant.

Institutional Review Board Statement: Not applicable.

Informed Consent Statement: Not applicable.

Data Availability Statement: When requested, the authors will make available all data used in this study.

Acknowledgments: This work is acknowledged by the Rwanda Agricultural Board (RAB) to avail the farming cooperatives operating in Rwanda and the coffee washing stations.

Conflicts of Interest: The authors declare no conflict of interest.

References

1. World Bank. Agricultural Development in Rwanda. Available online: https://www.worldbank.org/en/results/2013/01/23/agricultural-development-in-rwanda#:~:text=Agriculture%20is%20crucial%20for%20Rwanda\T1\textquoterights,of%20the%20country\T1\textquoterights%20food%20needs (accessed on 19 June 2023).
2. The Republic of Rwanda, Ministry of Trade, and Industry. Revised National Export Strategy. Available online: https://rwandatrade.rw/media/2015%20MINICOM%20National%20Export%20Strategy%20II%20(NES%20II).pdf (accessed on 19 June 2023).
3. Nzeyimana, I.; Hartemink, A.E.; de Graaff, J. Coffee Farming and Soil Management in Rwanda. *Outlook Agric.* **2013**, *42*, 47–52. [CrossRef]
4. Nurihun, B.A. The Relationship between Climate, Disease and Coffee Yield: Optimizing Management for Smallholder Farmers. Ph.D. Thesis, Ecology and Evolution at Stockholm University, Stockholm, Sweden, 2023. Available online: https://su.diva-portal.org/smash/get/diva2:1749585/FULLTEXT01.pdf (accessed on 19 June 2023).
5. Balodi, R.; Bisht, S.; Ghatak, A.; Rao, K.H. Plant Disease Diagnosis: Technological Advancements and Challenges. *Indian Phytopathol.* **2017**, *70*, 275–281. [CrossRef]
6. World Bank Group. Agriculture Global Practice Note, Rwanda Agricultural Sector Risk Assessment. December 2015. Available online: https://documents1.worldbank.org/curated/en/514891468197095483/pdf/102075-BRI-P148140-Box394821B-PUBLIC-Rwanda-policy-note-web.pdf (accessed on 21 June 2023).
7. Kifle, B.; Demelash, T. Climatic Variables and Impact of Coffee Berry Diseases (Colletotrichum Kahawae) in Ethiopian Coffee Production. *J. Biol. Agric. Healthc.* **2015**, *5*, 7.
8. Flood, J.; Cabi, U. Coffee wilt disease. *Burleigh Dodds Ser. Agric. Sci.* **2021**, *96*, 319–342. [CrossRef]
9. Phiri, N.; Baker, P.S.; CAB International. The Status of Coffee Wilt Disease in Africa. 2009. Available online: https://assets.publishing.service.gov.uk/media/57a08b4040f0b652dd000bb4/Coffee_CH02.pdf (accessed on 21 June 2023).
10. The Abundance of Pests and Diseases in Arabica Coffee Production Systems in Uganda—Ecological Mechanisms and Spatial Analysis in the Face of Climate Change. 2017. Available online: https://agritrop.cirad.fr/584976/1/PhD_Thesis_TL_2017.pdf (accessed on 21 June 2023).
11. Bigirimana, J.; Uzayisenga, B.; Gut, L.J. Population distribution and density of Antestiopsis thunbergia (Hemiptera: Pentatomidae) in the coffee growing regions of Rwanda in relation to climatic variables. *Crop. Prot.* **2019**, *122*, 136–141. [CrossRef]
12. Wikifarmer. Coffee Major Pest and Diseases and Control Measures. Available online: https://wikifarmer.com/coffee-major-pest-and-diseases-and-control-measures/ (accessed on 21 June 2023).
13. Riley, M.; Williamson, M.; Maloy, O. Plant Disease Diagnosis. *Plant Health Instr.* **2002**, *10*, 193–210. [CrossRef]
14. Miller, S.A.; Beed, F.D.; Harmon, C.L. Plant Disease Diagnostic Capabilities and Networks. *Annu. Rev. Phytopathol.* **2009**, *47*, 15–38. [CrossRef]
15. Badel, J.L.; Zambolim, L. Coffee bacterial diseases: A plethora of scientific opportunities. *Plant Pathol.* **2018**, *68*, 411–425. [CrossRef]
16. Food and Agriculture Organization of the United Nationals. Climate Change and Food Security: Risks and Responses. 2015. Available online: https://www.fao.org/3/i5188e/I5188E.pdf (accessed on 19 June 2023).
17. Dawod, R.G.; Dobre, C. Upper and Lower Leaf Side Detection with Machine Learning Methods. *Sensors* **2022**, *22*, 2696. [CrossRef]
18. Vu, D.L.; Nguyen, T.K.; Nguyen, T.V.; Nguyen, T.N.; Massacci, F.; Phung, P.H. HIT4Mal: Hybrid image transformation for malware classification. *Trans. Emerg. Telecommun. Technol.* **2019**, *31*, e3789. [CrossRef]
19. Shaikh, R.P.; Dhole, S.A. Citrus Leaf Unhealthy Region Detection by Using Image Processing Technique. In Proceedings of the IEEE International Conference on Electronics, Communication and Aerospace Technology, Coimbatore, India, 20–22 April 2017; pp. 420–423.
20. Yu, K.; Lin, L.; Alazab, M.; Tan, L.; Gu, B. Deep Learning-Based Traffic Safety Solution for a Mixture of Autonomous and Manual Vehicles in a 5G-Enabled Intelligent Transportation System. *IEEE Trans. Intell. Transp. Syst.* **2020**, *22*, 4337–4347. [CrossRef]
21. Khan, M.A.; Akram, T.; Sharif, M.; Javed, K.; Raza, M.; Saba, T. An automated system for cucumber leaf diseased spot detection and classification using improved saliency method and deep features selection. *Multimed. Tools Appl.* **2020**, *79*, 18627–18656. Available online: https://link.springer.com/article/10.1007/s11042-020-08726-8 (accessed on 24 June 2023). [CrossRef]
22. Karthik, R.; Hariharan, M.; Anand, S.; Mathikshara, P.; Johnson, A.; Menaka, R. Attention embedded residual CNN for disease detection in tomato leaves. *Appl. Soft Comput.* **2019**, *86*, 105933. [CrossRef]
23. Giraddi, S.; Desai, S.; Deshpande, A. Deep Learning for Agricultural Plant Disease Detection. In *Lecture Notes in Electrical Engineering*; Kumar, A., Paprzycki, M., Gunjan, V., Eds.; ICDSMLA 2019; Springer: Singapore, 2020; Volume 601. [CrossRef]
24. Scientist, D.; Bengaluru, T.M.; Nadu, T. Rice Plant Disease Identification Using Artificial Intelligence. *Int. J. Electr. Eng. Technol.* **2020**, *11*, 392–402.

25. Dubey, S.R.; Jalal, A.S. Adapted Approach for Fruit Disease Identification using Images. In *Image Processing: Concepts, Methodologies, Tools, and Applications*; IGI Global: Hershey, PA, USA, 2013; pp. 1395–1409. [CrossRef]

26. Yun, S.; Xianfeng, W.; Shanwen, Z.; Chuanlei, Z. PNN-based crop disease recognition with leaf image features and meteorological data. *Int. J. Agric. Biol. Eng.* **2015**, *8*, 60–68.

27. Harakannanavar, S.S.; Rudagi, J.M.; Puranikmath, V.I.; Siddiqua, A.; Pramodhini, R. Plant leaf disease detection using computer vision and machine learning algorithms. *Glob. Transit. Proc.* **2022**, *3*, 305–310. [CrossRef]

28. Li, G.; Ma, Z.; Wang, H. Image Recognition of Grape Downy Mildew and Grape. In Proceedings of the International Conference on Computer and Computing Technologies in Agriculture, Beijing, China, 29–31 October 2011; pp. 151–162.

29. Hitimana, E.; Gwun, O. Automatic Estimation of Live Coffee Leaf Infection Based on Image Processing Techniques. *Comput. Sci. Inf. Technol.* **2014**, *19*, 255–266. [CrossRef]

30. Manso, G.L.; Knidel, H.; Krohling, R.A.; Ventura, J.A. A smartphone application to detection and classification of coffee leaf miner and coffee leaf rust. *arXiv* **2019**, arXiv:1904.00742.

31. Rauf, H.T.; Saleem, B.A.; Lali, M.I.U.; Khan, M.A.; Sharif, M.; Bukhari, S.A.C. A citrus fruits and leaves dataset for the detection and classification of citrus diseases through machine learning. *Data Brief* **2019**, *26*, 104340. [CrossRef]

32. Sujatha, R.; Chatterjee, J.M.; Jhanjhi, N.; Brohi, S.N. Performance of deep learning vs machine learning in plant leaf disease detection. *Microprocessor. Microsyst.* **2021**, *80*, 103615. [CrossRef]

33. Barbedo, J.G.A. Factors influencing the use of deep learning for plant disease recognition. *Biosyst. Eng.* **2018**, *172*, 84–91. [CrossRef]

34. Vardhini, P.H.; Asritha, S.; Devi, Y.S. Efficient Disease Detection of Paddy Crop using CNN. In Proceedings of the 2020 International Conference on Smart Technologies in Computing, Electrical and Electronics (ICSTCEE), Bengaluru, India, 9–10 October 2020; pp. 116–119.

35. Mohanty, S.P.; Hughes, D.P.; Salathé, M. Using Deep Learning for Image-Based Plant Disease Detection. *Front. Plant Sci.* **2016**, *7*, 1419. [CrossRef]

36. Panigrahi, K.P.; Sahoo, A.K.; Das, H. A CNN Approach for Corn Leaves Disease Detection to Support Digital Agricultural System. In Proceedings of the 4th International Conference on Trends in Electronics and Information, Tirunelveli, India, 15–17 June 2020; pp. 678–683.

37. Tan, C.; Sun, F.; Kong, T.; Zhang, W.; Yang, C.; Liu, C. A Survey on Deep Transfer Learning. In Proceedings of the 27th International Conference on Artificial Neural Networks, Rhodes, Greece, 4–7 October 2018; pp. 270–279. [CrossRef]

38. Andrew, J.; Fiona, R.; Caleb, A.H. Comparative Study of Various Deep Convolutional Neural Networks in the Early Prediction of Cancer. In Proceedings of the 2019 International Conference on Intelligent Computing and Control Systems (ICCS), Madurai, India, 15–17 May 2019; pp. 884–890. [CrossRef]

39. Onesimu, J.A.; Karthikeyan, J. An Efficient Privacy-preserving Deep Learning Scheme for Medical Image Analysis. *J. Inf. Technol. Manag.* **2020**, *12*, 50–67. [CrossRef]

40. Too, E.C.; Yujian, L.; Njuki, S.; Yingchun, L. A comparative study of fine-tuning deep learning models for plant disease identification. *Comput. Electron. Agric.* **2019**, *161*, 272–279. [CrossRef]

41. Devaraj, P.; Arakeri, M.P.; Kumar, B.P.V. Early detection of leaf diseases in Beans crop using Image Processing and Mobile Computing techniques. *Adv. Comput. Sci. Technol.* **2017**, *10*, 2927–2945.

42. Qin, F.; Liu, D.; Sun, B.; Ruan, L.; Ma, Z.; Wang, H. Identification of Alfalfa Leaf Diseases Using Image Recognition Technology. *PLoS ONE* **2016**, *11*, e0168274. [CrossRef]

43. Geetharamani, G.; Arun Pandian, J. Identification of plant leaf diseases using a nine-layer deep convolutional neural network. *Comput. Electr. Eng.* **2019**, *76*, 323–338. [CrossRef]

44. Azimi, S.; Kaur, T.; Gandhi, T.K. A deep learning approach to measure stress levels in plants due to Nitrogen deficiency. *Measurement* **2020**, *173*, 108650. [CrossRef]

45. Gadekallu, T.R.; Rajput, D.S.; Reddy, M.P.K.; Lakshmana, K.; Bhattacharya, S.; Singh, S.; Jolfaei, A.; Alazab, M. A novel PCA–whale optimization-based deep neural network model for classification of tomato plant diseases using GPU. *J. Real-Time Image Process.* **2020**, *18*, 1383–1396. Available online: https://link.springer.com/article/10.1007/s11554-020-00987-8 (accessed on 24 June 2023). [CrossRef]

46. Sinha, A.; Shekhawat, R.S. Olive Spot Disease Detection and Classification using Analysis of Leaf Image Textures. *Procedia Comput. Sci.* **2020**, *167*, 2328–2336. [CrossRef]

47. Raikar, M.M.; Meena, S.M.; Kuchanur, C.; Girraddi, S.; Benagi, P. Classification and Grading of Okra-ladies finger using Deep Learning. *Procedia Comput. Sci.* **2020**, *171*, 2380–2389. [CrossRef]

48. Joshi, R.C.; Kaushik, M.; Dutta, M.K.; Srivastava, A.; Choudhary, N. VirLeafNet: Automatic analysis and viral disease diagnosis using deep learning in Vigna mungo plant. *Ecol. Inform.* **2020**, *61*, 101197. Available online: https://linkinghub.elsevier.com/retrieve/pii/S1574954120301473 (accessed on 24 June 2023). [CrossRef]

49. Kaur, P.; Harnal, S.; Tiwari, R.; Upadhyay, S.; Bhatia, S.; Mashat, A.; Alabdali, A.M. Recognition of Leaf Disease Using Hybrid Convolutional Neural Network by Applying Feature Reduction. *Sensors* **2022**, *22*, 575. [CrossRef] [PubMed]

50. Innocent Nzeyimana, Optimizing Arabica Coffee Production Systems in Rwanda. 2018. Available online: https://www.researchgate.net/publication/325615794_Optimizing_Arabica_coffee_production_systems_in_Rwanda (accessed on 24 June 2023).

51. Parraga-Alava, J.; Cusme, K.; Loor, A.; Santander, E. RoCoLe: A robusta coffee leaf images dataset for evaluation of machine learning based methods in plant diseases recognition. *Data Brief* **2019**, *25*, 104414. [CrossRef]

52. Coffee Farming in Rwanda: Savoring Success. Contribution to Newsletter 02/2015 of the SDC Agriculture and Food Security Network. Available online: https://www.shareweb.ch/site/Agriculture-and-Food-Security/news/Documents/2015_02_coffee_rwanda_fromm.pdf (accessed on 24 June 2023).

53. Hitimana, E.; Bajpai, G.; Musabe, R.; Sibomana, L.; Kayalvizhi, J. Implementation of IoT Framework with Data Analysis Using Deep Learning Methods for Occupancy Prediction in a Building. *Future Internet* **2021**, *13*, 67. [CrossRef]

54. Kuradusenge, M.; Hitimana, E.; Hanyurwimfura, D.; Rukundo, P.; Mtonga, K.; Mukasine, A.; Uwitonze, C.; Ngabonziza, J.; Uwamahoro, A. Crop Yield Prediction Using Machine Learning Models: Case of Irish Potato and Maize. *Agriculture* **2023**, *13*, 225. [CrossRef]

55. Koo, K.-M.; Cha, E.-Y. Image recognition performance enhancements using image normalization. *Hum. Cent. Comput. Inf. Sci.* **2017**, *7*, 33. [CrossRef]

56. Szegedy, C.; Ioffe, S.; Vanhoucke, V.; Alemi, A. Inception-v4, inception-resnet, and the impact of residual connections on learning. In Proceedings of the AAAI Conference on Artificial Intelligence 2017, San Francisco, CA, USA, 4–9 February 2017; Volume 31.

57. Joshi, K.; Tripathi, V.; Bose, C.; Bhardwaj, C. Robust Sports Image Classification Using InceptionV3 and Neural Networks. *Procedia Comput. Sci.* **2020**, *167*, 2374–2381. [CrossRef]

58. Ramcharan, A.; Baranowski, K.; McCloskey, P.; Ahmed, B.; Legg, J.; Hughes, D.P. Deep learning for image-based cassava disease detection. *Front. Plant Sci.* **2017**, *8*, 1852. [CrossRef]

59. Szegedy, C.; Vanhoucke, V.; Ioffe, S.; Shlens, J. Rethinking the Inception Architecture for Computer Vision. Available online: https://www.cv-foundation.org/openaccess/content_cvpr_2016/papers/Szegedy_Rethinking_the_Inception_CVPR_2016_paper.pdf (accessed on 24 June 2023).

60. Inception V3 Model Architecture. Available online: https://iq.opengenus.org/inception-v3-model-architecture/ (accessed on 19 July 2023).

61. Deep Learning. Deep Residual Networks (ResNet, ResNet50)—2023 Guide. Available online: https://viso.ai/deep-learning/resnet-residual-neural-network/ (accessed on 26 June 2023).

62. He, K.; Zhang, X.; Ren, S.; Sun, J. Deep residual learning for image recognition. In Proceedings of the 2016 IEEE Conference on Computer Vision and Pattern Recognition (CVPR), Las Vegas, NV, USA, 27–30 June 2016; pp. 770–778. [CrossRef]

63. Al-Gaashani, M.S.; Samee, N.A.; Alnashwan, R.; Khayyat, M.; Muthanna, M.S.A. Using a Resnet50 with a Kernel Attention Mechanism for Rice Disease Diagnosis. *Life* **2023**, *13*, 1277. [CrossRef]

64. Giuseppe, G.; Celano, A. A ResNet-50-based Convolutional Neural Network Model for Language ID Identification from Speech Recordings. In Proceedings of the Third Workshop on Computational Typology and Multilingual NLP, Online, 20 July 2023; pp. 136–144. Available online: https://aclanthology.org/2021.sigtyp-1.13.pdf (accessed on 26 June 2023).

65. Simonyan, K.; Zisserman, A. Very Deep Convolutional Networks for Large-Scale Image Recognition. *arXiv* **2014**, arXiv:1409.1556. [CrossRef]

66. Tammina, S. Transfer learning using VGG-16 with Deep Convolutional Neural Network for Classifying Images. *Int. J. Sci. Res. Publ.* **2019**, *9*, 9420. [CrossRef]

67. Ter-Sarkisov, A. Network of Steel: Neural Font Style Transfer from Heavy Metal to Corporate Logos. *Comput. Sci.* **2020**, *1*, 621–629. [CrossRef]

68. Francois Chollet. Xception: Deep Learning with Depthwise Separable Convolutions. Available online: https://openaccess.thecvf.com/content_cvpr_2017/papers/Chollet_Xception_Deep_Learning_CVPR_2017_paper.pdf (accessed on 26 June 2023).

69. Sutaji, D.; Yıldız, O. LEMOXINET: Lite ensemble MobileNetV2 and Xception models to predict plant disease. *Ecol. Inform.* **2022**, *70*, 101698. [CrossRef]

70. Huang, G.; Liu, Z.; Van Der Maaten, L.; Weinberger, K.Q. Densely Connected Convolutional Networks. In Proceedings of the IEEE Conference on Computer Vision and Pattern Recognition, Honolulu, HI, USA, 21–26 July 2017; pp. 4700–4708.

71. Zhou, T.; Ye, X.; Lu, H.; Zheng, X.; Qiu, S.; Liu, Y. Dense Convolutional Network, and Its Application in Medical Image Analysis. *Microsc. Image Anal. Histopathol.* **2022**, *2022*, 2384830. [CrossRef] [PubMed]

72. Powers, D.M.W. Evaluation from precision, recall, and F-measure to ROC, informedness, markedness & correlation. *J. Mach. Learn. Technol.* **2011**, *2*, 37–63.

73. Chicco, D. Ten quick tips for machine learning in computational biology. *BioData Min.* **2017**, *10*, 1–17. [CrossRef]

technologies

A Foreign Object Detection Method for Belt Conveyors Based on an Improved YOLOX Model

Rongbin Yao [1], Peng Qi [2], Dezheng Hua [2], Xu Zhang [2], He Lu [1] and Xinhua Liu [2,*]

[1] Lianyungang Normal College, Lianyungang 222006, China; rby1972@126.com (R.Y.); luhe119879@126.com (H.L.)
[2] School of Mechatronic Engineering, China University of Mining and Technology, Xuzhou 221116, China; ts22050044a31@cumt.edu.cn (P.Q.); hua_dezheng@cumt.edu.cn (D.H.); ts20050081a31@cumt.edu.cn (X.Z.)
* Correspondence: liuxinhua@cumt.edu.cn

Abstract: As one of the main pieces of equipment in coal transportation, the belt conveyor with its detection system is an important area of research for the development of intelligent mines. Occurrences of non-coal foreign objects making contact with belts are common in complex production environments and with improper human operation. In order to avoid major safety accidents caused by scratches, deviation, and the breakage of belts, a foreign object detection method is proposed for belt conveyors in this work. Firstly, a foreign object image dataset is collected and established, and an IAT image enhancement module and an attention mechanism for CBAM are introduced to enhance the image data sample. Moreover, to predict the angle information of foreign objects with large aspect ratios, a rotating decoupling head is designed and a MO-YOLOX network structure is constructed. Some experiments are carried out with the belt conveyor in the mine's intelligent mining equipment laboratory, and different foreign objects are analyzed. The experimental results show that the accuracy, recall, and mAP^{50} of the proposed rotating frame foreign object detection method reach 93.87%, 93.69%, and 93.68%, respectively, and the average inference time for foreign object detection is 25 ms.

Keywords: belt conveyor; foreign object detection; YOLOX; image enhancement; rotation detection

Citation: Yao, R.; Qi, P.; Hua, D.; Zhang, X.; Lu, H.; Liu, X. A Foreign Object Detection Method for Belt Conveyors Based on an Improved YOLOX Model. *Technologies* **2023**, *11*, 114. https://doi.org/10.3390/technologies11050114

Academic Editor: Valeri Mladenov

Received: 20 July 2023
Revised: 7 August 2023
Accepted: 8 August 2023
Published: 26 August 2023

1. Introduction

With the development of intelligent mines and the improvement of machine vision technology, real-time detection systems for belt conveyors have become an important research topic in recent years [1]. The transportation belt plays a pivotal role and is prone to severe accidents such as deviation, slipping, belt breakage, and longitudinal belt tearing during production [2]. An in-depth analysis revealed that non-coal foreign objects entering the belt conveyor system are accountable for 61% of belt tearing and breakage incidents, amounting to a total of 21 cases [3]. The accurate and rapid identification of foreign objects in the belt transportation system, followed by removal, can substantially mitigate damage to the belt system and ensure the safe and stable operation of the belt transportation system [4].

Intelligent detection for the production of safe underground coal mines has become a hot research topic [5,6]. By utilizing video surveillance images, combined with image processing and machine-vision-related technologies, the theory of mine image monitoring has been applied to multiple aspects of automatic safety detection in coal mines, such as the automatic identification of spontaneous combustion fires [7], coal production monitoring [8], face detection and recognition methods for underground miners [9], and the automatic recognition of coal rock interfaces in coal faces [10]. However, traditional belt conveyor foreign object systems still rely on cameras to transmit collected video data to the central control room, where the staff can monitor the coal transportation area and the surrounding environment in real time. This practice is associated with significant drawbacks, including the duplication of work and fatigue-induced misjudgments. At the same time, staff cannot address foreign objects in a timely manner, which can easily cause foreign objects to block the transportation belt or sharp parts to scratch the belt, resulting in belt tearing and causing major safety accidents.

Nowadays, many systems have been proposed for the detection of foreign objects on the belt, and some results have been achieved. With an understanding of the characteristics of remote sensing targets with different dimensions, including their dense distribution and complex background, Xu et al. [11] applied YOLO-V3 to the field of remote sensing to detect remote sensing targets at different scales. The lightweight, fast Illumination Adaptive Transformer (IAT) was proposed by Cui et al. [12] to restore a normally lit sRGB image from either low-light or under-exposed conditions. Wang et al. [13] introduced the attention mechanism into YOLO-V5 to detect Solanum rostratum Dunal seeds, and it was found that the CBAM attention mechanism can effectively improve accuracy during model recognition. However, for the underground environment of a coal mine, there is no special model for the detection of foreign bodies on a coal flow belt, and the performance of other models still needs to be further improved.

In this work, a foreign object detection method for belt conveyors is proposed, and the remainder of this paper is organized as follows. In Section 2, the status and deficiencies of current foreign object detection methods are introduced. In Section 3, the improved algorithm and model architecture are proposed. In Section 4, comparative experiments are carried out on the proposed model, and the effectiveness of the improved algorithm is verified using a self-made foreign object dataset. The conclusions and future works are summarized in Section 5.

2. Literature Review

2.1. Foreign Object Detection Methods Based on Image Processing

The foreign object detection methods for belt conveyors based on images work with images of coal and non-coal foreign objects by obtaining the shallow or deep abstract features of the object, and use image processing to detect foreign objects. This approach has the advantages of simple installation and maintenance processes and low application costs, and has become one of the research focuses of foreign object detection for belt conveyors. Due to the diversity of the types of foreign objects (such as anchor rods and wood) causing belt tearing, many scholars have begun to extract the color, texture, shape, spatial relationship, and other features of foreign objects from the image features, achieving the automatic detection of foreign objects through image processing. Jiang et al. [14] used extreme median filtering to perform image noise processing and improve the traditional Canny edge detection algorithm to obtain an improved edge detection method for a Canny operator. The algorithm is used to perform image edge detection, and the image gray histogram is used for enhanced foreign object image processing. Zhang et al. [15] proposed a new image segmentation algorithm for belt conveying. A multi-scale linear filter composed of a Hessian matrix and Gaussian function forms the core of the algorithm, which can effectively obtain the edge intensity image, form a good seed area for watershed segmentation, and segment the background between the coal pile and foreign objects. Saran et al. [16] developed an image-processing-based foreign object detection solution to detect foreign objects such as concrete boulders and iron bars that often occur in the conveyor belts used for G furnace raw coal. The solution uses a multi-mode imaging (polarization camera)-based system to distinguish foreign objects. Tu et al. [17] proposed a new moving target detection method to solve the difficulties caused by the intermittent motions, temperatures, and dynamic background sequences of moving targets. By further comparing the similarities of edge images, ghosts and real static objects can be classified. Lins et al. [18] developed a system based on the concept of machine vision, which aims to realize the automation of the crack measurement process. Using the above method, a series of images can be processed and the crack size can be estimated as long as a camera is installed on a truck or robot.

2.2. Foreign Object Detection Method Based on Deep Learning

With the rapid development of deep learning, using this data learning method to learn image data features and perceive the surrounding environment has good research value in foreign object detection, to obtain a foreign object detection model that is more adaptable to a complex and changeable environment [19,20]. Deep learning is achieved by establishing and simulating the information processing neural structure of the human brain to extract low-level to high-level features from external input data, enabling machines to understand the learning data and obtain useful information [21]. Pu et al. [22] used CNN to identify coal and gangue images and to help separate coal and gangue, and introduced transfer learning to solve the problems of massive trainable parameters and limited computing power faced by the model. In order to apply CNN to the field of target detection, Ren et al. [23] put forward the RCNN method, which uses a selective search to obtain pre-selected regions, and completes image recognition through CNN combined with SVM. Because the multi-stage implementation of the algorithm led to its huge time cost, Girshick et al. [24] further put forward the concept of a ROI (region of interest) pooling layer, and replaced SVM with fully connected neural network, and proposed the Fast-RCNN algorithm. In order to solve the problem of foreign objects on the belt conveyor in the coal mine damaging the belt conveyor, Wang et al. [25] proposed a video detection method of foreign objects on the surface of the belt conveyor based on SSD. Firstly, the deep separable convolution method was adopted to reduce the number of parameters of the SSD algorithm and improve the calculation speed. Then, the GIOU loss function was used to replace the position loss function in the original SSD, which improves detection accuracy. Finally, the extraction position of the feature map and the proportion of the default frame were optimized, which improves the detection accuracy. Considering the fast-running speed of the belt and the influence of background and light source on foreign object targets, Ma et al. [26] proposed an improved Center-Net algorithm, which improved detection efficiency. The normalization method was optimized to reduce computer memory consumption, and a weighted feature fusion method was added to fully utilize the features of each layer, improving detection accuracy. In the experimental environment, the average detection rate was about 20fps, and met the demand for the real-time detection of foreign objects. Xiao et al. [27] used a median filtering method to preprocess images with foreign objects, removed the influence of dust, improved the clarity of ore edges, and established a dataset to train the YOLOv3 belt foreign object detection algorithm. Finally, after sparse training based on the BN layer, the YOLOv3 model was lightweight, and its parameters were fine tuned. Compared with the original YOLOv3 model, the model achieved smaller calculations, faster processing, and a smaller size.

2.3. Discussion

However, although many approaches for detecting foreign objects have been developed in the above literature, some common disadvantages of them are summarized as follows. Firstly, due to the specific coal mine environment, with a lot of dust, noise, and a complex background for foreign objects, it is difficult to achieve accurate detection of foreign objects on the belt conveyor. Therefore, the general target detection algorithms cannot be easily migrated to the coal mine environment. At the same time, the robustness of the traditional foreign object detection algorithms is poor, and the extraction of foreign object features requires a wealth of experience. Finally, the current public foreign object detection dataset lacks coal-mine-belt foreign object detection data, so it cannot flexibly adapt to different scenarios in different mining areas.

In this paper, the image dataset of belt foreign bodies in the coal mine environment is collected and established, and a target detection algorithm based on improved YOLO-V5 is used to detect non-coal foreign bodies on the coal belt.

3. The Proposed Foreign Object Detection Method

3.1. Target Detection of YOLO Model

The YOLO series target detection algorithm is a supervised learning target detection algorithm [28]. Its basic principle is to divide the input image into several grids, then extract the features of each part of the image through the convolutional neural network, and finally output the predicted bounding box, which is the center coordinates of the predicted object; the length and width of the detected object; and the confidence of the object category.

As shown in Figure 1, the input foreign object image is divided into S × S squares, and features are extracted from each grid through the convolutional neural network, then, features are fused and analyzed to output the confidence degree of the foreign object target, the boundary box coordinate information, and the foreign object category. In order to improve the accuracy of foreign object detection, a fixed number of anchor boxes are used for each grid to assist in learning position information. Clustering analysis is performed on the known labels of the target detection object in the image to obtain the initial size of the anchor box.

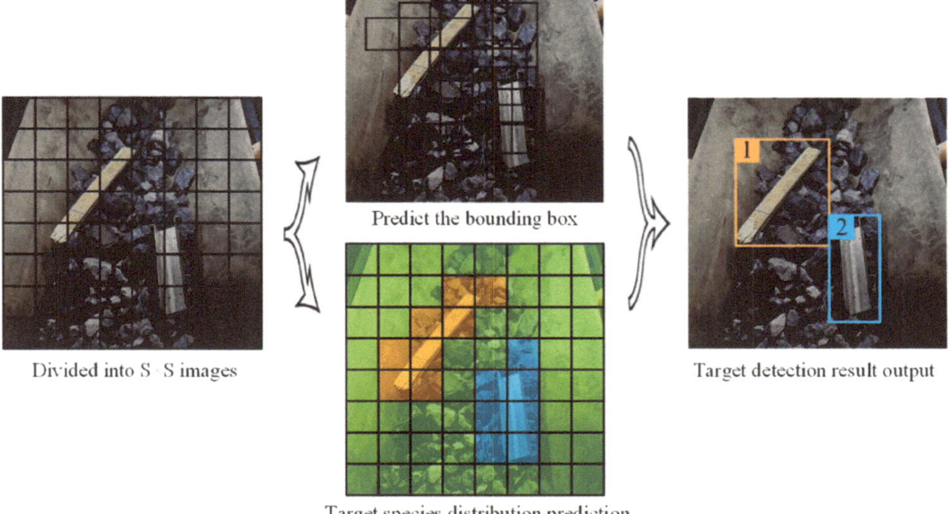

Figure 1. YOLO target detection process.

The framework of the YOLO series detection models has always been composed of three parts: the backbone feature extraction network, feature fusion layer, and detection decoupling head, as shown in Figure 2. The feature extraction network mainly extracts features from the input image data, then the feature fusion layer fuses the low-dimensional and high-dimensional features of the image to provide richer image information. Finally, the detection decoupling head outputs and predicts the position and category information of objects of interest. The YOLO series of algorithms all use a three branch detection head algorithm to predict objects of different scales, such as large, medium, and small.

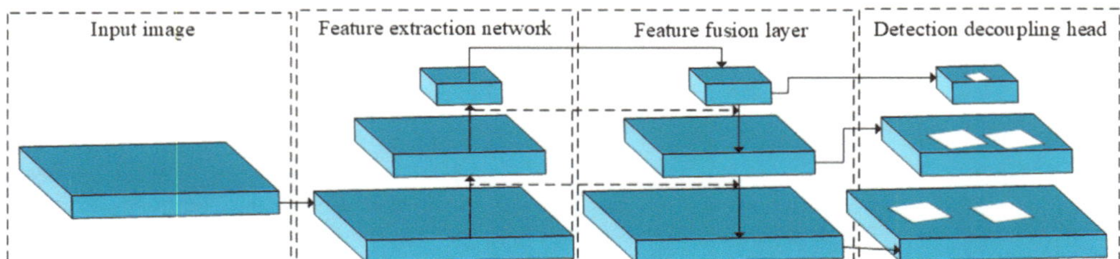

Figure 2. YOLO series network architecture.

In the actual target detection process, directly predicting the central coordinates, width, and height of the bounding box will result in too large a solution space for the predicted target, which will seriously waste computing resources. Therefore, an anchor frame mechanism is designed to accelerate the convergence of the model and improve the target detection accuracy, and the prediction principle of the bounding box is shown in Figure 3.

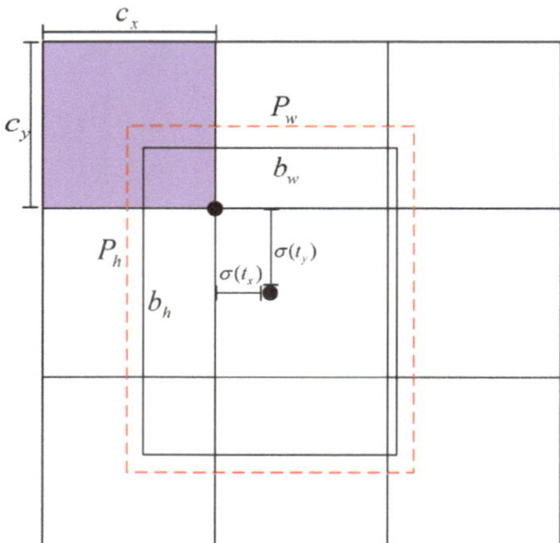

Figure 3. The prediction principle of YOLO's bounding box.

$$b_x = \sigma(t_x) + c_x \tag{1}$$

$$b_y = \sigma(t_y) + c_y \tag{2}$$

$$b_w = p_w e^{t_w} \tag{3}$$

$$b_h = p_h e^{t_h} \tag{4}$$

where p_w and p_h are the width and height of the anchor frame; b_w and b_h are the width and height of the prediction box; t_x and t_y are the offset from the anchor frame to the center of the prediction box; c_x and c_y are the coordinates of the upper left corner of the bounding box; $\sigma()$ is the normalized function.

3.2. Established Foreign Object Image Dataset

At present, the publicly available large-scale datasets do not include non-coal foreign objects. Therefore, it is necessary to establish an actual foreign object engineering dataset for belt conveyors to solve the problem of foreign object detection in practical engineering. This self-made dataset is named the belt conveyor foreign object detection dataset, and the sample categories of the dataset mainly include the following three types of foreign objects: iron, wood, and large gangue, as shown in Figure 4.

Figure 4. Different kinds of foreign object samples.

This work selected laboratory and belt-conveyor work scenarios for foreign object image collection. At the same time, foreign object image datasets were captured under different natural light conditions, and directional foreign objects such as iron and wood were offset to increase the information of image angles. In the laboratory environment, for the same foreign object, a foreign object dataset can be established that includes images of areas without coal flow, areas with coal flow, and areas obstructed by coal flow. Photos of foreign objects in different directions were collected in the laboratory environment to increase the diversity of foreign object dataset samples, as shown in Figure 5.

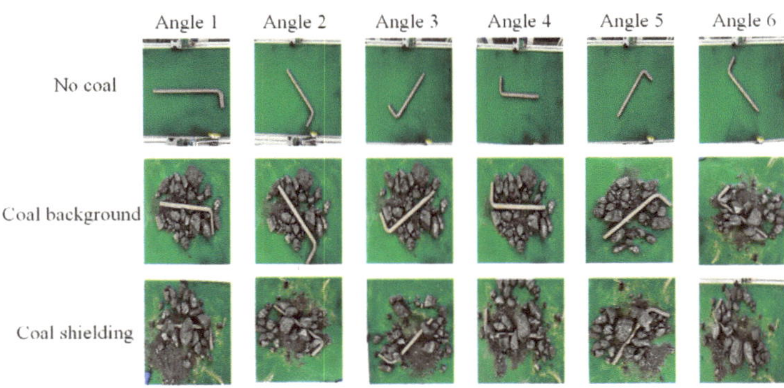

Figure 5. Single foreign material data samples.

In order to ensure the diversity of perspective in the collected dataset and better simulate the different shooting angles of cameras installed in actual working conditions, the top view images were collected by using a DJI drone with a pan tilt camera. The heights from the ground during the collection weare 1 m, 2 m, and 4 m, respectively, to ensure the diversity of perspective in the collected data, as shown in Figure 6.

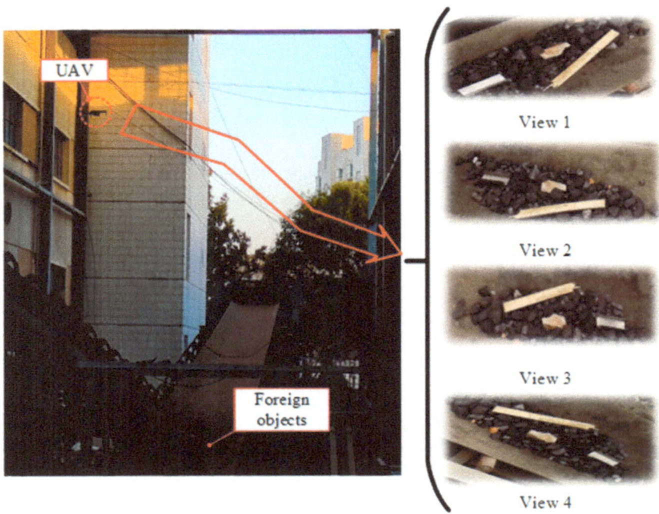

Figure 6. Multi-view image acquisition by UAV.

In order to improve the robustness and generalization of the model, the Mosaic multi-samples data augmentation method proposed by YOLOv4 was adopted. During the training process, four images in the training set were randomly selected, and the images were randomly scaled, cropped, and arranged for image combination. The sample size of the images during the training process was expanded, as shown in Figure 7.

Figure 7. Mosaic multi-sample data enhancement.

As shown in Figure 8, the images were expanded by means of horizontal flipping, random occlusion, random scaling, motion blur, random scaling and filling, and salt and pepper noise. A total of 1105 foreign object image datasets were collected for the belt conveyor foreign object image dataset, including 303 large gangue datasets, 401 iron tools datasets, 301 wood datasets, and 100 mixed target images. The belt conveyor foreign object image dataset was labeled with horizontal and rotating boxes, and the horizontal and rotating box foreign object detection datasets were constructed. Finally, the dataset was expanded to 8100 datasets through geometric expansion, and a complete dataset of foreign object images for belt conveyors had been constructed.

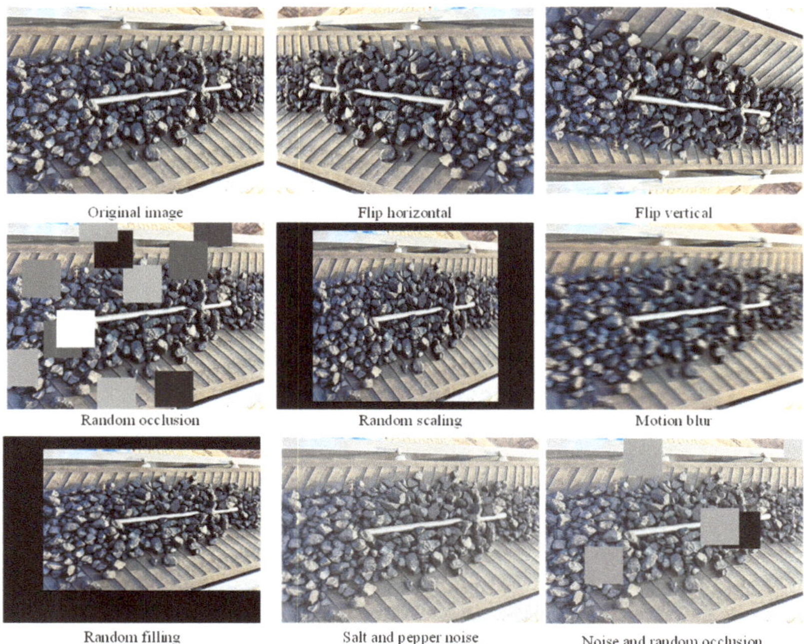

Figure 8. Geometry data enhancement.

3.3. Improved Depthwise Separable Convolution Block

Depthwise separable convolution breaks down the operations of standard convolution into depthwise convolution and point by point convolution [29]. Depthwise convolution performs separate spatial convolutions on each input channel, while point by point convolution combines the convolution results of each channel, which can greatly reduce the size and complexity of the model while maintaining high accuracy. The specific operations are shown in Figure 9.

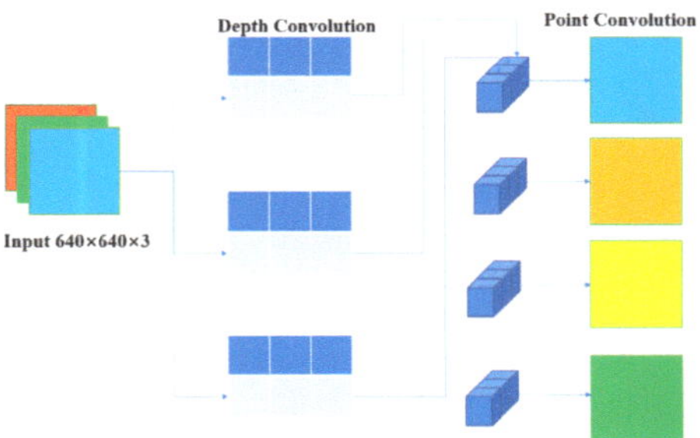

Figure 9. Depthwise separable convolution.

Assuming that the input image size is $640 \times 640 \times 3$ and the expected output size is $640 \times 640 \times 4$, and the ordinary convolution uses a convolution kernel of $3 \times 3 \times 3 \times 4$, then the parameter stc of the ordinary convolution is:

$$stc = H \times W \times chanel_{in} \times chanel_{out} = 3 \times 3 \times 3 \times 4 \tag{5}$$

Depthwise separable convolution is used for depth-by-depth convolution, and then point-by-point convolution of the channel relationship is carried out. First, convolution is performed by depth, and the number of parameters is as follows:

$$d_w = H \times W \times chanel_{in} = 3 \times 3 \times 3 \tag{6}$$

Then, through the point-by-point convolution operation, the total number of parameters for the depth-separable convolution is:

$$p_w = 1 \times 1 \times chanel_{in} \times chanel_{out} = 1 \times 1 \times 3 \times 4 \tag{7}$$

$$dsc = d_w + p_w = 3 \times 3 \times 3 + 1 \times 1 \times 3 \times 4 \tag{8}$$

The ratio of the number of parameters for the two convolution operations is:

$$\frac{dsc}{stc} = \frac{H \times W \times chanel_{in} + 1 \times 1 \times chanel_{in} \times chanel_{out}}{H \times W \times chanel_{in} \times chanel_{out}} \approx 0.36 \tag{9}$$

Using deep separable convolution for convolution operations can effectively reduce the number of parameters in the model, ensuring the feature extraction ability of the convolution and facilitating the light weight of the model. In addition, the Hard-Swish activation function was selected as the activation function of the belt conveyor foreign object detection model, as shown in Figure 10.

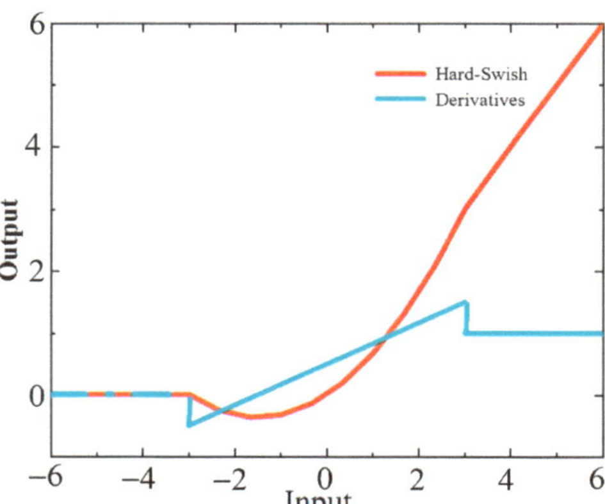

Figure 10. Hard-Swish activation function and its derivative.

The Hard-Swish activation function is a smooth function with no upper bound or lower bound. The activation function makes the model non-linear, which can effectively reduce the calculation cost in the embedded environment, and the expression is as follows:

$$\text{Hard-Swish}(x) = \begin{cases} 0 & if\ x \leq -3 \\ x & if\ x \geq +3 \\ x \times (x+3)/6 & otherwise \end{cases} \tag{10}$$

The basic module of the improved foreign object detection model is shown in Figure 11. Replacing the ordinary convolution at the end of the merge channel in the CSP1_ X and CSP2_X module with depthwise separatable convolution reduces the number of parameters in the convolution process and accelerates the inference speed of the model.

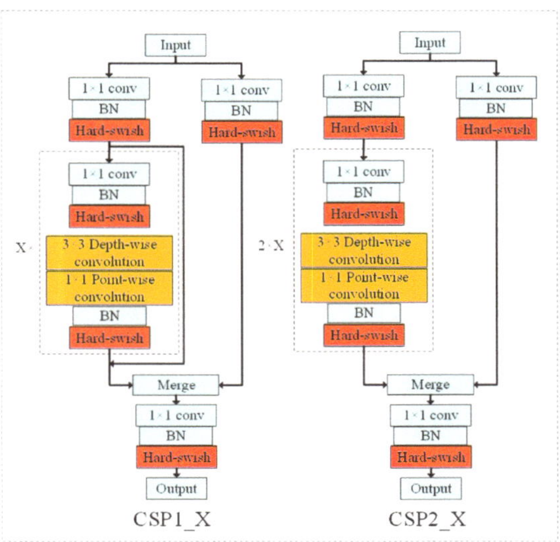

Figure 11. Improved CSP1_ X and CSP2_ X Structural blocks.

3.4. IAT Image Enhancement Module

In order to ensure the end-to-end output characteristics of deep learning, the IAT image enhancement module [30] is introduced to achieve image enhancement, and the network structure is shown in Figure 12. The color matrix in the IAT architecture represents the pixel weight weighted by a self-attention mechanism, in which the different colors are used to distinguish different patches from the original image. The IAT module can enhance the brightness of the image, restore the relevant details, improve the image quality, reduce the noise, and enhance the image contrast.

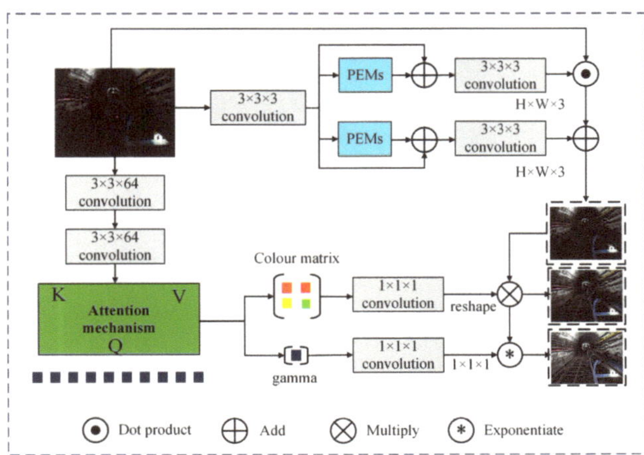

Figure 12. IAT image enhancement structure.

At the same time, the objective evaluation index Peak Signal to Noise Ratio (PSNR) for image enhancement is used as the specific evaluation index for image enhancement, and the formula is as follows:

$$PSNR = 10 \log_{10} \left[\frac{(2^n - 1)^2}{MSE} \right] \tag{11}$$

$$MSE = \frac{1}{H \times W} \sum_{i=1}^{H} \sum_{j=1}^{W} (X(i,j) - Y(i,j))^2 \tag{12}$$

where $X(i,j)$ are the pixel values of the original image, $Y(i,j)$ are the pixel values of the enhanced image, and H and W are the length and width of the image, respectively.

3.5. Improved CBAM Attention Block

CBAM [31] is a convolutional neural network module based on an attention mechanism, which is used to improve the overall performance of the model. Its essence is to inhibit the expression of redundant features by increasing the weight of non-redundant features. It is composed of a channel attention module (CAM) and a spatial attention module (SAM), and the specific network structure is shown in Figure 13.

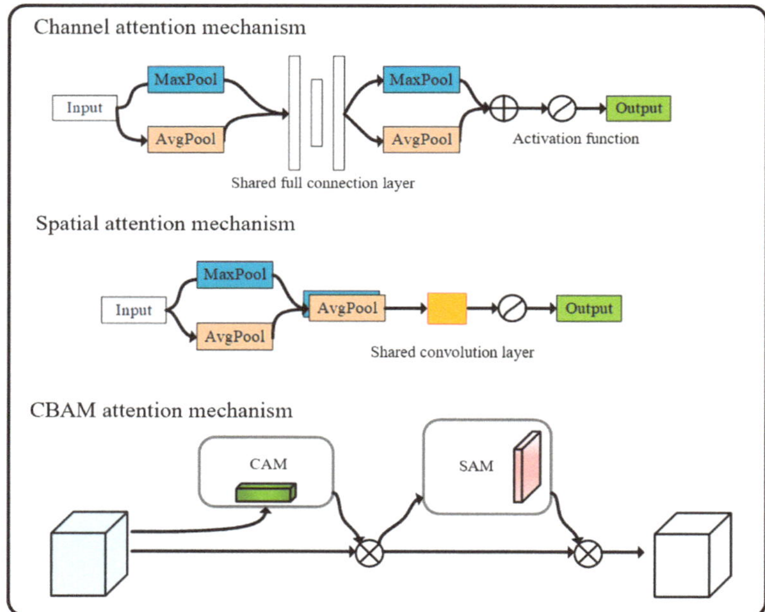

Figure 13. CBAM attention mechanism.

Therefore, in order to suppress redundant features and obtain attention feature maps that pay more attention to channels and spaces, the CBAM attention mechanism was introduced into the network structure, and the specific location of the addition is shown in Figure 14.

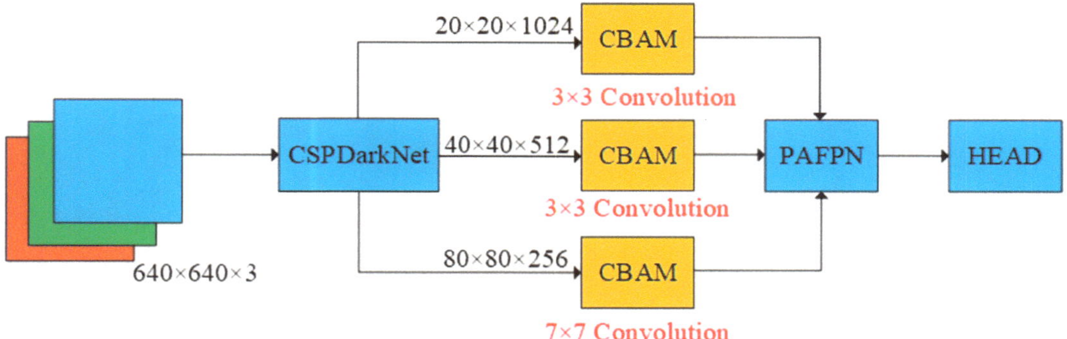

Figure 14. Improved addition location of CBAM in the network.

3.6. Designed Rotating Decoupling Head and MO-YOLOX Network

The detection boxes of the YOLO series object detection algorithms are all horizontal boxes, which is not conducive to the detection of foreign objects with diverse distribution directions such as ironware. Therefore, angle regression prediction was added to the head network of YOLOX, and a branch decoupling head based on the angle regression was constructed to accurately locate directional foreign objects. The structure of the rotary decoupling head is shown in Figure 15, where CBS*2 is an acronym for having two CBS modules. The overall network structure of MO-YOLOX is shown in Figure 16.

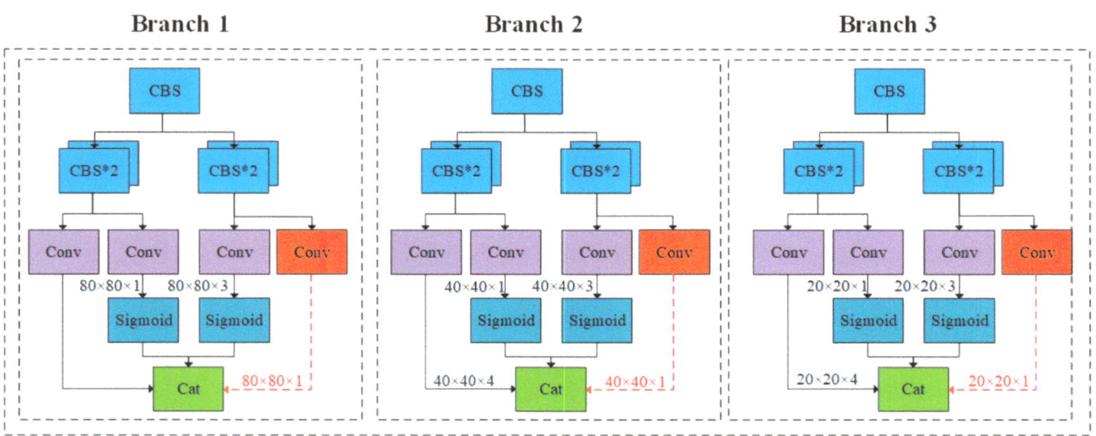

Figure 15. MO-YOLOX rotary decoupling head structure.

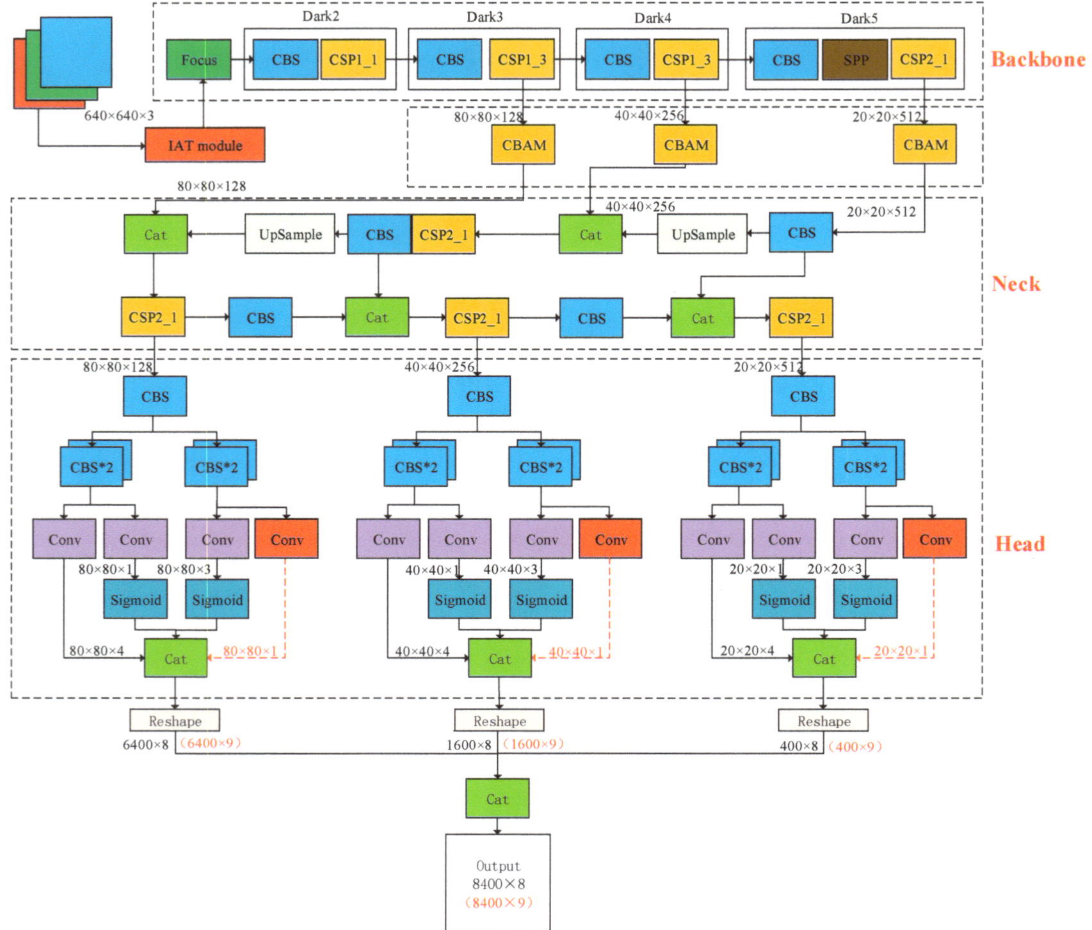

Figure 16. MO-YOLOX network model.

4. Experimental Example and Analysis

4.1. Experimental Platform

The proposed MO-YOLOX network model was trained in the GPU environment, and the environment configuration is shown in Table 1 below.

Table 1. Model training environment configuration.

Name	Parameter
CPU	Intel Core i9-10980XE
Hard disk	2 T
GPU	NVIDIA RTX A4000
Memory	16 G
Deep learning framework	Pytorch1.8.0
OS	Window10
Programming Language	Python3.8
CUDA	11.2

4.2. Experimental Comparisons

In order to verify the comprehensive performance of the proposed MO-YOLOX, comparison experiments of horizontal target detection and rotating target detection were carried out on the PASCALVOC dataset and the DOTA dataset, respectively. The PASCALVOC dataset is marked with horizontal boxes, including Bird, Dog, Cat, Person, Soft, Car, Bottle, and House. There are 312 image data, including a total of 1623 objects. The mainstream horizontal object detection models YOLOX-small and SSD300 [32] were selected for the comparative experiments. The experimental results are shown in Table 2, where excellent results are shown in bold.

Table 2. VOC dataset detection accuracy test results.

Metrics	MO-YOLOX	YOLOX-Small	SSD300
mAP^{50}	**71.9%**	70.6%	68.7%
Bird AP^{50}	**74.2%**	74.2%	71.2%
Dog AP^{50}	**78.7%**	75.2%	72.8%
Cat AP^{50}	78.5%	**80.3%**	73.9%
Person AP^{50}	**72.8%**	72.5%	69.8%
Sofa AP^{50}	71.1%	71.8%	**71.9%**
Car AP^{50}	**79.1%**	70.2%	70.8%
Bottle AP^{50}	49.2%	**50.9%**	49.9%
Average inference time	**21 ms**	28 ms	27 ms

In addition, in order to verify the effectiveness of MO-YOLOX in rotating target detection, the DOTA dataset marked with the rotating frame was selected, including plane (PL), ship (SH), large vehicle (LV), harbor (HA), small vehicle (SV), and baseball diamond (BD). Mainstream rotating object detection models, such as S2A-Net [33] and CFA [34], were selected for comparative experiments. The experimental results are shown in Table 3.

Table 3. DOTA dataset detection accuracy test results.

Metrics	MO-YOLOX	S2ANet	CFA
mAP^{50}	79.44%	79.26%	**79.57%**
PL AP^{50}	85.61%	**86.12%**	85.21%
SH AP^{50}	83.23%	82.23%	**83.82%**
LV AP^{50}	75.22%	76.32%	**80.91%**
HA AP^{50}	**76.51%**	75.41%	73.21%
SV AP^{50}	**78.29%**	75.23%	76.25%
BD AP^{50}	77.78%	**79.25%**	78.00%
Average inference time	**27 ms**	65 ms	66 ms

Compared with YOLOX-Small, the detection accuracy and reasoning speed of the MO-YOLOX target detection model with its attention mechanism and depthwise separable convolution are better than the original model. The average detection accuracy of the proposed model in the VOC test data set is higher than that of the original YOLOX-small and SSD300 model, and its detection accuracy in the DOTA data set is the same as that of S2ANet and CFA, but the reasoning time of MO-YOLOX is better than the above comparison algorithms. Therefore, the proposed foreign body detection network can meet the requirements of both detection accuracy and reasoning speed in the target detection task.

4.3. Experimental Testing and Analysis

The training dataset is the foreign object detection dataset of the belt conveyor, including the horizontal frame labeling dataset and the rotating frame marking dataset, and the relevant parameters are shown in Table 4. After 300 rounds of model training iterations, the proposed model can converge to relatively stable positions, and the loss values during the training process are shown in Figure 17. The models obtained from the above training were used as the optimal model for experimental comparison, and comparison experiments were conducted.

Table 4. Model training parameters.

Training Parameters	Setting Values
Activation function	Hard-Swish
Pooling method	Max-Pooling
Optimization algorithm	Adams, Batch-size = 8,
Loss function	Cross-entropy Loss function, KLD
Epoch	300
Data enhancement	Mosaic
Learning rate	Initial Learning rate $\alpha_0 = 0.01$, Nature Index attenuation
Dataset partitioning ratio	Training set:Verification set:Test set = 0.6:0.3:0.1

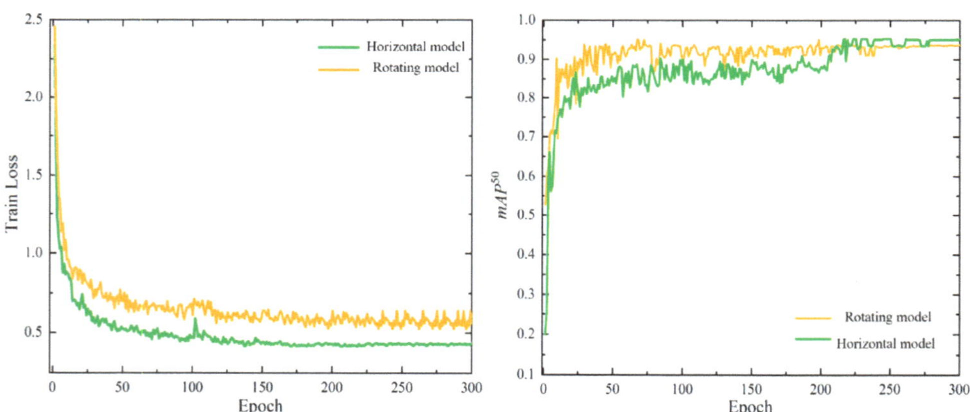

Figure 17. Model training curve.

In order to verify the detection performance of our foreign object detection model on the dataset, the self-made belt conveyor foreign object detection dataset was used for testing, with a total of 1070 images (including 253 background images). The confusion matrixes during the foreign object detection process were obtained, as shown in Figure 18. Ten-fold cross-validation was used to comprehensively evaluate the performance of the model, and the results of the cross-validation are shown in Figure 19.

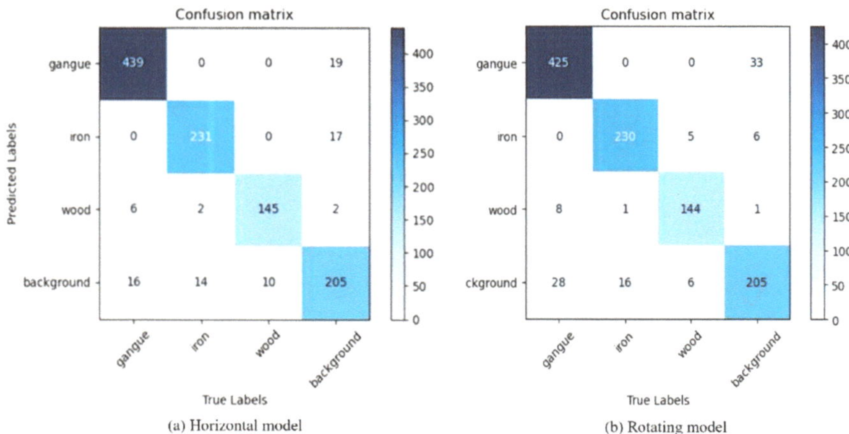

Figure 18. Confusion matrix results of the self-made dataset.

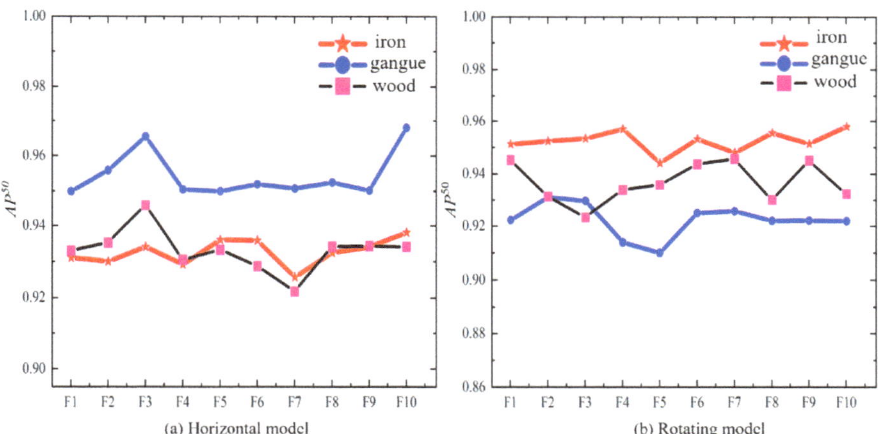

Figure 19. Cross-validation results of foreign object detection models.

It can be seen from Figures 20 and 21 that the proposed foreign object detection model can effectively detect foreign objects in the case of background coal flow. The rectangle in the figures is the target result predicted by the foreign object detection model, and different colors represent different categories. In Figure 21, the predicted angle information is represented by the long side of the rotating rectangular box, with angle values of 36.8, −30.3, and 65.1, which can verify the effectiveness of the rotation decoupling head in angle regression prediction. Figures 22 and 23 show the results of foreign object detection under coal-flow occlusion and the multi-angle detection results of the same foreign object, respectively. The proposed model can locate the foreign object in the image more accurately, and the performance indicators of the foreign object detection model are shown in Tables 5 and 6.

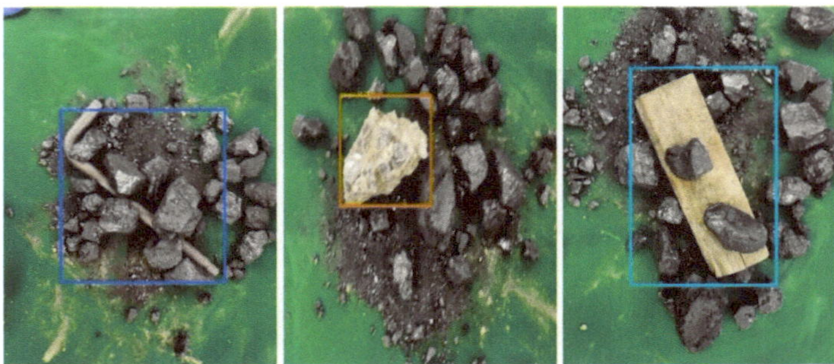

Figure 20. Test results of the foreign object detection of horizontal frames.

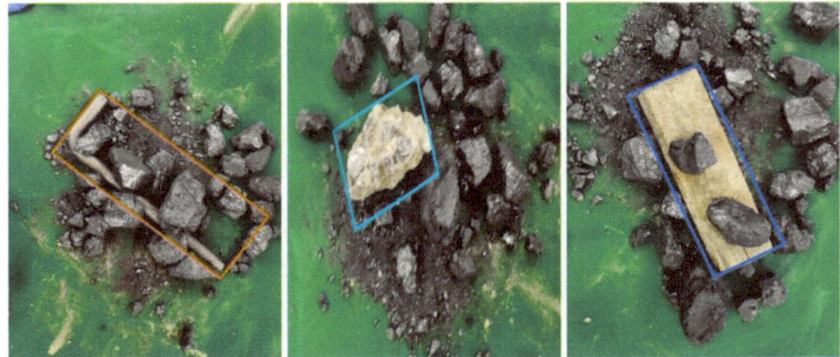

Figure 21. Test results of the foreign object detection of rotating frames.

(**a**) Horizontal box detection   (**b**) Rotation box detection

Figure 22. Detection of foreign objects under the shelter of coal flow.

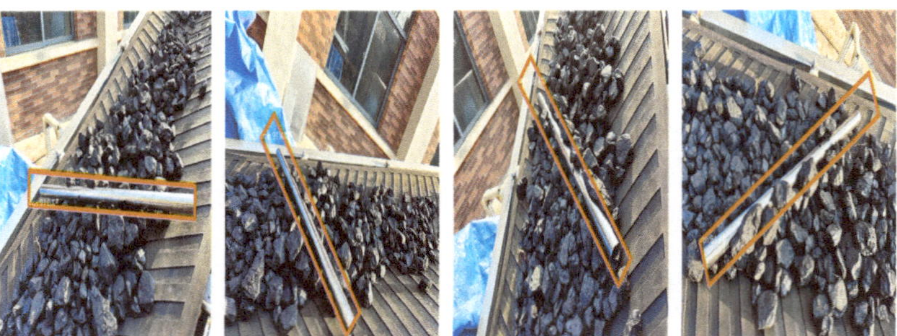

Figure 23. Multi-angle foreign object detection with the same foreign object sample.

Table 5. MO-YOLOX horizontal foreign object detection performance index parameters.

	Precision	Recall	AP^{50}	F2-Score	Inference Time/ms
iron	93.71%	93.20%	93.27%	93.30%	21
wood	93.12%	93.62%	93.30%	95.80%	23
large gangue	95.32%	95.92%	95.45%	93.52%	22
average value	94.05%	94.25%	94.01%	94.20%	22

Table 6. MO-YOLOX rotating frame foreign object detection performance index parameters.

	Precision	Recall	AP^{50}	F2-Score	Inference Time/ms
iron	95.11%	95.32%	95.25%	95.28%	28
wood	92.17%	92.51%	92.23%	92.44%	26
large gangue	94.32%	93.25%	93.56%	93.46%	29
average value	93.87%	93.69%	93.68%	93.73%	27.7

From the experimental results, it can be seen that the performance of the proposed foreign object detection model of the belt conveyor is superior to similar mainstream algorithms on both horizontal foreign object datasets and rotating foreign object datasets. Specifically, when the target foreign object and the coal mine stone have obvious differences in shape and texture, such as large gangue, the proposed horizontal frame foreign object detection model has very excellent performance. However, for slender ironware, the detection effect of the proposed model is slightly poor. It is exciting that the proposed rotating frame foreign body detection model has good detection sensitivity for targets with large length and width, such as slender iron bars, and the cost of angle prediction is a 5.7 ms increase in the reasoning time. There is an obvious difference in length and width between iron and wood, and the proposed model can effectively predict the angle of a foreign body. However, with the irregular gangue, its characteristics are quite different, and the angle information of the data label is irregular, which causes great difficulties in the angle regression prediction of the network. In addition, in the case of slight occlusion from the coal background, both horizontal frame foreign body detection and rotating foreign body detection can accurately detect foreign bodies and determine the types of foreign bodies. Therefore, the experimental results show that the proposed model meets the design requirements.

5. Conclusions and Future Works

In this paper, a foreign object image dataset for the belt conveyor is collected and established, and the IAT image enhancement module and CBAM attention mechanism are introduced. Secondly, a novel rotating decoupling head is designed to predict the angle information of foreign objects, and a MO-YOLOX network structure is constructed. The experimental results show that the proposed algorithm has a performance of 71.9% and 73.2% on the VOC and DOTA test datasets, respectively, with an average inference time of around 26 ms, which can meet the requirements of real-time inference. Ten-fold cross-validation is conducted on the self-built foreign object dataset of the belt conveyor, and the accuracy, recall, and mAP^{50} of horizontal frame foreign object detection are 94.05%, 94.25%, and 94.01%, respectively. Moreover, the accuracy, recall, and mAP^{50} of the rotating frame foreign object detection reaches 93.87%, 93.69%, and 93.68%, and the average inference time of foreign object detection is 25 ms.

However, the proposed foreign object detection method for belt conveyors we have designed has not yet considered embedded deployment as part of the industrial experiment. In the future, further research is needed on the pruning optimization of the model and embedded deployment.

Author Contributions: Conceptualization, R.Y. and P.Q.; methodology, D.H. and X.Z.; validation, H.L.; formal analysis, X.L.; data curation, P.Q.; writing—original draft preparation, D.H. and X.Z.; writing—review and editing, H.L.; funding acquisition, X.L. and D.H. All authors have read and agreed to the published version of the manuscript.

Funding: This research was funded by Lianyungang 521 High-level Talent Training Project grant number LYG06521202203, the Qing Lan Project for Excellent Teaching Team of Jiangsu Province grant number 2022, the National Natural Science Foundation of China grant number 51975568, the Natural Science Foundation of Jiangsu Province grant number BK20191341 and the Jiangsu Funding Program for Excellent Postdoctoral Talent grant number 2022ZB519.

Institutional Review Board Statement: Not applicable.

Informed Consent Statement: Not applicable.

Data Availability Statement: All the data that produce the results in this work can be requested from the corresponding author.

Conflicts of Interest: The authors declare that there are no conflicts of interest regarding the publication of this article.

References

1. Petrikova, I.; Marvalova, B.; Samal, S.; Cadek, M. Digital image correlation as a measurement tool for large deformations of a Conveyor Belt. *Appl. Mech. Mater.* **2015**, *732*, 77–80. [CrossRef]
2. Zimroz, R.; Stefaniak, P.K.; Bartelmus, W.; Hardygora, M. Novel techniques of diagnostic data processing for belt conveyor maintenance. In Proceedings of the 12th International Symposium Continuous Surface Mining—Aachen 2014; Springer International Publishing: Cham, Switzerland, 2015; pp. 31–40.
3. Cao, H.Q. Study and analysis on tear belt and break belt of belt conveyor in coal mine. *Coal Sci. Technol.* **2015**, *S2*, 130–134.
4. Hu, J.H.; Gao, Y.; Zhang, H.J. Recognition method of non-coal foreign matter in belt conveyor based on deep learning. *J. Mine Autom.* **2021**, *47*, 106653.
5. Chen, W.; Wang, X. Coal mine safety intelligent monitoring based on wireless sensor network. *IEEE Sens. J.* **2020**, *21*, 25465–25471. [CrossRef]
6. Gao, D.; Li, W.; Dai, K. Design of coal mine intelligent monitoring system based on ZigBee wireless sensor network. In Proceedings of the 2016 International Conference on Mechanics, Materials and Structural Engineering (ICMMSE 2016), Jeju Island, Republic of Korea, 18–20 March 2016; Atlantis Press: Amsterdam, The Netherlands, 2016; pp. 182–187.
7. Saydirasulovich, S.N.; Abdusalomov, A.; Jamil, M.K.; Nasimov, R.; Kozhamzharova, D.; Cho, Y.I. A YOLOv6-based improved fire detection approach for smart city environments. *Sensors* **2023**, *23*, 3161. [CrossRef]
8. Xu, P.; Zhou, Z.; Geng, Z. Safety monitoring method of moving target in underground coal mine based on computer vision processing. *Sci. Rep.* **2022**, *12*, 17899. [CrossRef]
9. Dai, L.; Qi, P.; Lu, H.; Liu, X.; Hua, D.; Guo, X. Image enhancement method in underground coal mines based on an improved particle swarm optimization algorithm. *Appl. Sci.* **2023**, *13*, 3254. [CrossRef]
10. Wei, W.; Li, L.; Shi, W.; Liu, J.P. Ultrasonic imaging recognition of coal-rock interface based on the improved variational mode decomposition. *Measurement* **2021**, *170*, 108728. [CrossRef]
11. Xu, D.; Wu, Y. Improved YOLO-V3 with DenseNet for multi-scale remote sensing target detection. *Sensors* **2020**, *20*, 4276. [CrossRef]
12. Cui, Z.; Li, K.; Gu, L.; Su, S.; Gao, P.; Jiang, Z.; Qiao, Y.; Harada, T. You Only Need 90K Parameters to Adapt Light: A Light Weight Transformer for Image Enhancement and Exposure Correction. *arXiv* **2022**, arXiv:2205.14871.
13. Wang, Q.; Cheng, M.; Huang, S.; Cai, Z.; Zhang, J.; Yuan, H. A deep learning approach incorporating YOLO v5 and attention mechanisms for field real-time detection of the invasive weed Solanum rostra-tum Dunal seedlings. *Comput. Electron. Agric.* **2022**, *199*, 107194. [CrossRef]
14. Jiang, L.; Peng, G.; Xu, B.; Lu, Y.; Wang, W. Foreign object recognition technology for port transportation channel based on automatic image recognition. *EURASIP J. Image Video Process.* **2018**, *2018*, 147. [CrossRef]
15. Zhang, Z.; Su, X.; Ding, L.; Wang, Y. Multi-scale image segmentation of coal piles on a belt based on the Hessian matrix. *Particuology* **2013**, *11*, 549–555. [CrossRef]
16. Saran, G.; Ganguly, A.; Tripathi, V.; Kumar, A.A.; Gigie, A.; Bhaumik, C.; Chakravarty, T. Multi-modal imaging-based foreign particle detection system on coal conveyor belt. *Trans. Indian Inst. Met.* **2022**, *75*, 2231–2240. [CrossRef]
17. Tu, L.; Zhong, S.; Peng, Q. Moving object detection method based on complementary multi resolution background models. *J. Cent. South Univ.* **2014**, *21*, 2306–2314. [CrossRef]
18. Lins, R.G.; Givigi, S.N. Automatic crack detection and measurement based on image analysis. *IEEE Trans. Instrum. Meas.* **2016**, *65*, 583–590. [CrossRef]

19. Ghasemi, Y.; Jeong, H.; Choi, S.H.; Park, K.B.; Lee, J.Y. Deep learning-based object detection in augmented reality: A systematic review. *Comput. Ind.* **2022**, *139*, 103661. [CrossRef]
20. Kaur, J.; Singh, W. Tools, techniques, datasets and application areas for object detection in an image: A review. *Multimed. Tools Appl.* **2022**, *81*, 38297–38351. [CrossRef] [PubMed]
21. LeCun, Y.; Bengio, Y.; Hinton, G. Deep learning. *Nature* **2015**, *521*, 436–444. [CrossRef]
22. Pu, Y.; Apel, D.B.; Szmigiel, A.; Chen, J. Image recognition of coal and coal gangue using a convolutional neural network and transfer learning. *Energies* **2019**, *12*, 1735. [CrossRef]
23. Girshick, R.; Donahue, J.; Darrell, T.; Malik, J. Rich feature hierarchies for accurate object detection and semantic segmentation. In Proceedings of the IEEE Conference on Computer Vision and Pattern Recognition, Columbus, OH, USA, 23–28 June 2014; pp. 580–587.
24. Ren, S.; He, K.; Girshick, R.; Sun, J. Faster r-cnn: Towards real-time object detection with region proposal networks. In Proceedings of the Advances in Neural Information Processing Systems, Montreal, QC, Canada, 7–12 December 2015; pp. 91–99.
25. Wang, Y.; Wang, Y.; Dang, L. Video detection of foreign objects on the surface of belt conveyor underground coal mine based on improved SSD. *J. Ambient. Intell. Humaniz. Comput.* **2020**, *14*, 5507–5516. [CrossRef]
26. Ma, G.; Wang, X.; Liu, J.; Chen, W.; Niu, Q.; Liu, Y.; Gao, X. Intelligent detection of foreign matter in coal mine transportation belt based on convolution neural network. *Sci. Program.* **2022**, *2022*, 9740622. [CrossRef]
27. Xiao, D.; Kang, Z.; Yu, H.; Wan, L. Research on belt foreign body detection method based on deep learning. *Trans. Inst. Meas. Control* **2022**, *44*, 2919–2927. [CrossRef]
28. Ge, Z.; Liu, S.; Wang, F.; Li, Z.; Sun, J. Yolox: Exceeding yolo series in 2021. *arXiv* **2021**, arXiv:2017.08430.
29. Chollet, F. Xception: Deep learning with depthwise separable convolutions. In Proceedings of the IEEE Conference on Computer Vision and Pattern Recognition, Honolulu, HI, USA, 21–26 July 2017; pp. 1251–1258.
30. Cui, Z.; Li, K.; Gu, L.; Su, S.; Gao, P.; Jiang, Z.; Qiao, Y.; Harada, T. Illumination adaptive transformer. *arXiv* **2022**, arXiv:2205.14871.
31. Yang, X.; Zhang, G.; Yang, X.; Tang, J.; He, T.; Yan, J. Detecting rotated objects as gaussian distributions and its 3-d generalization. *IEEE Trans. Pattern Anal. Mach. Intell.* **2022**, *45*, 4335–4354. [CrossRef]
32. Liu, W.; Anguelov, D.; Erhan, D.; Szegedy, C.; Reed, S.; Fu, C.Y.; Berg, A.C. SSD: Single Shot MultiBox Detector. In Proceedings of the Computer Vision–ECCV 2016: 14th European Conference, Amsterdam, The Netherlands, 11–14 October 2016; Springer International Publishing: Cham, The Netherlands; pp. 21–37.
33. Han, J.; Ding, J.; Li, J.; Xia, G.S. Align deep features for oriented object detection. *IEEE Trans. Geosci. Remote Sens.* **2021**, *60*, 5602511. [CrossRef]
34. Guo, Z.; Liu, C.; Zhang, X.; Jiao, J.; Ji, X.; Ye, Q. Beyond Bounding-Box: Convex-hull Feature Adaptation for Oriented and Densely Packed Object Detection. In Proceedings of the IEEE/CVF conference on Computer Vision and Pattern Recognition, Nashville, TN, USA, 20–25 June 2021; pp. 8792–8801.

Article

A Novel Approach to Quantitative Characterization and Visualization of Color Fading

Woo Sik Yoo [1,2,*], Kitaek Kang [1], Jung Gon Kim [1] and Yeongsik Yoo [3,*]

1 WaferMasters, Inc., Dublin, CA 94568, USA; kitaek.kang@wafermasters.com (K.K.); junggon.kim@wafermasters.com (J.G.K.)
2 Institute of Humanities Studies, Kyungpook National University, Daegu 41566, Republic of Korea
3 College of Liberal Arts, Dankook University, Yongin 16890, Republic of Korea
* Correspondence: woosik.yoo@wafermasters.com (W.S.Y.); ysyoophd@dankook.ac.kr (Y.Y.)

Abstract: Color fading naturally occurs with time under light illumination. It is triggered by the high photon energy of light. The rate of color fading and darkening depends on the substance, lighting condition, and storage conditions. Color fading is only observed after some time has passed. The current color of objects of interest can only be compared with old photographs or the observer's perception at the time of reference. Color fading and color darkening rates between two or more points in time in the past can only be determined using photographic images from the past. For objective characterization of color difference between two or more different times, quantification of color in either digital or printed photographs is required. A newly developed image analysis and comparison software (PicMan) has been used for color quantification and pixel-by-pixel color difference mapping in this study. Images of two copies of Japanese wood-block prints with and without color fading have been selected for the exemplary study of quantitative characterization of color fading and color darkening. The fading occurred during a long period of exposure to light. Pixel-by-pixel, line-by-line, and area-by-area comparisons of color fading and darkening between two images were very effective in quantifying color change and visualization of the phenomena. RGB, HSV, CIE L*a*b* values between images and their differences of a single pixel to areas of interest in any shape can be quantified. Color fading and darkening analysis results were presented in numerical, graphical, and image formats for completeness. All formats have their own advantages and disadvantages over the other formats in terms of data size, complexity, readability, and communication among parties of interest. This paper demonstrates various display options for color analysis, a summary of color fading, or color difference among images of interest for practical artistic, cultural heritage conservation, and museum applications. Color simulation for various moments in time was proposed and demonstrated by interpolation or extrapolation of color change between images, with and without color fading, using PicMan. The degree of color fading and color darkening over the various moments in time (past and future) can be simulated and visualized for decision-making in public display, storage, and restoration planning.

Keywords: color fading; color quantification; color difference; image analysis; statistical analysis; software; pixel-by-pixel color difference mapping

Citation: Yoo, W.S.; Kang, K.; Kim, J.G.; Yoo, Y. A Novel Approach to Quantitative Characterization and Visualization of Color Fading. *Technologies* **2023**, *11*, 108. https://doi.org/10.3390/technologies11040108

Academic Editor: Pietro Zanuttigh

Received: 12 July 2023
Revised: 1 August 2023
Accepted: 4 August 2023
Published: 8 August 2023

1. Introduction

The energy of photons has the potential to cause irreversible damage to objects [1–3]. The photon energy increases with the shortening of wavelength. For the visible wavelength range of 380–700 nm (violet to red), the photon energy corresponds to 3.26–1.77 eV [4]. The higher energy side of photons, from blue to ultraviolet (UV) light, can be especially deleterious to colorants, resulting in visible fading [5]. Conservators understand that light-sensitive objects have a finite life for display under illumination. Most collecting institutions often adopt guidelines to indicate the lighting conditions and duration of exhibitions based on the characteristics of substances and art media such as oil paintings, watercolors, metal sculptures, and so on. These guidelines assume that objects within a class share a similar known stability under the same illumination conditions. Some of those assumptions have been brought into question, and additional investigations use a micro fading tester (MFT) to predict a colorant's rate of fade over time [6].

The MFT is an analytical technique proposed and developed by Paul Whitmore of Carnegie Mellon University to determine the in situ light sensitivity of a piece of artwork [6,7]. The MFT was designed to rapidly induce and monitor color change in small areas of fugitive materials. This is achieved by exposing the sample surface to a stable, high-intensity focused light spot (typically 0.5 mm or less in diameter) and simultaneously examining the affected area using a spectrophotometer. The resulting color change may be perceived as photo-induced damage from the small area of exposure to very intense light. Once a minimum color change threshold is reached, even if the change is visually imperceptible, the test is terminated. The MFT allows the direct identification of the most fugitive colorants in an artwork by comparing the color change behavior of other colorants and light-sensitive standard materials. This predictive information for specific colorants used in an artwork is very valuable when deciding on a preservation lighting policy [1,7]. Characterization of color fading and its rate determination as a function of aging conditions are of strong interest in the fields of art, museum, cultural heritage, paint, fabrics, food, plant, materials, and so on [8–16].

There is a well-documented European Standard on the Conservation of cultural property—Test methods—Colour measurement of surfaces [17]. It describes a test method to measure the surface color of porous inorganic materials and their possible chromatic changes. However, no reference to the appearance of glossy surfaces is described. The method may be applied to porous inorganic materials either untreated or subjected to any treatment or aging. The method is suitable for the measurement of color coordinates of representative surfaces of objects, indoors or outdoors, and representative surfaces of specimens described in the document.

In this paper, we propose a novel approach to quantitative characterization visualization of color fading and color darkening by image comparisons using image analysis/processing software (PicMan) capable of color analysis from single pixels, lines, and areas of any shape and size up to the entire image. We demonstrate various display options for color analysis comparisons of color fading or color difference among images of interest for practical applications in the field of art, cultural heritage, and museums. Potential applications of color interpolation between images and the possibility of extrapolation beyond referenced images for predicting future color appearance simulation are discussed.

2. Images and Analysis Methods

2.1. Test Images

Two copies of a Japanese woodblock print by Utamaro used in a Microfading Workshop given in 2014 were selected as test images [1,18]. Figure 1 shows the two copies of the Japanese woodblock print by Utamaro (a) before and (b) after light exposure to cause significant color fading. According to the source, they were identical twins when they were printed. They were made from the same materials and printed in the same way. They were probably printed by the same people. Then the two copies were separated after printing, with one image experiencing significant color fading by light exposure. The light exposure has caused profound changes to a majority of colors used in the print. Some colors, such as the blue on the kimono (Japanese dress), are not very light-sensitive compared to the other colors. The purpose of the microfading experiment at the Getty workshop was to gain information on the light sensitivity of each colorant without the identical twins as a reference. Assuming the images were used as an example in a Microfading Workshop for professionals in the field, the images were digitized using the same equipment under the same conditions, including lighting conditions and environment.

(**a**)　　　　　　　　　　　　　　　　(**b**)

Figure 1. Two identical copies of a Japanese woodblock print by Utamaro before and after significant light exposure [17]: (**a**) as printed; (**b**) after light exposure.

2.2. Image Analysis/Processing Software (PicMan)

Image processing and analysis are very different from image editing (brightening, straightening, white balancing, contrast stretching, resizing, etc.) or image modification using commercially available software such as Photoshop, Illustrator, or Lightroom, as well as many other photo-editing software. More than one type of software is required to efficiently perform desired tasks on digital photographs, images, videos, and documents. Multiple image-editing application software requires tremendous data processing power and memory as resources. They are operating on PCs simultaneously in the background to perform very specific tasks of varying complexity. The demand for computing resources is increasing exponentially with the increase in the number of files and size of image files. The development of integrated image analysis/processing software suitable for frequently needed image editing, processing, and various analysis functions with image, video, and numerical data exporting capabilities is strongly desired. Ideally, all desired tasks should be able to be performed from a regular PC with reasonable computing power and without opening multiple applications and switching between applications.

As reported previously, the authors' group has been developing user-friendly software (PicMan, WaferMasters, Inc. Dublin, CA, USA) for image-based dimension (length, area, circumference, circularity, etc.) measurement/analysis, quantitative color analysis, statistical analysis, and various image-tailoring functions to address the deficiencies in the above-mentioned areas with existing software [19–21]. Pattern selection, area selection, highlighting, editing, coloring, and transparency application to any digital image can be easily done. A few applications of these functions have been successfully applied to archaeology, cultural heritage, conservation science, material science, biology, medical science, semiconductor research, and development studies [21–25]. PicMan can handle various formats of digital image and video files such as JPG, BMP, GIF, PNG, TGA, TIFF, WEBP, JPEG XR, PDF, CR2, DM3, ND2, MIRAX, MOV, MP4, AVI, WMV, etc. for image analysis and processing. Detailed application examples of PicMan can be found in previous reports from other study groups [19–25]. Several new functions, such as pixel-by-pixel color difference mapping, block comparison, color interpolation, and extrapolation, for color fading characterization, have been added to this study.

All digital images have a set of pixel information on the x and y coordinates and RGB brightness. Combinations of RGB brightness values at different coordinates determine the color, brightness, and shapes of interest. Combining the RGB brightness (8 bits per channel), values can generate more than 16 million colors (2^8 = 256 brightness values $\times$ 3 channels = $2^8 \times 3$ colors = 224 colors = 16,777,216 colors) in displays and monitor screens. However, it is more difficult to understand how we feel, recognize, and interpret colors and shapes from visual stimulation. The differences in the spectral intensity distribution of light (wavelength dependence of brightness in the visible wavelength range of 380–700 nm, 400–700 nm, or 400–780 nm) make us perceive corresponding colors. In the course of color model and theory development, many different concept customs of color spaces have been introduced to characterize and classify colors in quantitative and traceable manners suitable for the field of applications.

The most popular color spaces used across disciplines are RGB, HSV, CIE L*a*b*, and Munsell models. The RGB and CIE L*a*b* color spaces are built on cartesian coordinates, while the HSV and Munsell color spaces are based on cylindrical coordinates. The CIE L*a*b* coordinate is based on RGYB instead of RGB. The CIE L* value indicates lightness assuming grayscale, while the a* and b* values are determined by the balance between red–green and yellow–blue, respectively. For conversion between RGB and CIE L*a*b* color spaces, a new cartesian coordinate of XYZ is introduced to add Y (yellow) color component into account [20,25–27]. Then, the XYZ values were used for L*a*b* values in the CIE L*a*b* color spaces. All digital images are based on combinations of RGB channel brightness. The color information in the other color spaces can be calculated from the RGB values. Details of color space transformations and color conversion in different color spaces can be found elsewhere [25–27]. All 16,777,216 colors are assigned six-digit hexadecimal color codes for computer graphics. The capability of pixel-by-pixel color extraction, and the average color extraction of selected areas, can be very useful for quantitative color characterization for a variety of applications. All colors can be quantified as numerical values or corresponding hexadecimal color codes. It is extremely helpful for objective communication and reproduction of colors, hue, tint, tone, and shade.

3. Results

Color quantification and comparisons between the two images before and after color fading (shown in Figure 1) were made point-by-point, line-by-line, area-by-area, and block-by-block. Color information was extracted and exported as data files in CSV format for further analysis. Details of color characterization, statistical analysis, and comparison results are described in the following subsections.

3.1. Histogram Analysis

Figure 2 shows a screenshot image of image analysis software (PicMan) for color information extraction at 34 points of five-pixel diameter regions of interest (ROIs) and a 284×422 pixel (=119,848 pixels) rectangle (white border line) area per printed image. The selected rectangular ROIs are the identical location on individual images. RGB and L*a*b* histogram analysis was done for the rectangle (white border line) areas of the two images to gain insight into the overall color distribution and lightness values. Then, the identical 34 points were selected for point-by-point color comparison and color information extraction for further analysis.

The RGB and CIE L*a*b* histograms of two rectangle areas with white border lines on the images before and after color fading and color darkening (in Figure 2) were plotted in Figure 3a–d. The image before color fading showed wide RGB intensity distribution due to the vibrant colors of the print image, as shown in Figure 3a. The CIE L*a*b* histogram showed a high CIE L* value peak at around 78, corresponding to bright colors on the image. As seen in Figure 3b, the majority of colors show a slightly reddish color (i.e., low positive value for CIE a*) and mild yellowish color (i.e., low positive values for CIE b*). The presence of small portions of pixels with a greenish color component can be recognized from the small area under negative CIE a* values in the range of −5 to 0.

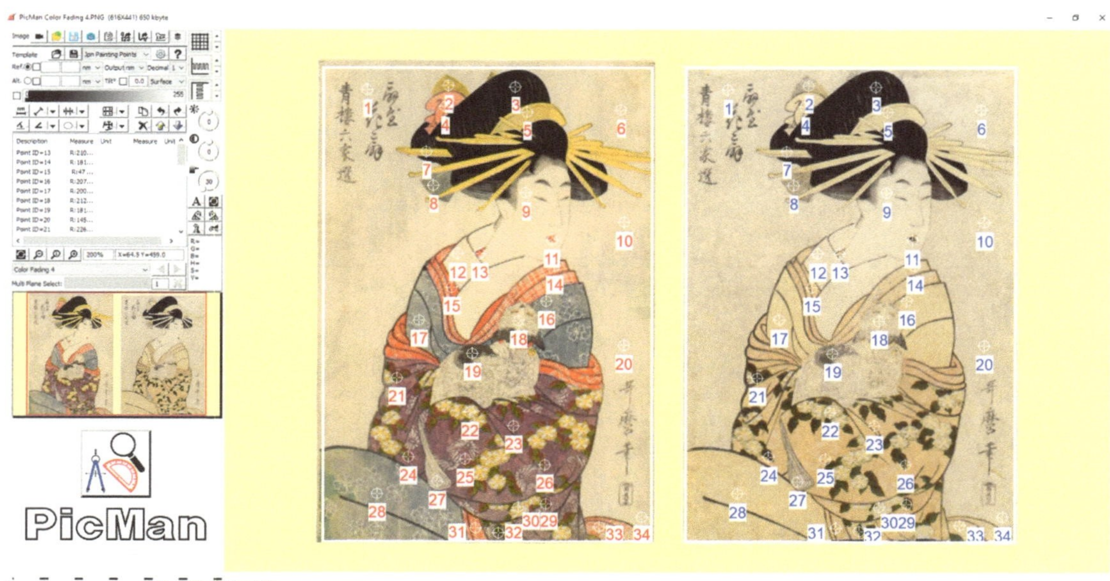

Figure 2. A screenshot image of image analysis software (PicMan) for color information extraction at 34 points of five-pixel diameter regions of interest (ROIs) and a 284 × 422 pixel (= 119,848 pixels) rectangle (white border line) area per print. For easy comparisons, different colors for the numbers of ROIs for two images were used.

The RGB and CIE L*a*b* histograms of the print after color fading were also plotted in the same vertical and horizontal scales for easy comparison (Figure 3c,d). The red peak became stronger, and small peaks below 192 became flatter, showing the brightening of the print after color fading (and darkening) after light exposure. The increase of RGB intensity and the loss of variation in RGB histogram are the results of discoloration and detailed pattern loss in print. The CIE L*a*b* histogram also showed a very sharp and intense positive CIE a* peak between 5–10 and a broadening of the positive side of the CIE b* histogram graph extending toward the higher positive CIE b* values toward yellowish color after color fading. These agree well with the RGB histogram and the effect of color fading.

As seen in Figure 3, a decrease of 'frequency (number of pixels with a given brightness in RGB channel)' and broadening of the green and blue channel histogram and the opposite trends of the green channel histogram are noticed after color fading or darkening of certain colors. Similar trends in L*a*b* color space (i.e., some values increased, others decreased) are observed. Most colors became lighter (color fading), and some became darker (color darkening) with light exposure. It should be noted that the pixels with the L* value in the range of 40–70 before color fading have increased their L* value to 60–74 after color fading. It is a good indicator for color fading in most printed areas by light exposure.

Table 1 shows the RGB and CIE L*a*b* statistics of a rectangle area before and after color fading (and darkening). As seen from the table, the average values of RGB increased while the standard deviation of RGB intensity decreased after color fading. The average color change between the two test images before and after color fading/darkening can be easily judged in L*, a*, b*, and RGB values. It clearly indicates that the color became brighter, and details are lost after color fading. The CIE L*a*b* statistics showed an increase in the average CIE L* value and a decrease in CIE a* and CIE b* values. This can be interpreted that most colors were lightened and approached natural tone colors. The standard deviations of the CIE L*a*b* values were decreased after color fading, a clear indicator of the loss of contrast after color fading.

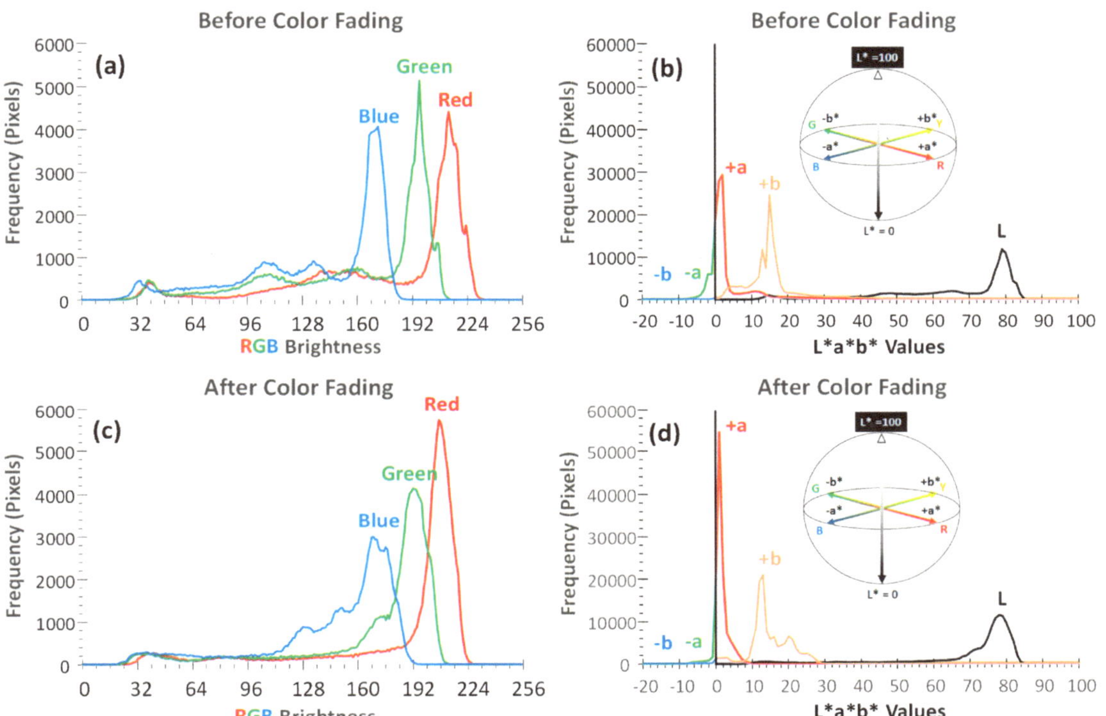

Figure 3. RGB and CIE L*a*b* histograms of a 284 × 422 pixel (=119,848 pixels) rectangle area per print: (**a**) RGB histogram of the original copy; (**b**) CIE L*a*b* histogram of the original copy; (**c**) RGB histogram of the faded copy; (**d**) CIE L*a*b* histogram of the faded copy.

Table 1. RGB and CIE L*a*b* statistics of a 284 × 422 pixel (= 119,848 pixels) rectangle area before and after color fading.

	Before Color Fading						After Color Fading					
	Red	**Green**	**Blue**	**L***	**a***	**b***	**Red**	**Green**	**Blue**	**L***	**a***	**b***
Count	119,848 Pixels						119,848 Pixels					
Minimum	18	22	10	7.7	−12.5	−5.4	21	13	10	4.9	−13.4	−8.3
Average	180.1	160.6	135.9	66.7	3.5	15.5	187.2	170.9	145.4	70.2	1.9	15.3
Maximum	242	215	190	86.4	51.5	52.9	231	219	199	87.8	15.3	33.6
Range	224	193	180	78.7	64.0	58.2	210	206	189	82.8	28.7	41.9
StdDev	47.2	46.0	41.2	17.7	6.6	7.2	44.4	41.6	38.8	16.6	1.9	5.4

3.2. Point-by-Point Color Extraction

Point-by-point color information was extracted from 34 points per image, as indicated in Figure 2. The identical locations on the two images, before and after color fading, were selected for color information extraction. The average colors of the five-pixel diameter circles were extracted and summarized in Table 2. Average RGB values, average HSV values, and average CIE L*a*b* values with average colors of 34 five-pixel diameter areas on the prints, before and after color fading (and darkening), are listed side-by-side for easy comparison.

Table 2. Average RGB values, average HSV values, average CIE L*a*b* values, and average colors of 34 five-pixel diameter areas at identical locations on the prints before and after color fading.

	Before Fading										After Fading									
No.	R	G	B	H	S	V	L*	a*	b*	Color	R	G	B	H	S	V	L*	a*	b*	Color
1	210	194	172	34.7	0.2	0.8	79.2	2.1	13.1		211	197	176	36.0	0.2	0.8	80.1	1.5	12.3	
2	181	153	95	40.5	0.5	0.7	64.5	2.7	34.2		187	173	151	36.7	0.2	0.7	71.3	1.4	13.1	
3	47	46	40	51.4	0.2	0.2	18.8	−0.9	4.0		49	47	43	40.0	0.1	0.2	19.5	0.0	2.9	
4	207	144	118	17.5	0.4	0.8	65.4	20.7	23.4		206	185	154	35.8	0.3	0.8	76.2	2.7	18.5	
5	200	175	116	42.1	0.4	0.8	72.4	0.9	33.5		207	190	156	40.0	0.3	0.8	77.6	0.7	19.4	
6	212	194	167	36.0	0.2	0.8	79.3	2.0	15.9		215	201	183	33.7	0.2	0.8	81.6	1.9	10.7	
7	181	155	96	41.6	0.5	0.7	65.1	1.7	34.4		179	162	136	36.3	0.2	0.7	67.4	2.0	15.8	
8	145	137	117	42.9	0.2	0.6	57.3	−0.5	11.8		172	156	133	35.4	0.2	0.7	65.2	2.1	14.2	
9	226	205	175	35.3	0.2	0.9	83.4	2.7	17.6		217	204	183	37.1	0.2	0.9	82.6	1.1	12.1	
10	216	198	171	36.0	0.2	0.9	80.7	2.0	15.8		210	194	171	35.4	0.2	0.8	79.2	1.9	13.6	
11	211	168	144	21.5	0.3	0.8	72.1	12.4	18.4		180	162	140	33.0	0.2	0.7	67.6	3.0	13.8	
12	218	201	171	38.3	0.2	0.9	81.6	1.2	17.2		213	197	176	34.1	0.2	0.8	80.3	2.2	12.6	
13	192	143	123	17.4	0.4	0.8	63.6	15.9	17.9		192	174	151	33.7	0.2	0.8	72.0	2.8	14.2	
14	209	157	134	18.4	0.4	0.8	69.0	16.3	19.6		208	185	154	34.4	0.3	0.8	76.3	3.4	18.8	
15	188	94	75	10.1	0.6	0.7	50.9	36.2	28.3		191	163	127	33.7	0.3	0.8	68.6	5.0	22.5	
16	157	154	134	52.2	0.2	0.6	63.3	−2.5	10.9		207	188	152	39.3	0.3	0.8	77.0	1.2	20.7	
17	161	159	141	54.0	0.1	0.6	65.2	−2.5	9.7		209	190	154	39.3	0.3	0.8	77.7	1.2	20.7	
18	215	198	167	38.8	0.2	0.8	80.5	1.1	17.7		207	189	162	36.0	0.2	0.8	77.4	2.1	16.0	
19	102	95	78	42.5	0.2	0.4	40.5	−0.3	10.8		140	129	110	38.0	0.2	0.6	54.5	0.9	11.8	
20	218	200	174	35.5	0.2	0.9	81.4	2.2	15.3		205	189	166	35.4	0.2	0.8	77.4	1.9	13.7	
21	117	88	64	27.2	0.5	0.5	39.8	8.6	18.5		94	89	66	49.3	0.3	0.4	37.7	−2.1	14.1	
22	201	176	126	40.0	0.4	0.8	72.9	1.9	28.9		193	176	142	40.0	0.3	0.8	72.5	0.8	19.7	
23	135	102	101	1.8	0.3	0.5	46.4	13.1	5.7		191	161	122	33.9	0.4	0.8	68.0	5.5	24.4	
24	125	96	78	23.0	0.4	0.5	43.2	9.1	14.8		122	110	81	42.4	0.3	0.5	46.8	−0.1	17.9	
25	139	104	98	8.8	0.3	0.6	47.3	13.1	8.8		190	158	115	34.4	0.4	0.8	67.0	5.8	26.9	
26	130	99	84	19.6	0.4	0.5	44.7	10.3	13.4		149	131	97	39.2	0.4	0.6	55.6	1.6	20.9	
27	190	175	151	36.9	0.2	0.8	72.1	1.5	14.3		180	164	141	35.4	0.2	0.7	68.2	2.1	14.1	
28	156	154	134	54.5	0.1	0.6	63.3	−2.9	10.8		206	186	150	38.6	0.3	0.8	76.3	1.6	20.9	
29	112	84	61	27.1	0.5	0.4	38.1	8.4	17.9		122	112	81	45.4	0.3	0.5	47.4	−1.2	18.7	
30	193	171	129	39.4	0.3	0.8	70.9	1.7	24.6		193	177	146	39.6	0.2	0.8	72.8	0.8	18.0	
31	205	137	111	16.6	0.5	0.8	63.4	23.0	24.5		183	156	121	33.9	0.3	0.7	65.9	4.8	22.1	
32	105	98	70	48.0	0.3	0.4	41.5	−2.0	17.0		131	122	92	46.2	0.3	0.5	51.3	−1.6	17.7	
33	205	155	127	21.5	0.4	0.8	68.0	14.9	21.9		207	186	152	37.1	0.3	0.8	76.5	2.2	20.0	
34	208	191	164	36.8	0.2	0.8	78.1	1.7	15.8		208	193	167	38.0	0.2	0.8	78.7	1.1	15.0	

All color information in a digital image file consists of a set of RGB intensity and pixel coordinate information. The RGB values can be translated into color values in other color space systems such as HSV, CIE L*a*b*, Munsell color, XYZ, and hexadecimal color codes. Table 2 shows HSV, CIE L*a*b*, and corresponding hexadecimal colors translated from RGB values of the images before and after color fading. As seen from the average colors for the 34 locations on the images, before and after color fading in Table 2, the majority of locations showed noticeable changes of colors after color fading (and darkening) by light illumination.

To make color differences before and after fading, the average colors at 34 locations were summarized in the order of $\Delta E_{L^*a^*b^*}$ (Table 3). The colors before and after color fading were shown side-by-side and provided differences in R, G, B, H, S, V, L*, a*, b*, and $\Delta E_{L^*a^*b^*}$. The color difference $\Delta E_{L^*a^*b^*}$ is defined as:

$$E_{L^*a^*b^*} = \sqrt{\left(L_1^* - L_2^*\right)^2 + \left(a_1^* - a_2^*\right)^2 + \left(b_1^* - b_2^*\right)^2} \tag{1}$$

where $\Delta E_{L^*a^*b^*} \approx 2.3$ (JND: a noticeable difference in the CIE76 formula. The $\Delta E_{L^*a^*b^*}$ values ranged from 1.2 to 36.3. Only three out of 34 locations have $\Delta E_{L^*a^*b^*} \approx 2.3$ (or JND) before and after color fading and color darkening. Background and hair colors were the only colors with negligible or unrecognizable fading. In other words, 31 out of 34 locations showed noticeable color differences.

Table 3. Difference between average RGB values, average HSV values, average CIE L*a*b* values, and their average colors of 34 five-pixel diameter areas at identical locations on the prints before and after color fading in the ascending order of $\Delta E_{L^*a^*b}^*$ rank.

ΔE Rank	No.	Color Before	After	ΔR	ΔG	ΔB	ΔH	ΔS	ΔV	ΔL*	Δa*	Δb*	$\Delta E_{L^*a^*b^*}$
1	34			0	−2	−3	−1.2	0.01	0	−0.6	0.62	0.8	1.2
2	1			−1	−3	−4	−1.3	0.01	−0.01	−0.9	0.6	0.81	1.4
3	3			−2	−1	−3	11.4	0.03	−0.01	−0.7	−0.89	1.15	1.6
4	10			6	4	0	0.6	0.02	0.03	1.5	0.13	2.22	2.7
5	18			8	9	5	2.8	0	0.03	3.1	−1	1.74	3.7
6	27			10	11	10	1.5	−0.01	0.04	3.9	−0.59	0.21	3.9
7	20			13	11	8	0.1	0.01	0.05	4	0.24	1.62	4.3
8	12			5	4	−5	4.2	0.05	0.01	1.3	−0.99	4.6	4.9
9	6			−3	−7	−16	2.3	0.06	−0.01	−2.3	0.14	5.15	5.6
10	9			9	1	−8	−1.8	0.07	0.04	0.8	1.61	5.53	5.8
11	30			0	−6	−17	−0.2	0.09	0	−1.9	0.92	6.61	6.9
12	8			−27	−19	−16	7.5	−0.04	−0.1	−7.9	−2.6	−2.38	8.7
13	22			8	0	−16	0	0.11	0.03	0.4	1.11	9.28	9.4
14	32			−26	−24	−22	1.8	0.03	−0.1	−9.8	−0.43	−0.73	9.8
15	24			3	−14	−3	−19.4	0.04	0.01	−3.6	9.15	−3.11	10.3
16	11			31	6	4	−11.5	0.1	0.12	4.5	9.38	4.56	11.4
17	21			23	−1	−2	−22.1	0.15	0.09	2.1	10.67	4.46	11.8
18	29			−10	−28	−20	−18.3	0.12	−0.04	−9.3	9.59	−0.75	13.4
19	19			−38	−34	−32	4.5	0.03	−0.15	−14	−1.19	−1.07	14.1
20	14			1	−28	−20	−16	0.1	0	−7.3	12.85	0.81	14.8
21	5			−7	−15	−40	2.1	0.17	−0.03	−5.2	0.2	14.17	15.1
22	33			−2	−31	−25	−15.6	0.11	−0.01	−8.5	12.7	1.94	15.4
23	26			−19	−32	−13	−19.6	0	−0.07	−10.9	8.69	−7.57	15.9
24	13			0	−31	−28	−16.3	0.15	0	−8.4	13.09	3.72	16.0
25	28			−50	−32	−16	15.9	−0.13	−0.2	−13	−4.43	−10.1	17.0
26	17			−48	−31	−13	14.7	−0.14	−0.19	−12.5	−3.67	−11	17.1
27	16			−50	−34	−18	12.9	−0.12	−0.19	−13.7	−3.64	−9.79	17.2
28	31			22	−19	−10	−17.3	0.12	0.08	−2.5	18.19	2.49	18.5
29	7			2	−7	−40	5.3	0.23	0.01	−2.3	−0.34	18.6	18.7
30	4			1	−41	−36	−18.3	0.18	0	−10.8	18.05	4.91	21.6
31	2			−6	−20	−56	3.8	0.29	−0.02	−6.8	1.27	21.11	22.2
32	25			−51	−54	−17	−25.6	−0.1	−0.2	−19.7	7.27	−18.1	27.7
33	23			−56	−59	−21	−32.1	−0.11	−0.22	−21.6	7.65	−18.7	29.6
34	15			−3	−69	−52	−23.6	0.26	−0.01	−17.7	31.17	5.76	36.3

Figure 4 shows the RGB value change before and after fading and their colors at 34 five-pixel diameter ROIs in the ascending order of CIE $\Delta E_{L^*a^*b^*}$ rank, from left to right. Thick colored lines and thin black lines are the intensity (or brightness) of each color channel before and after color fading. Color fading generally makes the intensity (or brightness) increase in all RGB color channels except for a few measurement locations indicated with arrows. It clearly indicates that most colors become lighter after color fading. However, a few colors became darker under light illumination, as seen in the locations with negative CIE ΔL* values in Table 3. The red and green channel intensity changed noticeably compared to the blue channel intensity after color fading by light illumination. In general, darker colors with lower red and green intensities tend to fade more and result in noticeable color changes. As the results changed in the CIE, $\Delta E_{L^*a^*b^*}\Delta$ was higher for the darker colors.

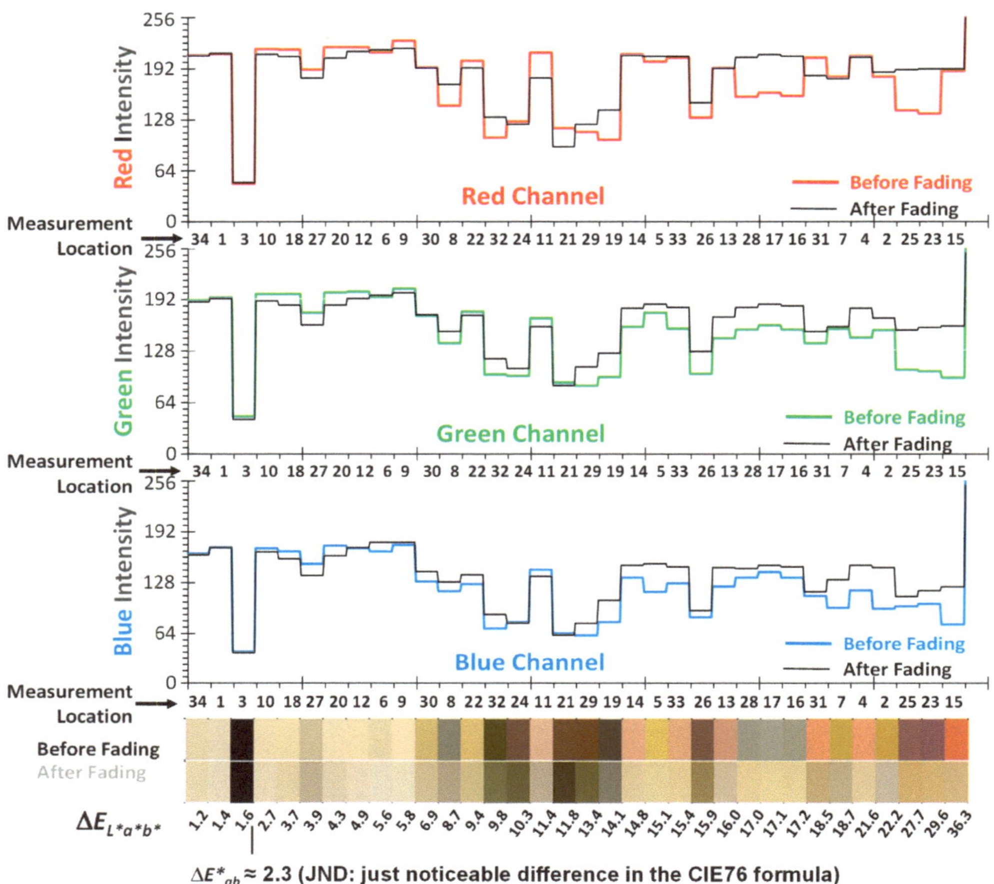

Figure 4. RGB value change before and after fading and their colors at 34 five-pixel diameter ROIs in the ascending order of CIE ΔE_{L*a*b*} rank, from left to right.

3.3. Line-by-Line Color Extraction

Line-by-line color information was extracted from 10 lines, as shown in Figure 5. Color information can be extracted from all lines in any direction, including free lines. Figure 6 shows the RGB intensity profile along 4 selected lines 2, 6, 8, and 10. Colors of a few locations per line showing noticeable color differences before and after color fading were shown together with RGB intensity line graphs.

Similar trends for color change by color fading, summarized in Tables 2 and 3, can be verified in Figure 6. Average RGB values and their ranges are very different between the two copies of the prints, before and after color fading. The average RGB value for the original copy of the print is smaller than the color faded copy. The RGB intensity range for the original copy of the print is much wider due to the large contrast of colors along the horizontal lines 6, 8, and 10. In contrast, the color-faded copy of the print shows a narrower range of RGB intensity values.

All colors can be quantified as RGB, HSV, CIE L*a*b*, XYZ, and Munsell color values, even with hexadecimal color codes. The extracted color information can be graphed for objective evaluation. Discoloration (or color fading) rate can be quantitatively determined per colorant as a function of the light exposure conditions. The MFT results can also be characterized by digital photography and used as a color fading database for individual colorants.

(**a**)   (**b**)

Figure 5. Ten numbered lines for color extraction and comparison between two identical copies of a Japanese woodblock print by Japanese artist Utamaro before and after significant light exposure: (**a**) as printed; (**b**) after light exposure.

3.4. Pixel-by-Pixel Image Comparison

Figure 7 shows a summary of the pixel-by-pixel image comparisons of the two copies of a print before and after color fading. For the alignment of two images, we overlapped the two images and translated one image in x and y coordinates pixel-by-pixel to minimize the color difference due to the misalignment. In fact, we found that the two images were digitized in the same magnification, and only two pixels in the y coordinate were off between the two images. Thus, the two images were suitable for pixel-by-pixel color comparisons. The pixel-by-pixel color differences were mapped in terms of the CIE ΔL^*, CIE Δa^*, CIE Δb^*, and CIE ΔE^* values at full scale (100% scale) and at quarter scale (25% of full scale) for easy recognition of changes by color fading. The full-scale images are shown in the top row, and the quarter-scale images are shown in the bottom row. Since the color differences along the CIE a^* and CIE b^* axes were small compared to the lightness of the CIE L^* axis, the full-scale pixel-by-pixel color difference mapping images for the CIE Δa^* and CIE Δb^* values did not show significant color differences. The light exposure induced color differences in lightness in CIE ΔL^*, color shift (in CIE Δa^* and CIE Δb^*), and color difference in CIE ΔE^* were four (4) times magnified in quarter scale images.

The CIE $L^*a^*b^*$ color space expresses color as three values. The lightness value CIE L^* defines black as 0 and white as 100. The CIE a^* axis is relative to the green–red opponent colors, green (negative CIE a^* values) to red (positive CIE a^* value). The b^* axis is relative to the blue–yellow opponent colors, blue (negative CIE b^* values) to yellow (positive CIE b^* value). The CIE a^* and CIE b^* axes are independent in the range of -128 to 127. In CIE $L^*a^*b^*$ color space, CIE L^* is for perceptual lightness, and CIE a^* and CIE b^* are for the four unique colors of human vision, red, green, blue, and yellow. The CIE ΔL^* and CIE ΔE^* values are in grayscale because they only have a magnitude in the brightness and overall color difference. The CIE Δa^* and CIE Δb^* values can be negative, zero (i.e., no color difference along the CIE a^* axis for red and green balance or CIE b^* axis for yellow and blue balance), or positive.

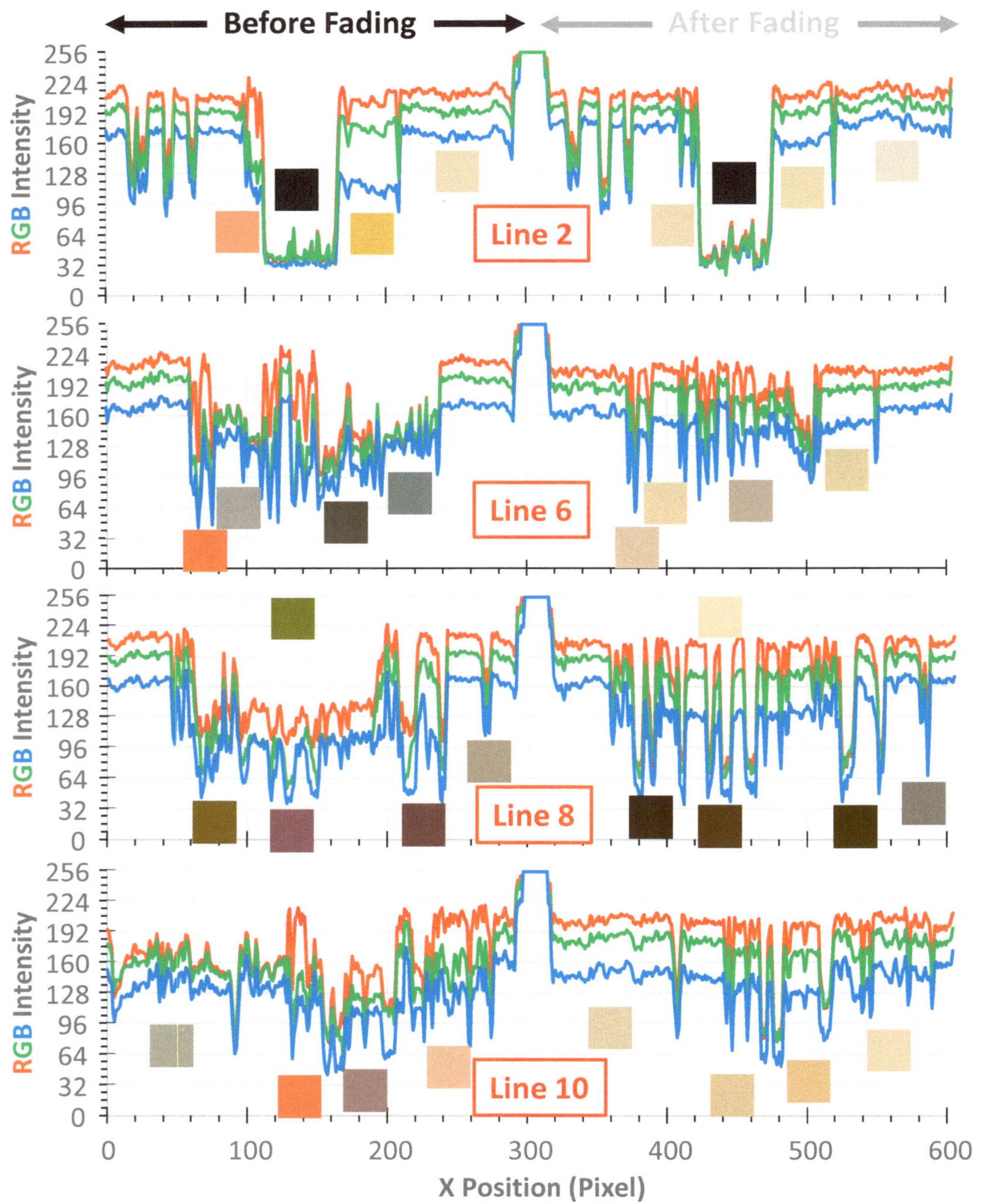

Figure 6. RGB values change across lines 2, 6, 8, and 10 on the image on the left. Colors of sections of noticeable color change before and after light illumination are shown for easier recognition and interpretation of RGB values.

The degrees of the color shifts along the CIE a* and CIE b* axes can be easily recognized, pixel by pixel, from Figure 7. The color difference CIE ΔE* can easily be determined and visualized pixel by pixel. This type of detailed pixel-by-pixel color difference characterization cannot be measured or visualized by conventional chroma meter measurements due to spatial resolution limits [19–21,28,29].

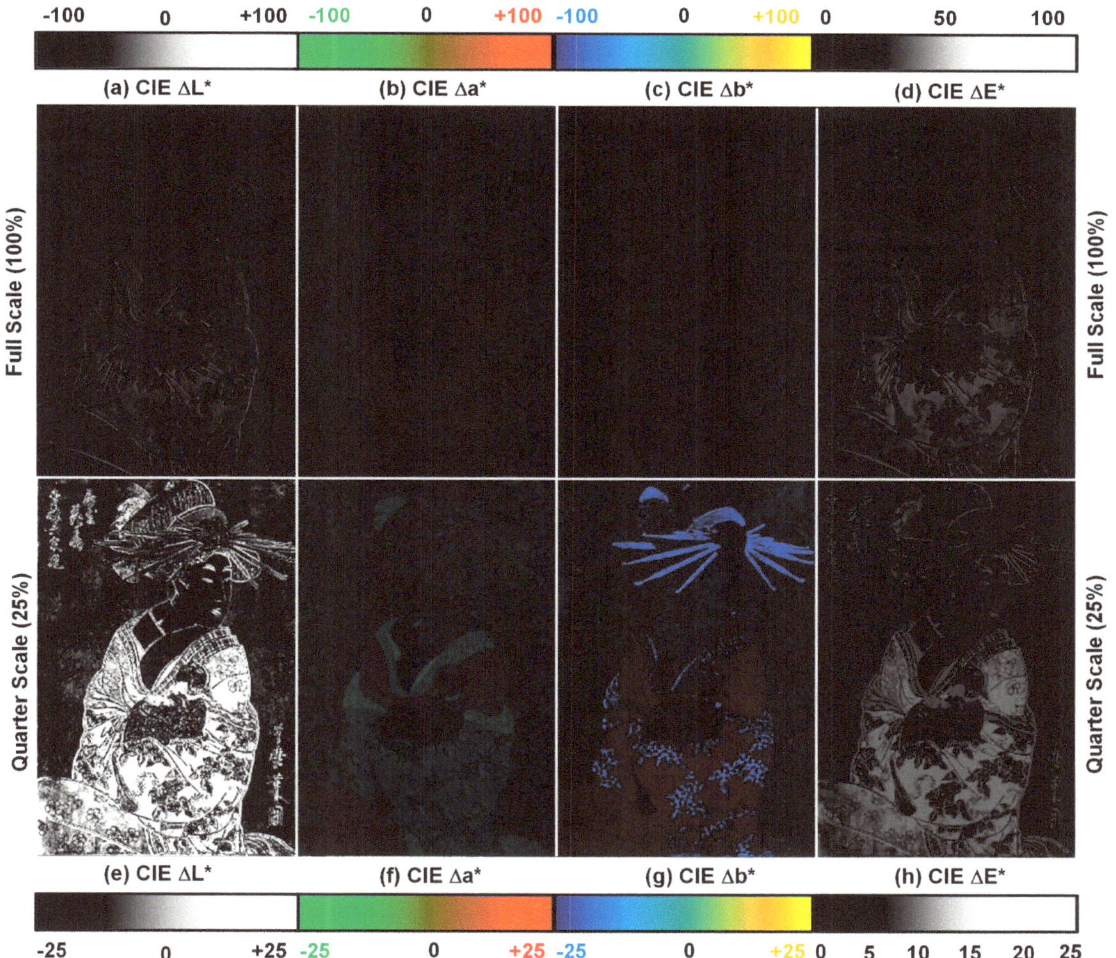

Figure 7. Pixel-by-pixel color difference in images in full scale (top row) and quarter scale (4× zoomed scale or 25% of full scale) (bottom row); (**a**) ΔL*; (**b**) Δa*; (**c**) Db*; (**d**) DE* in full scale; (**e**) DL*; (**f**) Δa*; (**g**) Δb*; (**h**) ΔE* in quarter scale (4× zoomed scale or 25% of full scale). Legends are given for easy recognition of pixel-by-pixel brightness, color shift, and color difference.

3.5. Block-by-Block Image Comparison

The two copies of a print, before and after color fading, were compared block-by-block using image processing/analysis software (PicMan). The block size was varied from 20 × 20 pixels to 50 × 50 pixels in 10-pixel increments (Figure 8). Upper-left and lower-right triangles show partial images before and after color fading for an easy comparison of the effect of color fading within areas of the same color patterns. The block-by-block image comparison technique, by alternating partial masking of the two comparing images with changing block size, provides intuitive synthesized images for side-by-side comparison on the reference image.

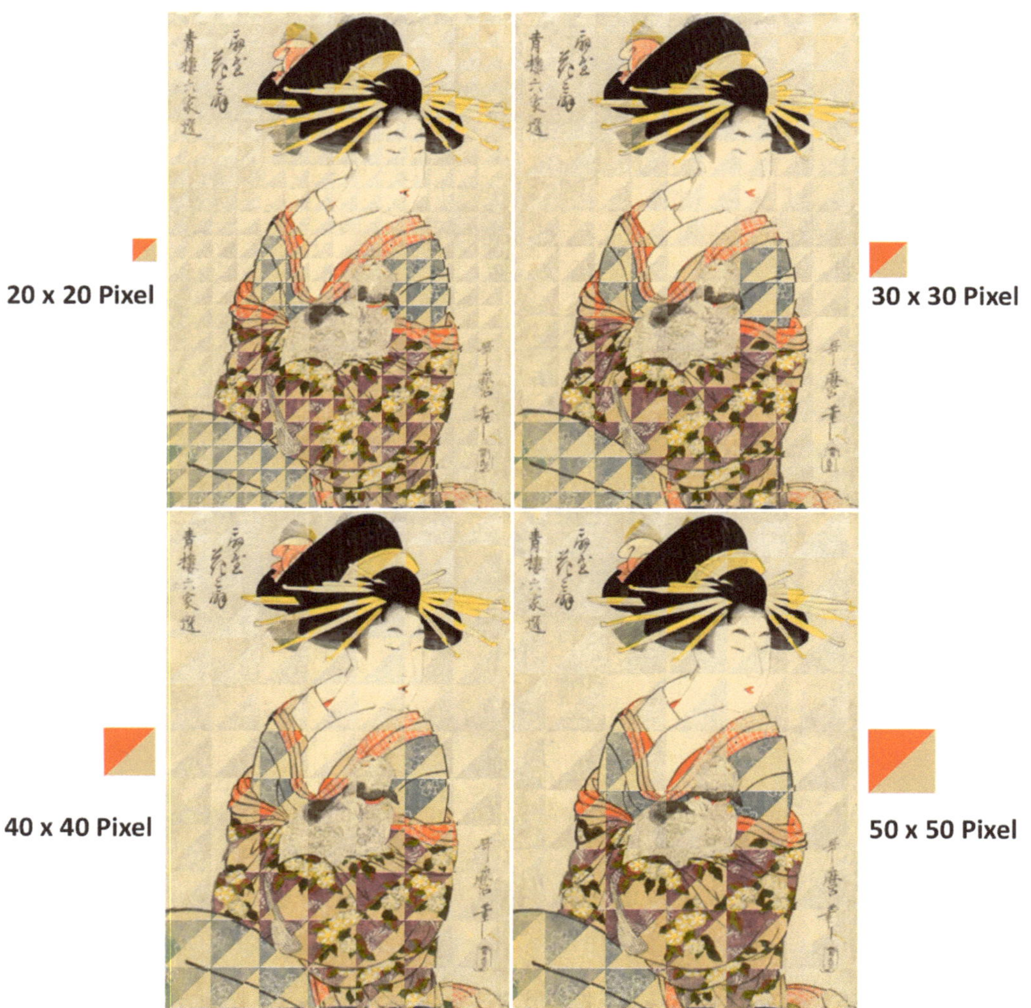

Figure 8. Block-by-block comparison images (top left triangle for the original image and bottom right triangle for color faded image) with four different block sizes of 20 × 20 pixel, 30 × 30 pixel, 40 × 40 pixel, and 50 × 50 pixel squared blocks.

The actual area of measurement can be calculated by the scale of an image in pixel/mm. The scale of an image can be easily calculated by measuring the actual dimensions of the print in mm and the size of the image in pixels. The image resolution of the scale is expressed in pixels/mm.

Background colors became darker after color fading by light exposure. The color difference in black hair and wig areas was hard to recognize by visual inspection due to their very low brightness (i.e., very low RGB values, low CIE L* value, low V value in a HSV color space). The red, pink, yellow, bluish, and purple colored areas showed noticeable color fading. The purple-colored area showed the most discoloration. The red, pink, and bluish-colored areas also had undergone significant discoloration and transformed into almost similar colors, as seen in Figures 1 and 8.

While the point-by-point, line-by-line, and area-by-area color characterization techniques are useful for quantifying color differences between two images, the block-by-block color comparison provides visual instinct on color differences between images at the desired block size.

3.6. Color Simulation by Interpretation and Extrapolation

In theory, if images of identical objects from different times are available, new images can be generated by image blending and/or image morphing. It is very similar to the interpolation between data sets acquired from two different times. The micro fading test data for discoloration under controlled light exposure conditions can be used as important data points for understanding color fading phenomena and mechanisms. If color extrapolation beyond the two data points is feasible, simulated images outside of the time frame for the two images can be generated.

Figure 9 shows image blending simulation results using two true images by interpolation. Figure 9a,b is the original image at time A, and the color faded image at time B. Figure 9c is the interpolated result image for color fading in progress in the middle of times A and B. The image was generated by 50:50 blending or interpolation of the two images. By changing the blending ratio, many simulated images can be generated at different times between times A and B, as shown in Figure 10. Furthermore, extrapolation can be done before time A and beyond time B assuming the rate of color change are constant.

(a)   (b)   (c)

Figure 9. Demonstration of image blending simulation using two true images by interpolation; (**a**) original image at time A; (**b**) interpolated image for color fading in progress; (**c**) fully faded image at time B.

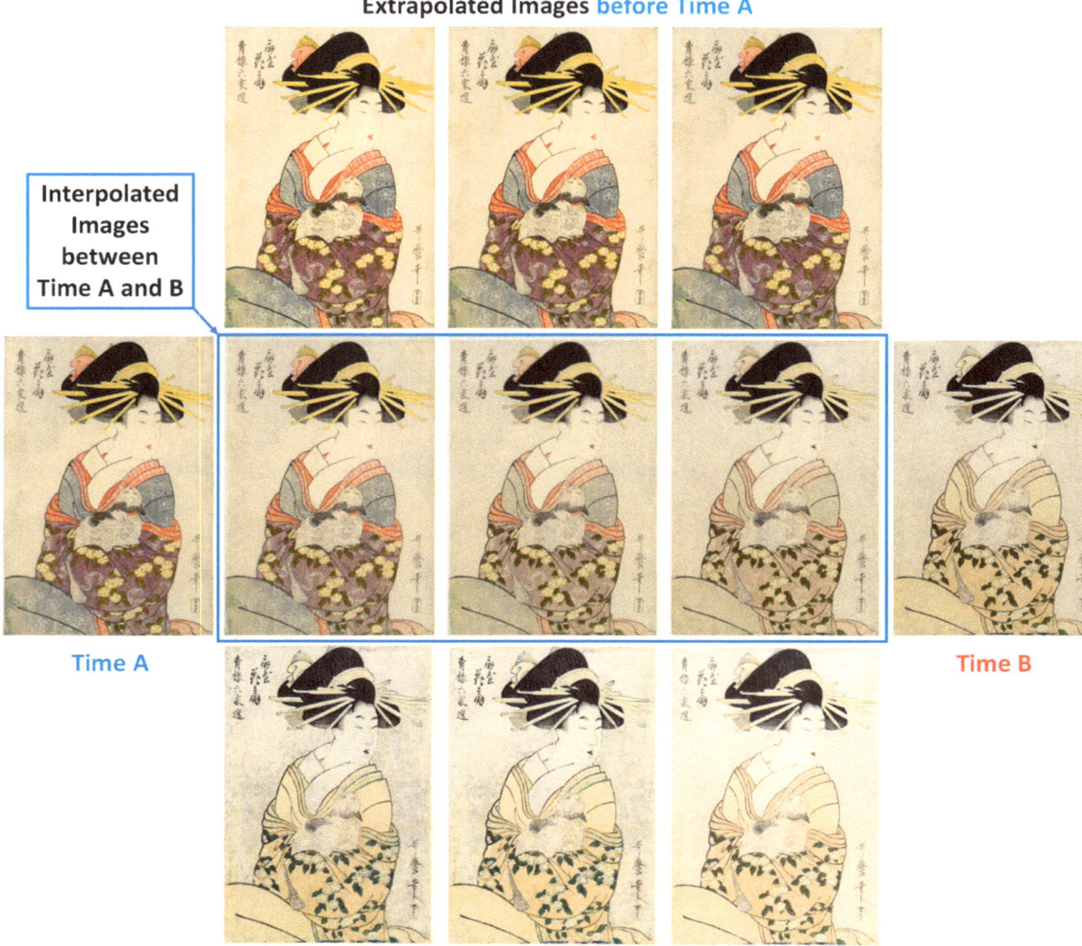

Figure 10. Demonstration of extrapolated images and interpolated images based on two images at time A (at the time of printing, before color fading occurs or any reference point of time) and B (at any point of time after time A for color comparison).

In the real world, the rate of color change of each colorant is neither linear nor infinite. The purpose of this paper is to demonstrate possible color change simulations using a simple color change model under a given light illumination condition. When the color change rate of individual colorants at a given illumination and time condition is characterized, a real color change rate can be applied for color change simulations for a realistic effect.

3.7. Potential Applications

Diagnoses of the conservation status of painting cultural heritages and color fading characteristics of pigments using this quantitative colorimetric analysis technique were also very promising [19–21]. It has been extremely useful for digital forensic studies of cultural heritage identification processes. In particular, printing techniques of ancient Korean books through ink tone analyses and character image comparisons with images of books with known printing techniques and printing dates [30–33]. Darkening by oxidation of *Hanji* (Korean paper) used in the ancient books was removed from photographed images for unbiased ink tone analysis and fair comparison between several versions of nearly identical ancient books. This led to the discovery of the world's oldest metal-type printed book (*The Song of Enlightenment* (南明泉和尚頌證道歌) in Korea in 1239). The world's oldest metal-type printed book from the Goryeo dynasty of Korea, from the 13th century, has been identified by comparing six nearly identical books from Korea from the 13th to 16th centuries, the *Jikji* (直指) printed in 1377 and the Gutenberg 42-line Bible, printed in 1455 using this quantitative image analysis technique [34]. Many other applications based on color/shape extraction and area measurement techniques using the newly developed image processing/analysis software (PicMan) are expanding. Image processing software applications combined with imaging devices such as USB cameras and professional digital cameras have been reported in previous papers [19–24].

4. Summary

Quantitative characterization of color from images is very important in the fields of color-related applications. In the fields of art, cultural heritage, and museum, color fading and color darkening under light exposure is one of many important and unavoidable natural processes. Color naturally fades with time under light illumination in an oxidizing environment. If the chemical environment of storage is altered, the color may fade or darken abnormally. Quantitative characterization of colors becomes very important, regardless of the conventional colorimetric approach or photometric approach. In this paper, a novel approach to the quantitative characterization of color and visualization of color fading and color darkening, using a newly developed image analysis and comparison software (PicMan), is proposed and demonstrated.

The rate of color fading depends on the substance, lighting condition, and chemical environment of storing facilities. Color fading is only observed as a result after the damage is done. The current color of objects of interest can only be compared with old photographs or the observer's perceived memory at the time of reference. There is no guarantee that the same individual will inspect or properly evaluate the color change. Color difference and color fading rate between two or more points of time in the past can only be determined using photographic images from the past unless the other colorimetric measurements were made at the time of inspection.

Quantification of color in either digital or printed photographs is required for objective characterization of color differences between two or more different times. Image analysis and comparison software-assisted color quantification and pixel-by-pixel color difference mapping were proposed and demonstrated in this study. Images of two copies of Japanese woodblock prints, with and without color fading, have been selected for exemplary study. Pixel-by-pixel, line-by-line, area-by-area, and block-by-block comparisons of color fading between two images were found to be very effective in the quantification of color fading and visualization of the phenomena. RGB, HSV, CIE L*a*b* values between images and the differences of single pixels to areas of interest in any shape and size can be quantified and exported as numerical and traceable data.

Color fading and color darkening analysis results were presented in numerical, graphical, and image formats for completeness. As demonstrated in this study, all formats have their own advantages and disadvantages over other formats in terms of data size, complexity, readability, and communication among parties of interest. Various display options for color analysis summary of color fading and color darkening were demonstrated using the images of interest for practical artistic, cultural heritage conservation, and museum applications. Color simulation for various moments in time was proposed and demonstrated by interpolation or extrapolation of color change between images, with and without color fading, using PicMan. The degree of color fading and color darkening in the various moments in the past and future can be simulated and visualized for decision-making in public display, storage, and restoration planning.

Author Contributions: All authors equally contributed to this study. Conceptualization, W.S.Y., K.K. and Y.Y.; material preparation, W.S.Y. and Y.Y.; methodology, Y.Y. and J.G.K.; software, W.S.Y. and K.K.; data acquisition and analysis, Y.Y., J.G.K. and W.S.Y.; writing—original draft preparation, review and editing, W.S.Y. and Y.Y. All authors have read and agreed to the published version of the manuscript.

Funding: This research received no external funding.

Conflicts of Interest: The authors declare no conflict of interest.

References

1. Special Post–Microfading Workshop! Available online: https://intheirtruecolors.wordpress.com/2014/09/26/special-post-microfading-workshop/ (accessed on 30 April 2023).
2. Van Gogh's Fading Colors Inspire Scientific Inquiry. Available online: https://cen.acs.org/articles/94/i5/Van-Goghs-Fading-Colors-Inspire.html (accessed on 30 April 2023).
3. Hofmann, M.; Hofmann-Sievert, R. Spectral Light Fading of Inkjet Prints. *Heritage* **2022**, *5*, 4061–4073. [CrossRef]
4. NASA Science, Tour of the Electromagnetic Spectrum. Available online: https://science.nasa.gov/ems/09_visiblelight#:~:text=Typically%2C%20the%20human%20eye%20can,from%20380%20to%20700%20nanometers (accessed on 30 April 2023).
5. Zhao, Y.; Wang, J.; Pan, A.; He, L.; Simon, S. Degradation of red lead pigment in the oil painting during UV aging. *Color Res. Appl.* **2019**, *44*, 790–797. [CrossRef]
6. Beltran, V.L.; Pesme, C.; Freeman, S.K.; Benson, M. *Microfading Tester: Light Sensitivity Assessment and Role in Lighting Policy Guidelines*; Getty Conservation Institute: Los Angeles, CA, USA, 2021; Available online: https://www.getty.edu/conservation/publications_resources/pdf_publications/pdf/Microfading_Tester.pdf (accessed on 30 April 2023).
7. Whitmore, P.M.; Pan, X.; Bailie, C. Predicting the Fading of Objects: Identification of Fugitive Colorants through Direct Nondestructive Lightfastness Measurements. *J. Am. Inst. Conserv.* **1999**, *38*, 395–409. [CrossRef]
8. Fieberg, J.E.; Knutås, P.; Hostettler, K.; Smith, G.D. "Paintings Fade Like Flowers": Pigment Analysis and Digital Reconstruction of a Faded Pink Lake Pigment in Vincent van Gogh's Undergrowth with Two Figures. *Appl. Spectrosc.* **2017**, *71*, 794–808. [CrossRef] [PubMed]
9. Amanatiadis, S.; Apostolidis, G.; Karagiannis, G. Consistent Characterization of Color Degradation Due to Artificial Aging Procedures at Popular Pigments of Byzantine Iconography. *Minerals* **2021**, *11*, 782. [CrossRef]
10. Eickhoff, A.J.; Hunter, R.S.; Measurement of the Fading Rate of Paints. RESEARCH PAPER RP1478, Part of Journal of Research of the National Bureau of Standards, Volume 28, June 1942. Available online: https://nvlpubs.nist.gov/nistpubs/jres/28/jresv28n6p773_A1b.pdf (accessed on 30 April 2023).
11. Kirby, J.; Saunders, D. 'Fading and Colour Change of Prussian Blue: Methods of Manufacture and the Influence of Extenders'. *Natl. Gallery Tech. Bull.* **2004**, *25*, 73–99. Available online: http://www.nationalgallery.org.uk/technical-bulletin/kirby_saunders2004 (accessed on 30 April 2023).
12. Melo, M.J.; Nabais, P.; Vieira, M.; Araújo, R.; Otero, V.; Lopes, J.; Martín, L. Between past and future: Advanced studies of ancient colours to safeguard cultural heritage and new sustainable applications. *Dye. Pigment.* **2022**, *208*, 110815. [CrossRef]
13. Zheng, L.; Wang, Z.; Shen, S.; Xia, Y.; Li, Y.; Hu, D. Blurring of ancient wall paintings caused by binder decay in the pigment layer. *Sci Rep.* **2020**, *10*, 21075. [CrossRef] [PubMed]
14. Kubik, M.A. Preserving the Painted Image: The Art and Science of Conservation. *Color Des. Creat.* **2010**, *5*, 1–8.
15. Lee, H.S.; Coates, G.A. Characterization of Color Fade during Frozen Storage of Red Grapefruit Juice Concentrates. *J. Agric. Food Chem.* **2002**, *50*, 3988–3991. [CrossRef] [PubMed]
16. Xia, Y.; Song, X.; Jia, Z.; Wang, X. Mechanism and quantitative characterization of color fading phenomenon of HTV composite insulators silicone rubber. In Proceedings of the 2018 12th International Conference on the Properties and Applications of Dielectric Materials (ICPADM) 2018, Xi'an, China, 20–24 May 2018. [CrossRef]
17. CSN EN 15886; Conservation of Cultural Property-Test Methods-Colour Measurement of Surfaces. CEN: Brussels, Belgium, 2010. Available online: https://www.en-standard.eu/csn-en-15886-conservation-of-cultural-property-test-methods-colour-measurement-of-surfaces/ (accessed on 31 July 2023).
18. Two Copies of a Japanese Woodblock Print by Utamaro. Available online: https://intheirtruecolors.files.wordpress.com/2014/09/fadeexample_woodblock.png (accessed on 30 April 2023).
19. Kim, G.; Kim, J.G.; Kang, K.; Yoo, W.S. Image-Based Quantitative Analysis of Foxing Stains on Old Printed Paper Documents. *Heritage* **2019**, *2*, 2665–2677. [CrossRef]
20. Yoo, W.S.; Kang, K.; Kim, J.G.; Yoo, Y. Extraction of Color Information and Visualization of Color Differences between Digital Images through Pixel-by-Pixel Color-Difference Mapping. *Heritage* **2022**, *5*, 3923–3945. [CrossRef]
21. Eom, T.H.; Lee, H.S. A Study on the Diagnosis Technology for Conservation Status of Painting Cultural Heritage Using Digital Image Analysis Program. *Heritage* **2023**, *6*, 1839–1855. [CrossRef]
22. Wakamoto, K.; Otsuka, T.; Nakahara, K.; Namazu, T. Degradation Mechanism of Pressure-Assisted Sintered Silver by Thermal Shock Test. *Energies* **2021**, *14*, 5532. [CrossRef]

23. Jo, S.-I.; Jeong, G.-H. Single-Walled Carbon Nanotube Synthesis Yield Variation in a Horizontal Chemical Vapor Deposition Reactor. *Nanomaterials* **2021**, *11*, 3293. [CrossRef] [PubMed]

24. Zhu, C.; Espulgar, W.V.; Yoo, W.S.; Koyama, S.; Dou, X.; Kumanogoh, A.; Tamiya, E.; Takamatsu, H.; Saito, M. Single Cell Receptor Analysis Aided by a Centrifugal Microfluidic Device for Immune Cells Profiling. *Bull. Chem. Soc. Jpn.* **2019**, *92*, 1834–1839. [CrossRef]

25. PicManTV. Available online: https://www.youtube.com/@picman-TV (accessed on 30 April 2023).

26. Kruschwitz, J.D.T. *Field Guide to Colorimetry and Fundamental Color Modeling*; SPIE Press: Washington, DC, USA, 2018.

27. Von Goethe, J.F. *Theory of Colours*; The MIT Press: Cambridge, MA, USA, 1970.

28. Rhyne, T.-M. *Applying Color Theory to Digital Media and Visualization*, 1st ed.; Association for Computing Machinery: New York, NY, USA, 2016.

29. Minolta, K. CR-400 Chroma Meter. Available online: https://sensing.konicaminolta.us/us/products/cr-400-chroma-meter-colorimeter/ (accessed on 30 April 2023).

30. Yoo, W.S. The World's Oldest Book Printed by Movable Metal Type in Korea in 1239: The Song of Enlightenment. *Heritage* **2022**, *5*, 1089–1119. [CrossRef]

31. Yoo, W.S. How Was the World's Oldest Metal-Type-Printed Book (The Song of Enlightenment, Korea, 1239) Misidentified for Nearly 50 Years? *Heritage* **2022**, *5*, 1779–1804. [CrossRef]

32. Yoo, W.S. Direct Evidence of Metal Type Printing in The Song of Enlightenment, Korea, 1239. *Heritage* **2022**, *5*, 3329–3358. [CrossRef]

33. Yoo, W.S. Ink Tone Analysis of Printed Character Images towards Identification of Medieval Korean Printing Technique: The Song of Enlightenment (1239), the Jikji (1377) and the Gutenberg Bible (~1455). *Heritage* **2023**, *6*, 2559–2581. [CrossRef]

34. American Printing History Association. History of Printing Timeline. Available online: https://printinghistory.org/timeline/ (accessed on 30 April 2023).

 *technologies*

Communication

Adapting the H.264 Standard to the Internet of Vehicles

Yair Wiseman 

Computer Science Department, Bar-Ilan University, Ramat-Gan 5290002, Israel; wiseman@cs.biu.ac.il

Abstract: We suggest two steps of reducing the amount of data transmitted on Internet of Vehicle networks. The first step shifts the image from a full-color resolution to only an 8-color resolution. The reduction of the color numbers is noticeable; however, the 8-color images are enough for the requirements of common vehicles' applications. The second step suggests modifying the quantization tables employed by H.264 to different tables that will be more suitable to an image with only 8 colors. The first step usually reduces the size of the image by more than 30%, and when continuing and performing the second step, the size of the image decreases by more than 40%. That is to say, the combination of the two steps can provide a significant reduction in the amount of data required to be transferred on vehicular networks.

Keywords: Internet of Vehicles; H.264; quantization tables

Citation: Wiseman, Y. Adapting the H.264 Standard to the Internet of Vehicles. *Technologies* **2023**, *11*, 103. https://doi.org/10.3390/technologies11040103

Academic Editor: Bang Wang

Received: 7 June 2023
Revised: 12 July 2023
Accepted: 25 July 2023
Published: 3 August 2023

1. Introduction

The H.264 video compression format is very old. H.264 was developed by the Moving Picture Experts Group (MPEG), a working group founded in 1988 by the International Electrotechnical Commission (IEC) and the International Organization for Standardization (ISO) [1]. The MPEG group finalized the last drafts of the H.264 standard in early 2003, and a short time afterward, H.264 was adopted with slight modifications by ISO [2].

Early H.264 variants were MPEG-1, MPEG-2, and MPEG-4 and were able to achieve a lower average compression ratio. However, since then, H.264 has been significantly improved, and contemporary versions of H.264 are much better. These improved compression ratios have made possible real-time and video applications using H.264. Additionally, the H.264 format's automatic error correction feature has contributed to the possibility of use by real-time and video applications [3].

Real-time decision making is very essential for the Internet of Things (IoT) and the Internet of Vehicles (IoV) [4]. Therefore, H.264 is used in many applications, including the Internet of Vehicles. H.264 is often employed to transmit not only videos but also still images between vehicles (V2V), vehicles to roadside infrastructure (V2I), and vehicles to central servers (V2X) with the aim of improving safety and efficiency by sharing information about traffic conditions and hazards with other road users and also facilitating the vehicle to observe traffic lights and signs. These features are essential requirements for autonomous vehicles; however, contemporary partially driverless vehicles also have at least some of these features [5].

H.264 can be a good option for IoV applications because it provides a high compression ratio while retaining reasonable image quality [6]. These attributes make H.264 suitable for applications where bandwidth might be restricted, such as vehicular networks and other IoT applications [7].

No distinct optimization for H.264 aimed at IoV with the purpose of better handling of such information has been proposed. In this paper, we suggest a scheme employing two steps of reducing the amount of data transmitted on Internet of Vehicle networks for enhancing H.264 aimed at IoV.

2. Background

H.264 represents images using chroma subsampling that provides lower resolution for chrominance information, whereas luminance information is represented by a better resolution [8]. Chroma subsampling exploits the smaller attention of typical human eyes to minute differences in chrominance information [9].

Usually, H.264 encodes an image employing YUV components [10]. The YUV compressed blocks are interleaved together within the compressed image; yet the amount of the Y data, the amount of the U data, and the amount of the V data are usually unequal. The amount of the Y data can be even four times more than the U data and the V data.

The Y data are for the luminance data within the block, whereas the U data and the V data are for the chrominance data within the block. As was mentioned above, there is smaller attention of typical human eyes to minute differences in chrominance information, so accordingly, there is a smaller amount of data of U and V in an H.264 compressed file [11].

A common practice is a quartered resolution of the U data and the V data, which is called 4:1:1 chroma subsampling [12]. This 4:1:1 denotes there are four times more luminance data than the chrominance data. In other words, for each block of 16×16 that can be divided to four blocks of 8×8, as shown in Figure 1, there are four Y blocks and just one U block and one V block. Each row in an H.264 frame consists of lines of Y, U, and V blocks. The lines are positioned from top to bottom, where the blocks of each line are from left to right.

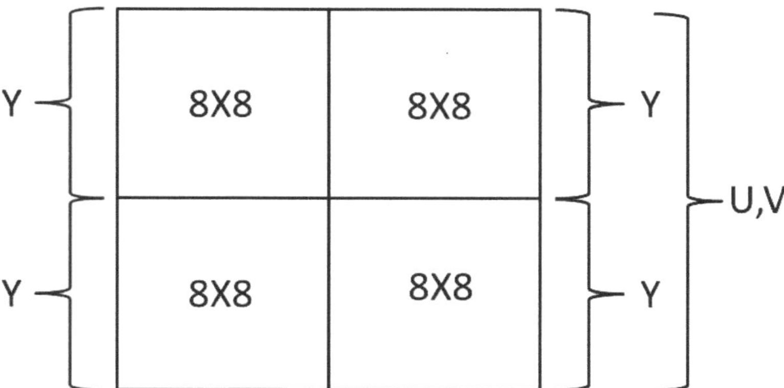

Figure 1. H.264 block of 16×16.

However, another chroma subsampling procedure of JPEG is very common. The 4:2:0 subsampling is very similar to 4:1:1. The 4:2:0 subsampling also keeps more information about the Y component and a smaller amount of information about the U and the V components. In 4:2:0 subsampling, the horizontal and the vertical sampling are different. The pixels of the U and the V components are sampled just on odd lines and not on even lines; however, in each sampled line, two pixels of each U and V component are sampled.

Figure 2 shows an example of the 4:2:0 subsampling. The six pixels marked with the letter X lose their original color and are replaced by the color of the upper-left pixel in the quad.

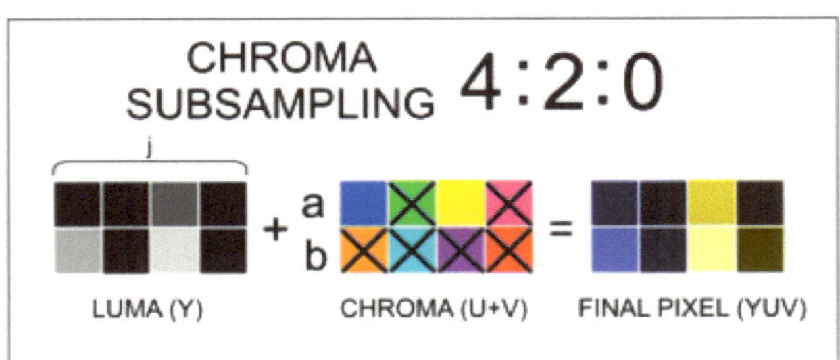

Figure 2. 4:2:0 subsampling layout [13].

The compression ratios of 4:1:1 and 4:2:0 are roughly the same, and both of them are popular when using H.264 [14,15].

H.264 has three types of frames: I-frames, P-frames, and B-frames [16]:

- I-frames (intracoded frames)—standalone frames containing all the data required to show the frame. Actually, this frame is very similar to a JPEG still image [17].
- P-frames (predictive frames)—predicted from another decoded frame.
- B-frames (bidirectional predicted frames)—predicted from both a previously decoded frame and a yet-to-come decoded frame.

H.264 applies three steps to each block in each frame:

1. Transform each block to a frequency space employing the forward discrete cosine transform (FDCT) [18].
2. Employ quantization; i.e., each value is divided by a predefined quantization coefficient and next rounded to an integer value [19].
3. Compress the file employing a version of canonical Huffman codes [20].

In order to decompress a frame, we have perform do the same operations as in compression, but in reverse order. That is, first decoding the Huffman codes, then multiplying by the constants by which we divided the coefficients in the quantization step, and finally performing the inverse discrete cosine transform (IDCT).

3. Motivation

Video is used for communication between vehicles for several reasons. First, there is information that cannot be conveyed in a still image, such as the direction of movement or speed, and must be added as additional information to the still image. In a video, on the other hand, the direction of movement and speed can be deduced from the video itself.

In addition, video is a more reliable format. Each frame in the video is also a kind of backup for its adjacent frames. If only one frame is damaged, the adjacent frames can compensate for the loss of one frame. Indeed, a large loss of frames will create a problem, but a small percentage of frame loss can be unnoticed.

When it comes to choosing the desired encoder, the H.264 format has several advantages. H.264 is a broadly supported codec [21]. H.264 is supported by numerous up-to-date video players and also by many modern-day devices. This feature means that V2V communication using H.264 can be employed by many brands of vehicles [22].

In addition, H.264 is a royalty-free codec [23]. That is, there is no need to pay licensing fees for using H.264. Therefore, also from an economic point of view, it is better to choose H.264, because, naturally, the public of vehicle users prefer to pay as little as possible.

Vehicular networks are still in their early stages, but their significant potential to upgrade transportation efficiency and safety is agreed upon. As this technology develops further, videos for vehicular networks are expected to become a standard element in these networks [24]. The suggested algorithm in this paper aimed at promoting the use of video in vehicular networks.

4. Reduced Image Quality

H.264 was designed for the human eye and the quality level that people are able to see [25]. The level of quality that a vehicle needs to provide information for safety and the efficiency of its use is much lower. Actually, H.264 dedicates 24 bits for each pixel in the image, which allows for a very large range of colors and a high-quality image. A vehicle is able to perform well even with a much lower level of image quality. Reducing the number of bits from 24 to 3 lowers the image quality significantly; however, for what a vehicle needs, it is a sufficient quality.

Therefore, an algorithm is proposed that, on the one hand, will reduce the number of colors to only 8 frequent colors and, on the other hand, will not choose colors that are almost the same, but colors with high variance and that will finely represent the information in the original image for IoV. This algorithm for selecting the 8 most frequent colors is

---

Algorithm 1 Selecting the 8 most frequent colors

Sort the colors by their count values.
Take the top 8 colors.
Calculate the variance of each pair of colors using Equation (1):

$$variance = \sqrt{(r_1 - r_2)^2 + (g_1 - g_2)^2 + (b_1 - b_2)^2} \tag{1}$$

While there is a pair (x_1, x_2) that holds (variance < δ)
 Begin
   Find from the pair (x_1, x_2) the color with the smaller occurrences and remove it from the
top 8 colors
   Add from the sorted list the next top color.
   Calculate the variances of the new color with the old colors
 End

In Algorithm 1, r_1, g_1, b_1 are the RGB (red, green, blue) values of the first color, and r_2, g_2, b_2 are the RGB values of the second color. δ is the Feigenbaum constant [26].

An example that shows the result of this algorithm is shown in Figures 3–5. Figure 3 contains the original image. Figure 4 contains the edited picture according to Algorithm 1. Therefore, this figure contains only 8 colors. Figure 5 shows the selected colors and their probabilities in the edited picture.

Figure 3. An original full-color image of a street.

Figure 4. The same image of a street but with only 8 colors.

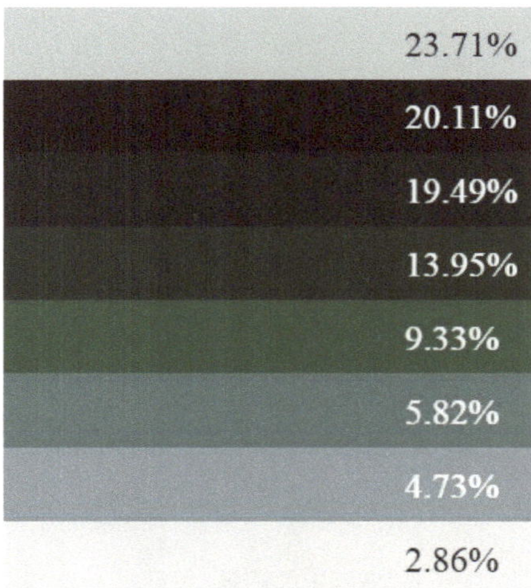

Figure 5. The distribution of the selected 8 colors.

For a vehicle, the essential data are observing other vehicles, the boundaries of the road, and recognizing the traffic signs. All of the essential tasks can be achieved with the 8-color image; nonetheless, the sending of the 8-color image on the vehicular network will be much faster because this image can be compressed better.

5. H.264 Quantization Tables

The reduced number of colors can be very helpful for a better compression ratio; however, H.264 assumes that the differences between the values within each frame are small, and the quantization tables of H.264 were formed according to this assumption. Hence, if we want to employ H.264, we must create different quantization tables that can fit frames with sharper changes.

In [27], the authors suggest quantization-optimized H.264 encoding for traffic video tracking applications. However, their suggestion does not support sharp changes, and it seems to be unsuitable for the approach in our paper.

The standard quantization tables of 90%, which are frequently used in vehicular networks, are shown in Tables 1 and 2.

Table 1. H.264 luminance quantization table of 90%.

3	2	2	3	5	8	10	12
2	2	3	4	5	12	12	11
3	3	3	5	8	11	14	11
3	3	4	6	10	17	16	12
4	4	7	11	14	22	21	15
5	7	11	13	16	21	23	18
10	13	16	17	21	24	24	20
14	18	19	20	22	20	21	20

Table 2. H.264 chrominance quantization table of 90%.

3	4	5	9	20	20	20	20
4	4	5	13	20	20	20	20
5	5	11	20	20	20	20	20
9	13	20	20	20	20	20	20
20	20	20	20	20	20	20	20
20	20	20	20	20	20	20	20
20	20	20	20	20	20	20	20
20	20	20	20	20	20	20	20

These default quantization tables were developed assuming that the possibility of sharp changes in the image will not often come out. This assumption is correct for regular images; however, the images of our application have a different kind of information containing more sharp changes. Therefore, although putting smaller numbers in the upper-left part of the quantization tables can be suitable for our application, the large values in the lower-right part of the quantization table can be harmful. The lower-right part of the quantization tables consists of data in relation to the sharper changes, and dividing this part's values by such large numbers will cause a significant amount of data to be lost. Therefore, in blocks of images that contain sharp changes, the difference between the values of the upper-left part and the values of the lower-right part in the quantization tables must be reduced.

The standard quantization tables of 90% have significant disparity between the highest value in the table and the smallest value in the table. In the luminance quantization table, the values' range is from 2 to 20, which represents a 10-fold difference. The chrominance quantization table has a smaller disparity, but still, the disparity is significant.

Other standard quantization tables of less than 90% can have even a larger disparity and can be even more than 10-fold. In our application, there are sharp changes in the images, and such disparity should be lowered.

Accordingly, attuning the quantization tables for our application that has sharp changes is essential because such an alteration of the quantization tables can enhance the compression ratio.

Typically, there is still disparity between the lower-right part and the upper-left part of H.264 blocks in our application; however, the disparity is slighter. Therefore, this formula is recommended for the luminance component:

$$f(x) = \frac{8 * \tan^{-1}\left(\frac{x}{16}\right)}{\pi} + 8 * DOTTIE \tag{2}$$

where $x = X^2 + Y^2$. X is the coefficient's index in the X-axis, and Y is the coefficient's index in the Y-axis; $f(x)$ is the coefficient's value in the quantization table.

DOTTIE is the Dottie number [28]. The Dottie number is the unique real fixed point of the cosine function. In other words, if the cosine function is repeatedly applied to a random number, it will generate a sequence that will converge to the Dottie number.

Similarly, this formula is recommended for the chrominance component:

$$f(x) = \pi * \tan^{-1}\left(\frac{x}{16}\right) + 8 * DOTTIE \tag{3}$$

where $x = X^2 + Y^2$. X is the coefficient's index in the X-axis, and Y is the coefficient's index in the Y-axis; $f(x)$ is the coefficient's value in the quantization table.

DOTTIE is the Dottie number.

The graphs of the recommended formulas are shown in Figures 6 and 7. These graphs were generated using [29].

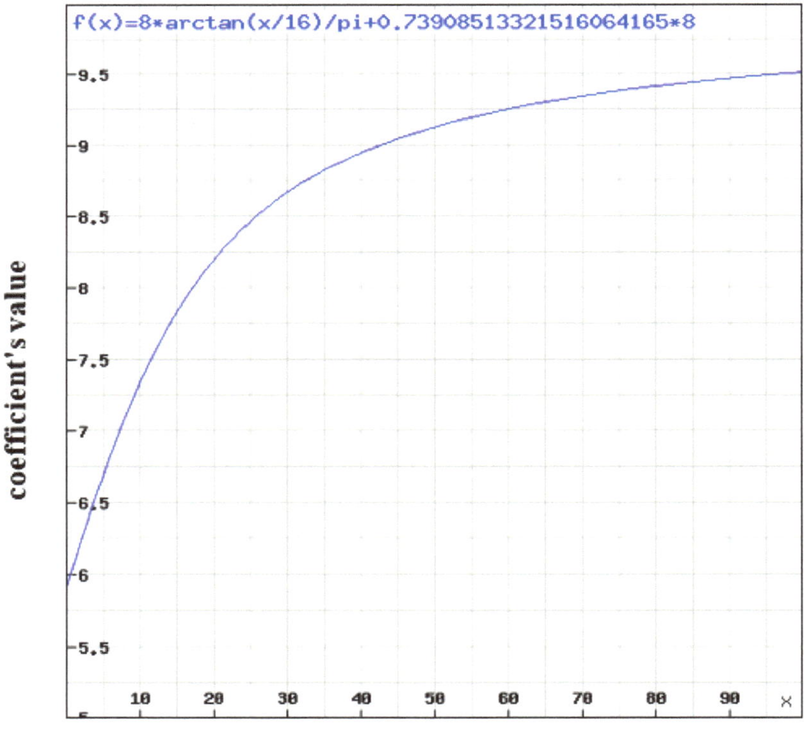

Figure 6. Luminance quantization table's graph.

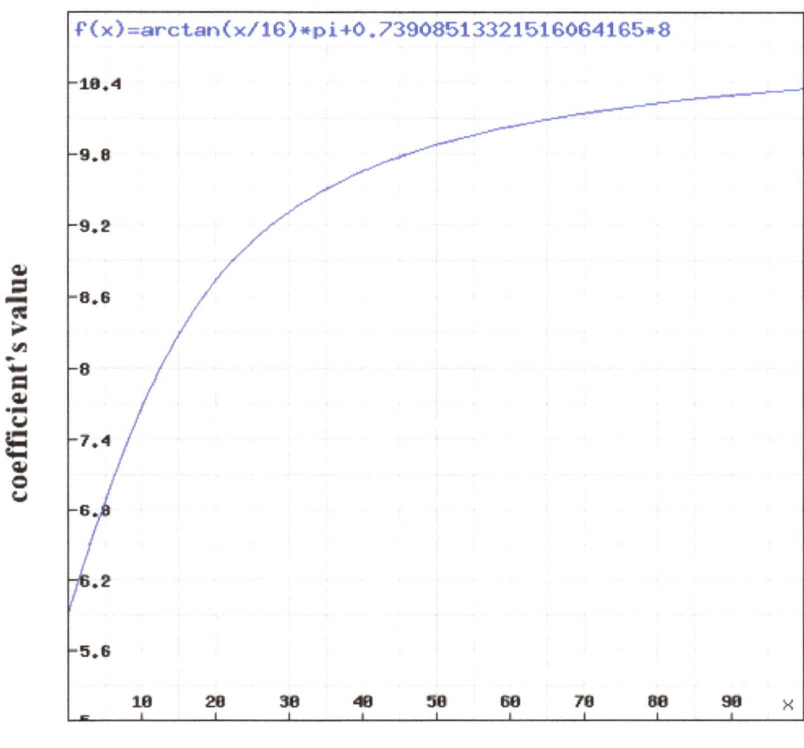

Figure 7. Chrominance quantization table's graph.

The quantization tables that were calculated according to these formulas are shown in Tables 3 and 4.

H.264 does not support real numbers for the quantization tables. The developers of H.264 decided about this because real numbers give a resolution that is good enough. There is no need for the real numbers' better precision that would result in less good compression to obtain a quality that the human eye cannot distinguish. In our application, we are even willing to further decrease the quality of the image, so clearly, we do not need the exactness of the real numbers. Therefore, we rounded the real numbers to natural numbers, and thus, the table we created contains only natural numbers.

Table 3. Luminance quantization table according to the suggested formula.

6	6	6	6	8	8	9	9
6	6	6	6	8	8	9	9
6	6	6	6	8	8	9	9
6	6	6	8	8	9	9	9
8	8	8	8	9	9	9	9
8	8	8	9	9	9	9	9
9	9	9	9	9	9	9	9
9	9	9	9	9	9	9	9

Table 4. Chrominance quantization table according to the suggested formula.

6	6	6	6	8	8	9	10
6	6	6	6	8	8	9	10
6	6	6	6	8	8	9	10
6	6	6	8	8	9	9	10
8	8	8	8	9	9	10	10
8	8	8	9	9	10	10	10
9	9	9	9	10	10	10	10
10	10	10	10	10	10	10	10

The curve of the graphs starts low and then increases because the functions are based on the tangent function, which starts low and then increases for the reason that the ratio of the opposite leg to the adjacent leg increases as the angle increases.

It can be easily seen that the disparity of the values in the suggested quantization tables is much smaller. From a disparity of 10-fold, we moved toward a disparity of less than 2-fold. This disparity might not be suitable for a regular frame of H.264; however, for this specific application, it is much more suitable.

6. Results

We used Figures 3 and 4 for assessing the effectiveness of the suggested technique. When shifting from a full-color image to an 8-color image, many components of the information were removed. Therefore, the image size was reduced from 382,874 bytes to 260,600 bytes, which is a reduction of 31.936%.

When we changed the quantization table to the suggested quantization tables that are suitable to an image with sharp changes, the size of the image was reduced to 204,798 bytes, which is a total decrease of 46.510%.

The new constructed image is shown in Figure 8. It is fine enough for a vehicle, and as mentioned, it is almost half in its size; therefore, this image can be transmitted much faster.

Figure 8. The image of a street with 8 colors and the suggested quantization tables.

We tried the suggested technique also on a larger image of 3,055,616 bytes. The image is of a parking lot and is shown in Figure 9.

Figure 9. Image of a parking lot with full color.

In this image, the moving to 8 colors made only minor changes to the image, mostly in the upper part of the image, which has some different colors. The size of the original picture was reduced by 35.154% to just 1,981,449 bytes when modifying to an 8-color image, because the original image had some minor changes that were removed in the 8-color image. The cases of no change are the best cases for the compression algorithm of H.264, which will generate the smaller files in such cases. Therefore, there is a large gain in Figure 10.

The shifting to the new suggested quantization tables also contributes an additional reduction in file size. The new file size is 1,755,971 bytes, which is a total decrease of 42.533%.

In this image, the contribution of the new suggested quantization tables was much smaller, because the file already had many zeros, which indicate "no change". It does not matter if we divide the zero in the quantization step by a small or a large number. Dividing zero by any number will yield zero, so the efficiency of the original quantization tables and the new suggested quantization tables will be the same in the cases of a zero value.

The new image, which has only unnoticed change, is shown in Figure 11.

The results of the two images' compression are summarized in Table 5.

Figure 10. Image of a parking lot with 8 colors.

Table 5. Compression's results of the images.

Image	Reduction after the First Step	Reduction after the Second Step
Road	31.936%	46.510%
Parking lot	35.154%	42.533%

Figure 11. Image of a parking lot with 8 colors and the suggested quantization tables.

7. Conclusions

Reducing data transmission in vehicle networks is imperative when many vehicles turn to smart driving [30]. We have proposed two steps to reduce the amount of data transmitted on vehicle networks.

Step 1: Reduce Color Resolution

The first step is to reduce the color resolution of images from full color to 8 colors. This may seem like a significant reduction, but it is still sufficient for most vehicle applications. For example, traffic lights and road signs can be easily distinguished in 8 colors.

Step 2: Modify Quantization Tables

The second step is to modify the quantization tables used by the H.264 video codec. Quantization tables are used to reduce the amount of data required to represent an image. By modifying the quantization tables, we can further reduce the amount of data required to transmit 8-color images.

The combination of these two steps can reduce the amount of data transmitted by almost 50%. This can significantly improve the efficiency of vehicle networks.

Funding: This research received no external funding.

Institutional Review Board Statement: Not applicable.

Informed Consent Statement: Not applicable.

Data Availability Statement: Not applicable.

Conflicts of Interest: The authors declare no conflict of interest.

References

1. Punchihewa, A.; Bailey, D. A Review of Emerging Video Codecs: Challenges and Opportunities. In Proceedings of the 2020 35th IEEE International Conference on Image and Vision Computing New Zealand (IVCNZ), Wellington, New Zealand, 25–27 November 2020; pp. 1–6.
2. Lu, Y.; Li, S. A review on developing status of stereo video technology. In Proceedings of the 2012 IEEE International Conference on Computer Science and Information Processing (CSIP), Xi'an, China, 24–26 August 2012; pp. 715–718.
3. Bahri, N.; Werda, I.; Grandpierre, T.; Ayed MA, B.; Masmoudi, N.; Akil, M. Optimizations for real-time implementation of H.264/AVC video encoder on DSP processor. *Int. Rev. Comput. Softw. (IRECOS)* **2013**, *8*, 2025–2035.
4. Kang, K.D. A Review of Efficient Real-Time Decision Making in the Internet of Things. *Technologies* **2022**, *10*, 12. [CrossRef]
5. Wiseman, Y. Autonomous vehicles. In *Research Anthology on Cross-Disciplinary Designs and Applications of Automation*; IGI Global: Hershey, PA, USA, 2022; pp. 878–889.
6. Liu, X.; Li, Y.; Dai, C.; Li, P.; Yang, L.T. An efficient H. 264/AVC to HEVC transcoder for real-time video communication in Internet of Vehicles. *IEEE Internet Things J.* **2018**, *5*, 3186–3197. [CrossRef]
7. Budati, A.K.; Islam, S.; Hasan, M.K.; Safie, N.; Bahar, N.; Ghazal, T.M. Optimized Visual Internet of Things for Video Streaming Enhancement in 5G Sensor Network Devices. *Sensors* **2023**, *23*, 5072. [CrossRef] [PubMed]
8. Zhou, J.; Zhou, D.; Zhang, H.; Hong, Y.; Liu, P.; Goto, S. A 136 cycles/MB, luma-chroma parallelized H. 264/AVC deblocking filter for QFHD applications. In Proceedings of the 2009 IEEE International Conference on Multimedia and Expo, New York, NY, USA, 28 June–July 2009; pp. 1134–1137.
9. Sharrab, Y.O.; Sarhan, N.J. Detailed comparative analysis of vp8 and h. 264. In Proceedings of the 2012 IEEE International Symposium on Multimedia, Irvine, CA, USA, 10–12 December 2012; pp. 133–140.
10. Ho, Y.H.; Lin, C.H.; Chen, P.Y.; Chen, M.J.; Chang, C.P.; Peng, W.H.; Hang, H.M. Learned video compression for yuv 4:2:0 content using flow-based conditional inter-frame coding. In Proceedings of the 2022 IEEE International Symposium on Circuits and Systems (ISCAS), Austin TX, USA, 28 May–1 June 2022; pp. 829–833.
11. Gunjal, B.L.; Mali, S.N. Comparative performance analysis of DWT-SVD based color image watermarking technique in YUV, RGB and YIQ color spaces. *Int. J. Comput. Theory Eng.* **2011**, *3*, 714. [CrossRef]
12. Sinha, A.K.; Mishra, D. Deep Video Compression using Compressed P-Frame Resampling. In Proceedings of the 2021 National Conference on Communications (NCC), Virtual Conference, 7–30 July 2021; pp. 1–6.
13. KVMGalore. 2023. Available online: http://kb.kvmgalore.com/lookup/ (accessed on 9 May 2023).
14. Hsia, S.C.; Chou, Y.C. VLSI implementation of high-throughput parallel H. 264/AVC baseline intra-predictor. *IET Circuits Devices Syst.* **2014**, *8*, 10–18. [CrossRef]
15. Dumic, E.; Mustra, M.; Grgic, S.; Gvozden, G. Image quality of 4:2:2 and 4:2:0 chroma subsampling formats. In Proceedings of the 2009 IEEE international symposium ELMAR, Zadar, Croatia, 28–30 September 2009; pp. 19–24.
16. Shahid, Z.; Chaumont, M.; Puech, W. Fast protection of H. 264/AVC by selective encryption of CAVLC and CABAC for I and P frames. *IEEE Trans. Circuits Syst. Video Technol.* **2011**, *21*, 565–576.
17. Park, J.H.; Lee, S.H.; Lim, K.S.; Kim, J.H.; Kim, S. A flexible transform processor architecture for multi-CODECs (JPEG, MPEG-2, 4 and H. 264). In Proceedings of the 2006 IEEE International Symposium on Circuits and Systems (ISCAS), Kos, Greece, 21–24 May 2006.
18. Okade, M.; Mukherjee, J. Discrete Cosine Transform: A Revolutionary Transform That Transformed Human Lives [CAS 101]. *IEEE Circuits Syst. Mag.* **2022**, *22*, 58–61.
19. Malvar, H.S.; Hallapuro, A.; Karczewicz, M.; Kerofsky, L. Low-complexity transform and quantization in H. 264/AVC. *IEEE Trans. Circuits Syst. Video Technol.* **2003**, *13*, 598–603.
20. Klein, S.T.; Wiseman, Y. Parallel Huffman decoding with applications to JPEG files. *Comput. J.* **2003**, *46*, 487–497. [CrossRef]
21. Vetro, A.; Wiegand, T.; Sullivan, G.J. Overview of the stereo and multiview video coding extensions of the H. 264/MPEG-4 AVC standard. *Proc. IEEE* **2011**, *99*, 626–642. [CrossRef]
22. Abou-Zeid, H.; Pervez, F.; Adinoyi, A.; Aljlayl, M.; Yanikomeroglu, H. Cellular V2X transmission for connected and autonomous vehicles standardization, applications, and enabling technologies. *IEEE Consum. Electron. Mag.* **2019**, *8*, 91–98.
23. Ravi, A.; Rao, K.R. Performance analysis comparison of the Dirac video codec with H 264/MPEG-4 Part 10 AVC. *Int. J. Wavelets Multiresolut. Inf. Process.* **2011**, *9*, 635–654. [CrossRef]

24. Jiang, X.; Yu, F.R.; Song, T.; Leung, V.C. Resource allocation of video streaming over vehicular networks: A survey, some research issues and challenges. *IEEE Trans. Intell. Transp. Syst.* **2021**, *23*, 5955–5975. [CrossRef]
25. Amor, M.B.; Kammoun, F.; Masmodi, N. A pretreatment using saliency map Harris to improve MSU blocking metric performance for encoding H. 264/AVC: Saliency map for video quality assessment. In Proceedings of the 2016 IEEE International Image Processing, Applications and Systems (IPAS), Hammamet, Tunisia, 5–7 November 2016; pp. 1–4.
26. Briggs, K. A precise calculation of the Feigenbaum constants. *Math. Comput.* **1991**, *57*, 435–439.
27. Soyak, E.; Tsaftaris, S.A.; Katsaggelos, A.K. Quantization optimized H. 264 encoding for traffic video tracking applications. In Proceedings of the 2010 IEEE International Conference on Image Processing, Hong Kong, 26–29 September 2010; pp. 1241–1244.
28. Wiseman, Y. JPEG Quantization Tables for GPS Maps. *Autom. Control. Comput. Sci.* **2021**, *55*, 568–576. [CrossRef]
29. Rechneronline. Draw Function Graphs. 2023. Available online: https://rechneronline.de/function-graphs/ (accessed on 9 May 2023).
30. Yaqoob, I.; Khan, L.U.; Kazmi, S.A.; Imran, M.; Guizani, N.; Hong, C.S. Autonomous driving cars in smart cities: Recent advances, requirements, and challenges. *IEEE Netw.* **2019**, *34*, 174–181. [CrossRef]

 *technologies*

Article

Comparative Analysis of Image Classification Models for Norwegian Sign Language Recognition

Benjamin Svendsen and Seifedine Kadry *

Department of Applied Data Science, Noroff University College, 4612 Kristiansand, Norway
* Correspondence: seifedine.kadry@noroff.no

Abstract: Communication is integral to every human's life, allowing individuals to express themselves and understand each other. This process can be challenging for the hearing-impaired population, who rely on sign language for communication due to the limited number of individuals proficient in sign language. Image classification models can be used to create assistive systems to address this communication barrier. This paper conducts a comprehensive literature review and experiments to find the state of the art in sign language recognition. It identifies a lack of research in Norwegian Sign Language (NSL). To address this gap, we created a dataset from scratch containing 24,300 images of 27 NSL alphabet signs and performed a comparative analysis of various machine learning models, including the Support Vector Machine (SVM), K-Nearest Neighbor (KNN), and Convolutional Neural Network (CNN) on the dataset. The evaluation of these models was based on accuracy and computational efficiency. Based on these metrics, our findings indicate that SVM and CNN were the most effective models, achieving accuracies of 99.9% with high computational efficiency. Consequently, the research conducted in this report aims to contribute to the field of NSL recognition and serve as a foundation for future studies in this area.

Keywords: Norwegian Sign Language recognition; machine learning; image classification; assistive technologies; fingerspelling

Citation: Svendsen, B.; Kadry, S. Comparative Analysis of Image Classification Models for Norwegian Sign Language Recognition. *Technologies* **2023**, *11*, 99. https://doi.org/10.3390/technologies11040099

Academic Editors: Pietro Zanuttigh, Gwanggil Jeon and Imran Ahmed

Received: 23 May 2023
Revised: 29 June 2023
Accepted: 12 July 2023
Published: 15 July 2023

1. Introduction

An integral part of every human's life is communication. It is the way they express themselves and understand each other. For most humans, communication comes naturally. They use their ears as input and mouths as output. A less fortunate minority of humans cannot use their hearing or voice to communicate. They must rely on sign language to communicate, where visual movements and hand gestures are used.

According to the World Health Organization, 5% of the world (430 million) struggle with hearing loss, which is likely to increase [1]. Seventy million out of those 430 million are completely deaf [2]. For the deaf and hard of hearing, it is often difficult to communicate with most of the population because a minority speaks sign language. This can cause issues in their daily lives that can be tough to deal with. Using machine learning models to create sign language translation models is one way to assist the deaf and hard of hearing [3].

Typically, when humans do not speak the same language, they use a translator to help them understand each other. The same goes for someone who communicates through sign language when trying to understand an individual who uses oral communication. Sign interpreters are a common way to help the deaf and hard of hearing by translating for them [4]. This is an excellent service where individuals can receive aid from professional sign language interpreters in real-time. The issue with such services is availability because there is a lack of interpreters worldwide for this service always to be available [5]. Machine learning classification models can be implemented to complement sign interprets by recognizing sign language to assist the deaf and hard of hearing.

Using machine learning models for sign language recognition is a well-researched topic. However, despite advancements in sign language recognition, there is a lack of research focused on NSL recognition. Therefore, further research is needed to determine the most effective classification model for NSL recognition to assist the deaf and hard of hearing in Norway and build a foundation for future work.

Our paper compares machine learning classification models to determine which approach is most suitable for NSL recognition to address this issue. The research will use a custom-made dataset containing 24,300 images of 27 NSL alphabet performed by two signers using different backgrounds and lightning.

However, it is essential to note some limitations of our study. The dataset includes images from only two signers, which may limit the model's ability to generalize across a diverse range of signers. Additionally, while our research provides valuable insights into recognizing NSL signs, it does not offer comprehensive guidelines for creating and implementing complete NSL recognition systems. The results obtained in this study should be seen as a foundation for further research and development.

2. Literature Review

Vision-based sign language recognition is a well-researched topic involving developing machine learning models to recognize sign language. These models use various inputs, such as cameras or sensors, to identify and interpret sign language [6]. There are several categories within sign language recognition, including fingerspelling, isolated words, lexicons of words, and continuous signs [7].

Despite the promising potential of this technology, there is a noticeable gap in research focusing on NSL recognition. This is surprising, considering an estimated 250,000 to 300,000 hearing-impaired individuals in Norway, with 3500 to 4000 being completely deaf [8].

We aim to research specific machine learning models, examining their strengths and weaknesses. We aim to highlight why these models are suitable in our comparative analysis of NSL recognition. We also reference results from previous research papers to provide context and results from research on other sign languages.

2.1. K-Nearest Neighbor

K-Nearest Neighbor (K-NN) is a simple yet fundamental non-parametric supervised classification model requiring little data distribution knowledge [9]. It is known as a lazy learner because there are no parameters for KNN [10]. The general steps to how the K-NN model works written by [11] goes as follows: Based on the input data, KNN finds the K most similar instances to an unseen observation by calculating the distance between them. Then, the algorithm picks the k points based on the training set closest to the new observations made and calls it set A. The last step is assigning the new observations to the most probability class. This is determined by searching for the match that closely looks like the new example among the learned training observations. Because KNN offers such simplicity, it is a commonly used classification algorithm, but when the dataset and its features are large, the model can lose much of its efficiency [12].

Because KNN is a classifier that employs the scikit-learn library in Python, the optimal hyperparameters for the model can be determined using GridSearchCV or Randomized-SearchCV methods. These methods explore all possible parameter combinations and samples from parameter distributions and can leverage cross-validation schemes and score functions provided by scikit-learn [13].

In [14], they created an ISL recognition system where the input image was first categorized as either a single-handed or two-handed sign language gesture using a HOG-SVM classifier. Then, they used feature extraction using SIFT and HOG, and finally, they used KNN to classify the images. They used a dataset containing 520 images to train their model and 260 images to test their model, where 60 of those images were from 6 single-handed sign language gestures and 200 images were divided into 20 double-handed sign language gestures. They achieved an accuracy of 98.33% on their single-handed data and 89.5% on their double-handed data.

In [11], they created a KNN ASL sign language recognition model where they used logarithmic plot transformation and histogram equalization for image reprocessing, then canny edges and L*a*b color space for segmentation. They used HOG for feature extraction on a dataset containing 26 alphabet signs, with 14 samples for training and eight for testing. Their KNN model achieved 94.23% accuracy on their test set.

2.2. Convolutional Neural Network

Convolutional Neural Network (CNN), inspired by the organization of the visual cortex in the human brain, is a popular, supervised machine learning model for image classification tasks due to its ability to generalize well and handle large-scale datasets [15]. CNN models generally consist of convolutional, pooling, and fully connected layers and are trained using backpropagation to recognize visual patterns from the pixels of images [16]. The process is as follows in [6]. First, the convolutional layer extracts the input data by a convolution operation, reduces feature dimensionality through a pooling layer, then passes the most significant features to the fully connected layer to categorize the images.

When creating CNN models, the creator has many options, and the hyper-parameters one chooses will significantly impact the model's accuracy [15]. A large dataset is crucial for training an effective deep learning model such as a CNN [17]. Dropout is also often an added technique used in CNN models to prevent overfitting. Randomly dropping out neurons during training with a specified probability makes the model more robust and generalized, reducing reliance on any single neuron [18].

In [19], they used CNN for their real-time ASL recognition model, skin color detection, and convex hull algorithm for segmentation of the hand location. Then resized that hand region to 28 × 28 pixels and converted it to grayscale as part of their feature extraction. They achieved 100% accuracy on their test data and 98.05% on their real-time system. Their model was trained on a static fingerspelling dataset created by [20], which consists of 900 images of 25 hand gestures. Reference [21] created a SLRNet-8 CNN model ASL recognition, which was trained and tested on four different datasets containing a combined size of 154,643 in 38 classes (alphabet, 0–9, delete, nothing, space). They grayscaled the images, performed normalization, and resized them to 64 × 64 pixels. Their model achieved an average accuracy of 99.92% on the mixed dataset. Reference [22] created a CNN model using a public dataset from MNIST to create a better-performing model. They performed augmentation to create a total of 34,627 28 × 28 static images of the alphabet (excluding J, Z). The paper was written to compete with state-of-the-art methodologies and achieved an accuracy of 99.67%, which beat other models such as SVM, DNN, and RNN.

2.3. Support Vector Machine

Support Vector Machine (SVM) is a supervised learning model used for pattern recognition and classification or regression analysis. It works by finding an optimal hyperplane that separates the data classes by maximizing the distance between the margin and the classes, reducing classification errors [6]. SVM is an effective machine learning algorithm for sign language prediction, particularly in high-dimensional spaces and when the number of samples exceeds the number of dimensions [23]. While SVM is excellent at classifying, it requires a lot of computational power if the dataset size is too large. Therefore, it can also have computational issues if implemented in real-time applications [24]. According to [24], having a good training set with quality labels and balanced data is essential to creating a successful SVM model. Such as the KNN classifier, SVM is a classifier that employs the scikit-learn library in Python to build the model. Therefore the optimal hyperparameters for the model can be determined using GridSearchCV or RandomizedSearchCV.

Ref. [25] used SVM to create a static PSL (Pakistan Sign Language) model. They used K-means clustering segmentation to separate the fingers and palm area (foreground) from the background. They then used multiple kernel learning to test which feature extraction method was the most effective. They tried EOH (Edge Orientation Histograms), LBP (Local Binary Patterns), and SURF, and the results showed that HOG combined with linear kernel function achieved the best accuracy of 89.52%. They used a dataset containing 6633 static images of 36 alphabet signs from 6 signers to build this model. In [26], they created an SVM model that performed pre-processing by converting the images to grayscale, normalizing the data, and applying gamma correction. For feature extraction, they used Multilevel-HOG. They performed this on an ISL complex background dataset containing 2600 images of 26 classes in grayscale and an ASL dataset containing 2929 images of 29 classes in grayscale. They achieved 92% accuracy on the ISL complex background dataset and 99.2% accuracy on the ASL dataset.

2.4. Key Findings

The finding in our literature search reveals that sign language recognition is a well-researched topic. Primarily for static sign language recognition models, there was a lot of research on KNN, CNN, and SVM models. Based on our summary in Table 1, we can see that the segmentation varied across all three classification models. Still, HOG was very popular for feature extraction across all the models. CNN seems to have a lot of research that does not include segmentation or feature extraction of their datasets. Still, they usually balance that out with quality pre-processing and large datasets, one of the most important aspects of CNN models. The accuracy of all the models is very good, and there is no clear indication of which model best suits sign language recognition.

Table 1. Summary of finding different machine learning models.

References	Segmentation	Feature Extraction	Classification	Dataset	Accuracy
(B. Gupta, Shukla, and Mittal 2016)	NaN	SIFT and HOG	HOG-SVM to categorize images as single or two-handed gestures, then KNN for classification	780 ISL images of 6 one-handed and 20 two-handed alphabets	98.33% for one-handed 89.5% for two-handed
(Mahmud et al. 2018)	Canny edges and L*a*b color space	HOG	KNN	572 ASL images on 26 alphabets	94.23%
(Taskiran, Killioglu, and Kahraman 2018)	Skin color detection and convex hull algorithm	Rezised hand region to 28 × 28 pixels and grayscaled it	CNN	900 ASL images on 25 alphabets	100% on test data 98.05% on real-time system
(Rahman et al. 2019)	NaN	NaN	SLRNet-8 CNN	154,643 ASL images on 38 gestures from four different datasets	99.92%
(Mannan et al. 2022)	NaN	NaN	CNN	34,627 ASL images on 26 alphabets	99.67%
(Shah et al. 2021)	K-means clustering to separate foreground and background	HOG	Linear SVM	6633 PSL images on 36 alphabets from 6 signers	89.52%
(Joshi, Singh, and Vig 2020)	NaN	Multilevel-HOG	SVM	2600 ISL images on 26 grayscale alphabets with complex background	92% accuracy on ISL
				2929 ASL images on 29 grayscale alphabets.	99.2% accuracy on ASL

During our research, we found some gaps in the literature that we hope we can assist. Upon studying the literature, there is no research provided on Norwegian Sign Language. We aim to close this gap by creating SVM, KNN, and CNN models for Norwegian Sign Language recognition. Another gap we identified was that many of the datasets the researchers used to train their models were simple images with non-complex backgrounds. While this will achieve great accuracy for the models, an important aspect of image recognition is to build robust models that can generalize. If the models are trained on simple images, it is tough to see how these can be used in real-life applications. We, therefore, aim to create our NSL model based upon a combination of non-complex and complex images with different lighting and backgrounds to develop models that would generalize well. While this might negatively affect our model's accuracy, we hope it is better suited for real-life situations to further assist the deaf and hard of hearing in Norway.

3. Methods

This section covers the methodology and implementation to find the most effective classification models for NSL recognition. All tests were performed on our local computer with the specs: RTX 3060ti GPU, AMD Ryzen 7 5700G CPU @ 3.8 GHz, and 16 GB of RAM. Figure 1 displays a summary and visualization of the entire data pipeline that was conducted for our methodology and implementation. We started by using our camera for image capturing. The images are then stored as a dataset. Once completed, the dataset is used for three different classification models using their distinct segmentation and feature extraction methods. Finally, the three classification models predict signs of the unseen test set to evaluate their performance.

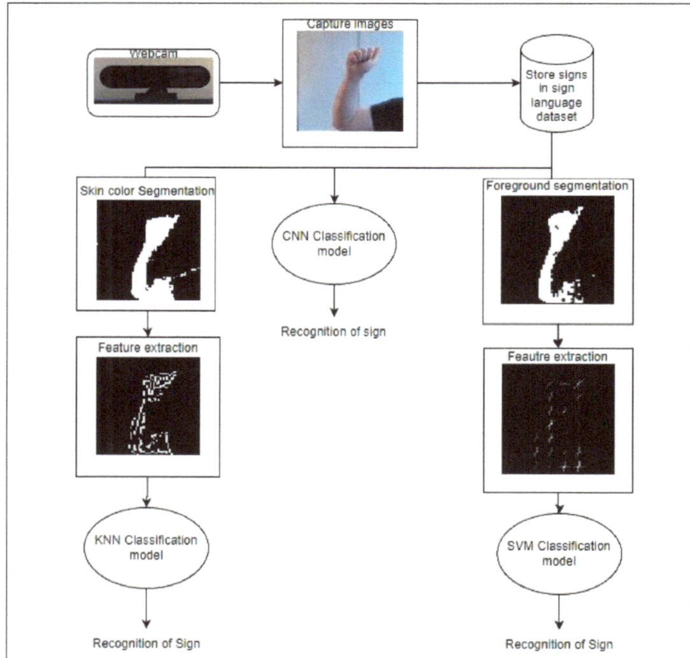

Figure 1. Flowchart illustrating the pipeline of our method.

3.1. Image Acquisition

To build our NSL (Norwegian sign language) recognition model, we had to acquire a dataset containing images or videos of NSL. During our extensive research, we could not find any public NSL dataset and therefore had to start creating our own from scratch. The first step in creating our NSL dataset was determining which sign language category we wanted to create. Because little to no research has been conducted on NSL recognition, we decided that the best approach would be to create a static fingerspelling dataset of the NSL alphabet. We chose this sign language category because languages are built from the ground up, starting with the alphabet. Therefore, we wanted to create a strong base for further work in NSL recognition by creating a baseline for NSL alphabet recognition.

The signs must be performed and statically read to develop a static fingerspelling sign language model. The Norwegian Sign Language (NSL) includes 29 signs from A to Å. However, the letters Å and H cannot be statically displayed, so they were excluded from our research. Static images can display the other 27 signs, most of them as one-handed, but J, P, Q, U, X, Z, and Ø had to be represented as two-handed as there is no static one-handed version. All information about the NSL alphabet and the drawn images in Figure 2 was gathered from the Norwegian Sign Language Dictionary, www.tegnordbok.no accessed on 11 July 2023, which was developed by Statped (State Special Needs Education Service) on behalf of the Norwegian Directorate for Education and Training [27].

Our dataset contains images captured using a Lenovo Performance FHD Webcam with a resolution of 1920 × 1080. However, because we focused on extracting the hand from the images, we implemented a technique called Range of Interest (ROI) to restrict the area of interest to 256 × 256 pixels. This allowed us to retain the high-definition quality of the webcam while minimizing the amount of noise and background the training algorithm had to deal with. In other words, by using an ROI, we ensured that the training process was optimized for accuracy and efficiency.

The Dataset

For our dataset, we created 900 images for each of the 27 NSL alphabet signs, resulting in 24,300 images. The images of one female and one male were taken to ensure that the model could generalize better. We captured 600 images inside with natural and artificial lighting on a white background. Of those, 300 images were taken by the female and 300 by the male. We also captured 300 more images with a complex background to help the model better handle real-world scenarios. These images were taken by the male participant indoors with various objects in the background. Table 2 shows the distribution of the dataset, and Figure 3 illustrates the different scenarios. We also ensured that the dataset was balanced and that all the classes had the same number of pictures, as this is an essential step in making recognition models that are good at generalizing and aim for a high accuracy [28].

While different backgrounds and people are important steps in creating a dataset for a well-generalized model, we also ensured that the signs we performed during our image acquisition consisted of different angles and rotations of the different signs. In the same way that people pronounce words differently with their voice, signers have different hand orientations when they perform sign language gestures. By incorporating rotated images in the dataset, we can help the model learn to recognize the gestures regardless of the orientation of the signer's hands. Examples of signs with different angles are shown in Figure 4, and different rotations are illustrated in Figure 5.

Based on the literature we have gathered on sign language recognition, a dataset consisting of 24,300 images represents a substantial baseline for most machine learning models. It should be noted that specific machine learning models may not optimally perform when trained on excessively large datasets. Hence, we decided not to create a larger dataset, recognizing that various machine learning models have different limitations and requirements. In the case of machine learning models such as CNN, which require a considerable amount of data for effective training, data augmentation techniques can be employed to expand the dataset size as needed.

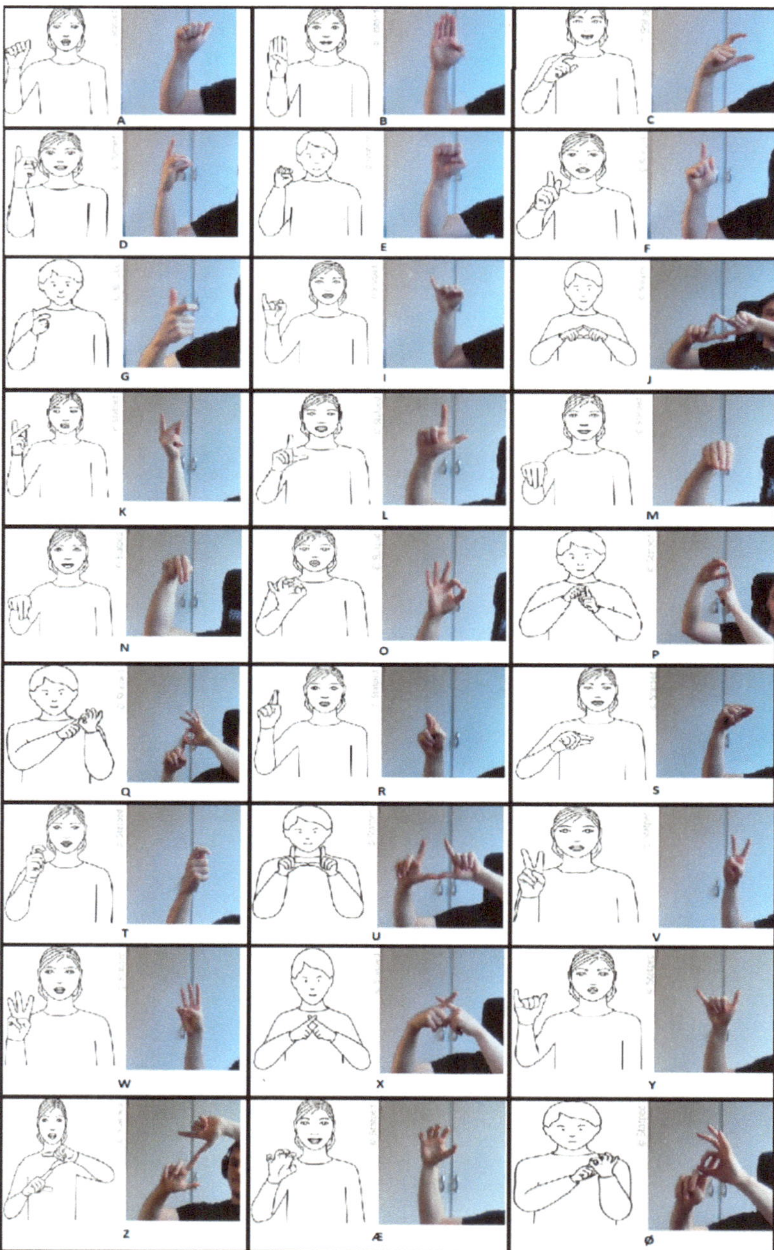

Figure 2. Drawn images and captured images of each sign were used in our comparative analysis.

Table 2. Distribution of the NSL dataset.

Alphabet	Male Complex Background	Male Artificial and Natural Lighting, White Background	Female Artificial and Natural Lighting, White Background	Total
A–Ø (Å and H Excluded)	300	300	300	24,300

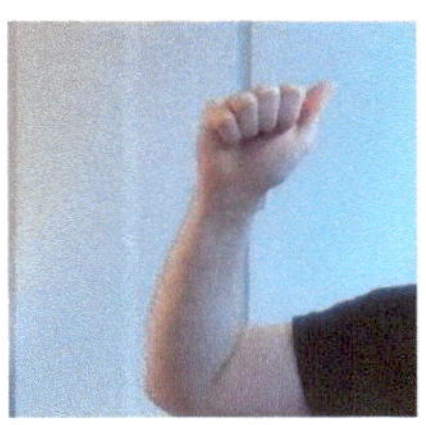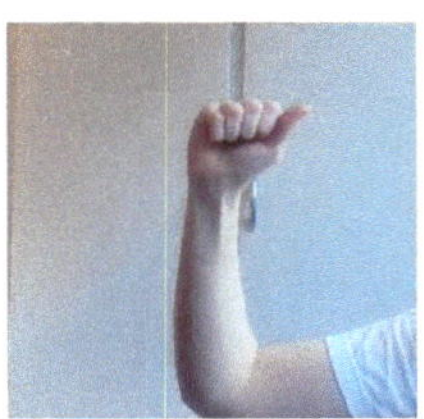

Figure 3. From left to right, a male hand with artificial and natural lighting on white background, a female hand with artificial and natural lighting on white background, and a male hand with a complex background.

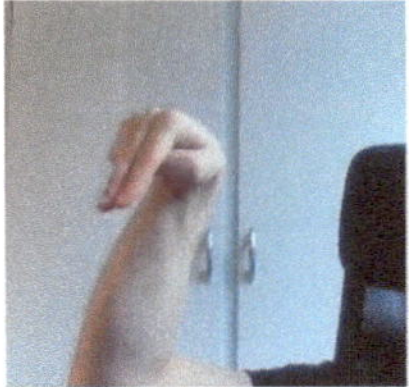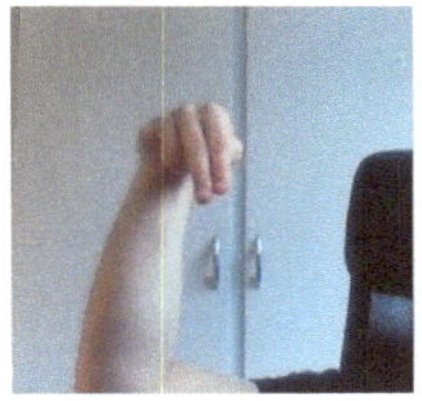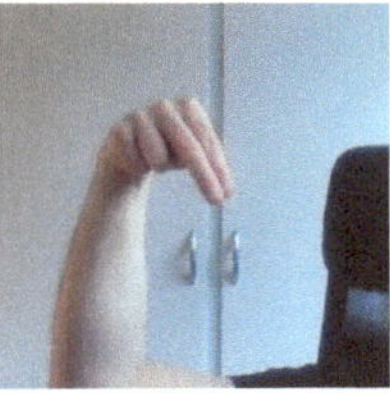

Figure 4. Images show different angles of the sign N.

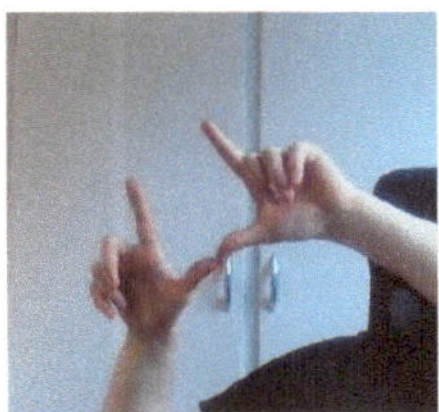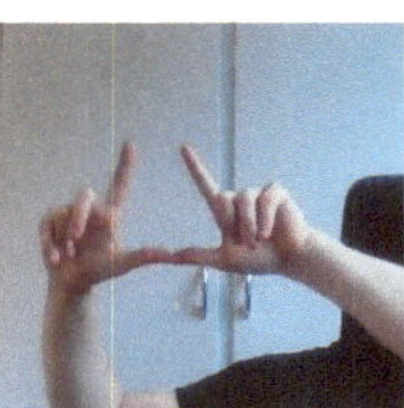

Figure 5. Images show different rotations of the sign U.

Based on the size of the dataset and the measures taken during the image acquisition process, we are very confident in the quality and diversity of our dataset for NSL recognition. The dataset includes a variety of lighting conditions, backgrounds, and hand orientations. By capturing images from male and female singers and incorporating different angles and rotations, we sought to enhance the generalizability of the dataset for use across various machine learning models. Overall, our dataset can serve as a strong foundation for our own and future research in NSL recognition.

3.2. KNN Model

After creating the dataset, we decided that the first recognition model to create was KNN. We chose this model because it is a simple yet respected model for classification. Additionally, many sign language recognition research studies have utilized KNN, and we wanted to provide similar research for NSL recognition to compare our results. Due to

KNN's struggle with large datasets, we decided not to perform augmentation and use the dataset as is.

3.2.1. Preparing the Data

The KNN classifier is sensitive to noise and large complex datasets. Therefore, the data preparation process plays a crucial role in ensuring the optimal performance of the model. The process we have implemented consists of three primary steps: pre-processing, segmentation, and feature extraction. These steps are designed to transform the raw image data into a structured and standardized format, which can be efficiently utilized to train and evaluate the KNN model while mitigating the impact of noise and complexity. Figure 6 illustrates the different steps described below.

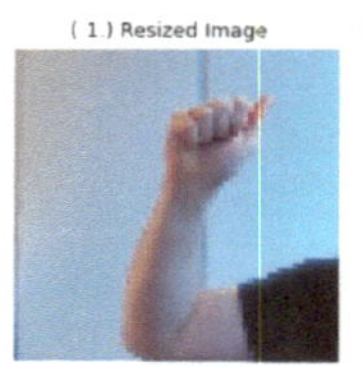
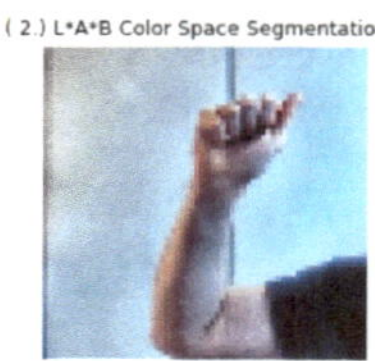
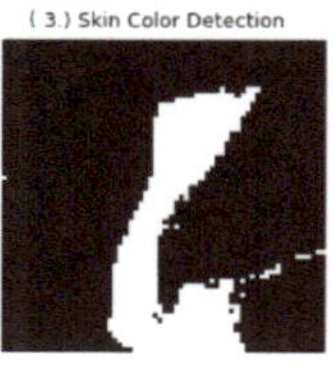
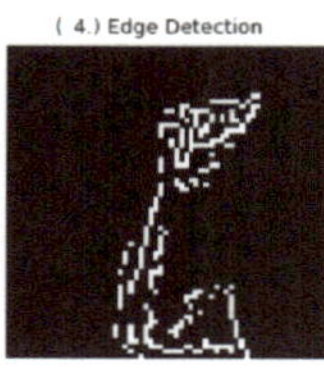
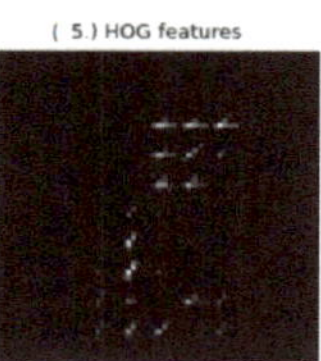

Figure 6. Data steps for our KNN model.

3.2.2. Pre-Processing

The pre-processing step ensures consistency among the images in the dataset. We achieved this by resizing the images from 256×256 pixels to a uniform dimension of 64×64 pixels. This guarantees that all images have the same dimensions, which is crucial for comparing and processing the images during the later stages of the process. Additionally, it reduces the computational time of the model. This step is illustrated in Step 1 in Figure 6.

3.2.3. Segmentation

The segmentation process isolates specific regions of interest within the images, reducing noise and allowing for more accurate classification. We achieved this using LAB color space segmentation, which provides a more uniform representation of color differences than the RGB color space. The resized images are first converted to the LAB color space, and the Contrast Limited Adaptive Histogram Equalization (CLAHE) technique is applied to the L channel to enhance the image's contrast. After that, the images are segmented into skin and non-skin regions based on a predefined color range, focusing on the regions containing skin color and effectively removing irrelevant areas within the images. Illustrated are Steps 2 and 3 in Figure 6.

3.2.4. Feature Extraction

Once the images have been segmented, feature extraction provides the KNN model with the most relevant features. We applied the Canny edge detection technique to the grayscale version of the segmented images to achieve this. Subsequently, the skin color mask is applied to the edge image, further refining the features by focusing on the regions of interest. The HOG features are then computed for the masked edge image, generating a feature vector that effectively captures the essential shape and characteristics of the images. L2 normalization is employed to ensure that the HOG features have the same scale and can be effectively compared across images. Illustrated are Steps 4 and 5 in Figure 6.

3.2.5. Model Training and Hyperparameter Tuning

After performing all these steps on the images, the normalized HOG features are flattened and added to a 'data' list. The corresponding labels for the images are added to the 'labels' list. Finally, the 'data' and 'labels' lists are converted to numpy arrays. The 'data' array is reshaped, maintaining the number of samples (rows) and arranging the HOG features into a 2D array with the appropriate number of columns. This structured format is crucial to train and evaluate the KNN model.

Having prepared the images and extracted their HOG features, we found the best parameters for the KNN model. We began by dividing the dataset into training and testing sets, using a 30–70% split to ensure that the model's performance could be evaluated on previously unseen data. Following this, we conducted a randomized search combined with cross-validation to find the optimal hyperparameters for the model.

To carry out this process, we constructed a parameter grid containing various combinations of the number of neighbors (K values ranging from 1 to 100), weights (uniform or distance), and distance metrics (Euclidean, Manhattan, and Minkowski). Using cross-validation, we then instantiated a RandomizedSearchCV object to search for the best hyperparameters. To enhance the search efficiency, we set the number of iterations to 50 and employed 10-fold cross-validation. The search process was further expedited by running it in parallel, utilizing all available CPU cores.

Because KNN is already a high computational time model, we decided to create two identical models, one with a 64 × 64 resized image and one with 32 × 32, to compare computational time and accuracy trade-offs.

3.2.6. Experiment 1: 64 × 64 Images

Figure 7 displays the accuracy for the different values of K and the different metrics. Upon completion of the randomized search and cross-validation process, the optimal hyperparameters for the KNN model with a 64 × 64 resolution were identified as follows: 'metric': 'Manhattan', 'n_neighbors': 4, and 'weights': 'uniform'. By training the KNN model with these parameters, we achieved an outstanding accuracy of 99.9% on the unseen test dataset. This accuracy illustrates the effectiveness of the methods we have implemented in this model.

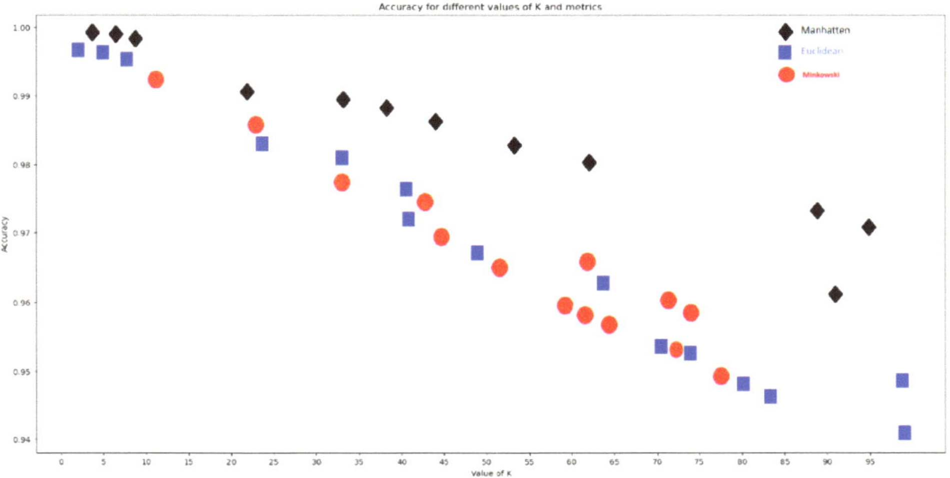

Figure 7. Result of RandomizedSearchCV for our 64 × 64 KNN model.

In Table 3, we present the performance of the 64 × 64 KNN model. The classification report from sklearn metrics provides a comprehensive view of the precision, recall, F1-score, and support for each sign, as well as the number of errors made by the model.

Table 3. Classification report of the 64 × 64 KNN model.

Sign	Precision	Recall	f1-Score	Support	Number of Errors
A	1.00	1.00	1.00	271	0
B	1.00	1.00	1.00	271	0
C	1.00	1.00	1.00	272	0
D	1.00	1.00	1.00	267	0
E	1.00	1.00	1.00	280	0
F	1.00	1.00	1.00	261	0
G	1.00	1.00	1.00	272	0
I	1.00	1.00	1.00	280	0
J	1.00	1.00	1.00	247	0
K	1.00	1.00	1.00	262	0
L	1.00	1.00	1.00	295	0
M	1.00	1.00	1.00	255	0
N	1.00	1.00	1.00	285	0
O	1.00	1.00	1.00	263	0
P	1.00	1.00	1.00	252	0
Q	1.00	1.00	1.00	276	0
R	1.00	1.00	1.00	256	0
S	1.00	1.00	1.00	244	0
T	1.00	1.00	1.00	281	0
U	1.00	1.00	1.00	268	0
V	1.00	1.00	1.00	281	0
W	1.00	1.00	1.00	274	0
X	1.00	1.00	1.00	279	0
Y	1.00	1.00	1.00	271	0
Z	1.00	1.00	1.00	283	1
Æ	1.00	1.00	1.00	256	0
Ø	1.00	1.00	1.00	296	0
Accuracy	1.00	1.00	1.00	7298	1

Precision, which measures the model's ability to identify positive instances correctly, is at 1.00 for all signs, indicating that the model has a high degree of accuracy when it predicts a sign. Recall, which measures the model's ability to find all the positive instances, is also at 1.00 for all signs, showing that the model is highly effective at identifying all instances of a sign in the dataset. The F1-score, the harmonic mean of precision and recall, is also at 1.00 for all signs, indicating a balance between precision and recall in the model's performance.

Despite the model's high accuracy of 99.99%, it is important to note that it took the model 188 s to make predictions on the unseen data. This trade-off between accuracy and time is a consideration for the practical implementation of the model. However, the overall performance of the model demonstrates its robustness and effectiveness in NSL recognition.

To further confirm the model's accuracy, we aimed to display correctly and wrongly predicted images. Figure 8 shows randomly selected correctly predicted images, where the actual image is displayed as the original image input, and the predicted image is presented as the image version of the HOG features. Figure 9 displays the wrongly predicted image found in the entire model, highlighting the exceptional overall accuracy achieved by the model.

3.2.7. Experiment 2: 32 × 32 Images

The second part of the KNN model result testing involved resizing the images to 32 × 32 to see if the computational time could be reduced without sacrificing too much accuracy. Because all the steps were the same as in Experiment 1, only the results are summarized here. Firstly, a RandomizedSearchCV was performed on the 32 × 32 images, which resulted in the optimal hyperparameters for the KNN model as 'metric': 'Manhattan', 'n_neighbors': 4, and 'weights': 'distance', with an accuracy of 97.2%. This model only used

18 s to predict the signs of the unseen test data. The result of the RandomizedSearchCV is shown in Figure 10.

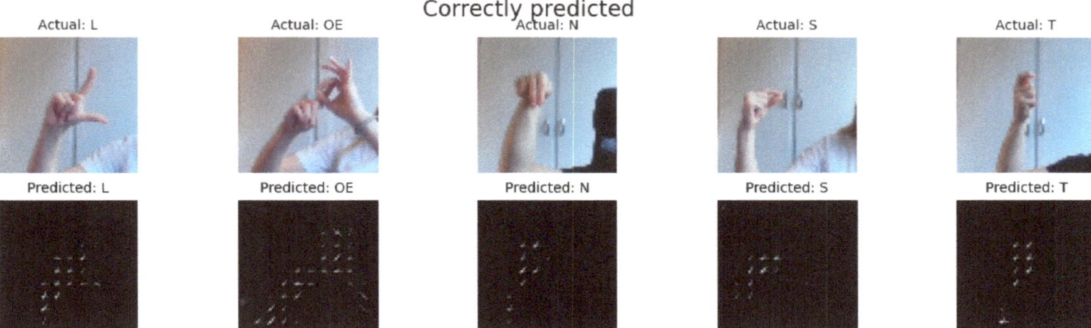

Figure 8. Correctly predicted signs of the 64 × 64 KNN model.

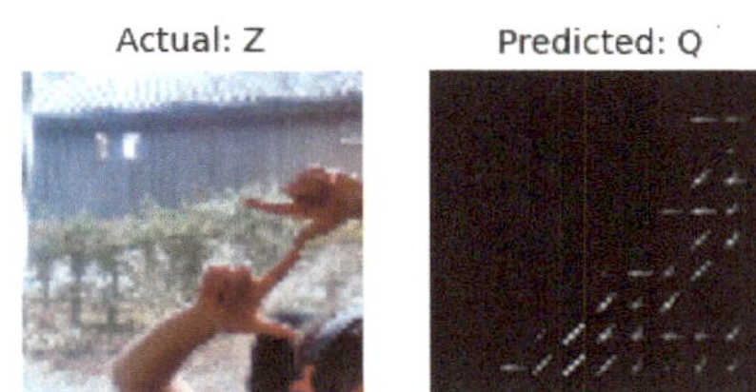

Figure 9. Wrongly predicted signs of the 64 × 64 KNN model.

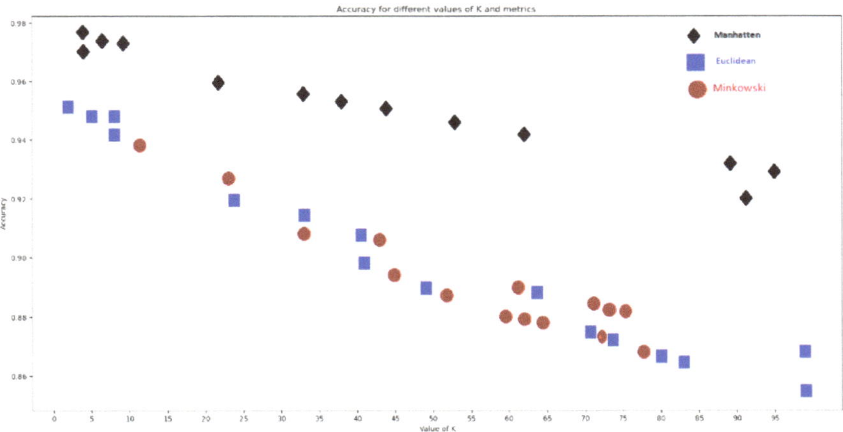

Figure 10. KNN 32 × 32 Gridsearch result.

Table 4 provides a detailed performance report of the model for each sign. While the model performs well overall, with an accuracy of 97%, certain signs such as W, V, F, and L have shown sub-optimal performance.

Table 4. Classification report of the 32 × 32 KNN model.

Sign	Precision	Recall	f1-Score	Support	Number of Errors
A	1.00	1.00	1.00	265	0
Æ	0.99	1.00	0.99	273	3
B	0.99	1.00	0.99	270	5
C	0.98	0.98	0.98	264	10
D	0.96	0.95	0.96	276	5
E	0.98	0.97	0.98	247	18
F	0.91	0.98	0.94	278	2
G	0.96	0.97	0.97	255	12
I	0.94	0.98	0.96	293	1
J	0.98	0.99	0.99	270	8
K	0.98	0.97	0.97	265	10
L	0.97	0.91	0.94	278	16
M	0.96	0.97	0.96	287	7
N	0.98	0.95	0.96	267	10
O	0.96	0.96	0.96	259	0
Ø	1.00	1.00	1.00	258	5
P	0.99	0.98	0.99	282	4
Q	0.99	1.00	0.99	269	6
R	0.99	0.99	0.99	258	11
S	0.95	1.00	0.97	263	4
T	0.97	0.98	0.98	299	21
U	0.98	0.99	0.99	247	19
V	0.91	0.95	0.93	266	9
W	0.91	0.84	0.88	269	10
X	0.99	0.99	0.99	283	0
Y	0.99	0.95	0.97	280	1
Z	1.00	0.99	0.99	277	0
Accuracy	0.97	0.97	0.97	7298	159

Precision varies across signs, with the lowest being 0.92 for the sign 'V'. This indicates that the model has a slightly lower accuracy when predicting this sign. Recall also varies, with the lowest being 0.88 for the sign 'W', indicating that the model has slightly lower completeness when identifying this sign. The F1-score is lowest for the sign 'W' at 0.90, indicating a slight imbalance between precision and recall for this sign.

The confusion matrix in Figure 11 provides further insights into errors, showing that 'W' was wrongly predicted as 'V' 18 times, 'V' was predicted as 'W' 9 times, 'L' was wrongly predicted as 'F' 7 times, and 'Y' was wrongly predicted as 'I' 6 times.

Despite these errors, the model's overall performance demonstrates its effectiveness, highlighting areas for improvement.

3.3. CNN Model

CNNs are known for their ability to handle noise and perform well on image data, so we opted to create the model without employing any segmentation or feature extraction. This is also the approach most of the CNN literature we came across took, such as [21,22].

We chose TensorFlow as the framework for developing our CNN model due to its support for GPU acceleration, enabling faster training. TensorFlow is an open-source platform for large-scale machine learning, which we utilized for importing, pre-processing, splitting, training, evaluating, visualizing, and testing our dataset [29].

3.3.1. Preparing the Data

Although we decided against segmentation or feature extraction, data preparation remains essential for achieving optimal performance with CNN models. We used Tensor-Flow's image_dataset_from_directory function to load and pre-process the dataset. The images were resized to 64 × 64 pixels, and the pixel values were normalized to a range of 0–1. We shuffled the data during loading to avoid any biases during training and set a batch size of 32. Once the dataset was prepared, we divided the data into training 70%, validation 15%, and testing 15% sets.

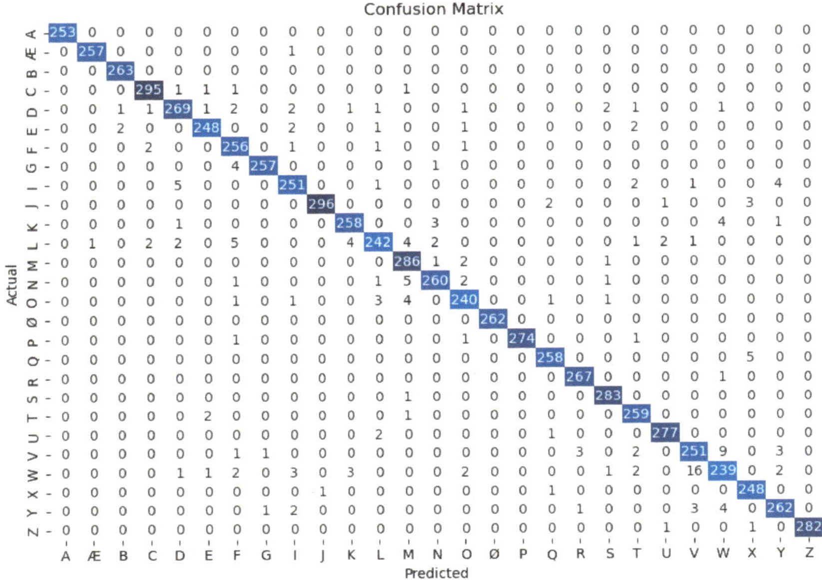

Figure 11. KNN 32 × 32 confusion matrix.

3.3.2. Building the Model

Figure 12 shows the architecture of our CNN model. The CNN model consists of three convolutional layers, each followed by a max pooling layer. The model is then flattened to a 1D array, which is used for the connected layer consisting of two dense layers. Before both dense layers, a dropout layer is added to prevent overfitting. For the convolutional layers, we use ReLU because it is a commonly used activation function.

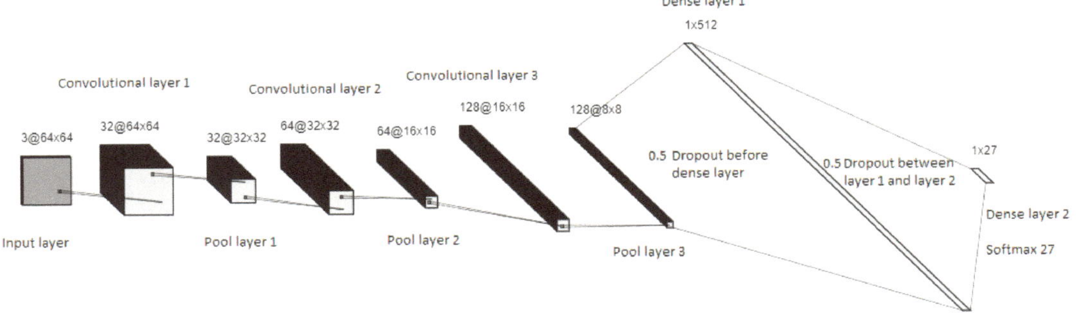

Figure 12. Visualization of the CNN architecture.

3.3.3. Model Training and Hyperparameters Tuning

Once the model was built, we started finding the best hyperparameters. Using standard procedures in CNN, we compiled the model using the Adam optimizer, as it is a popular choice for its computational efficiency and low memory requirements [30]. We selected a learning rate of 0.0001, which balances convergence speed and accuracy. The sparse categorical cross-entropy loss function was used, as it is appropriate for multi-class classification problems where the target classes are mutually exclusive. To prevent overfitting during training, we employed an early stopping callback with a patience of 3, which halts the training process if there is no improvement in validation loss after three consecutive epochs.

We initially trained the model for up to 50 epochs so the model could train until it no longer improved and used a batch size of 32. We did this to find the optimal number of epochs for our model to create an accurate model without overfitting it to the training data. The image in Figure 13 displays the results of our initial model, displaying training loss versus validation loss and training accuracy versus validation accuracy. As we can see, the model stopped at epoch 10, meaning the validation loss did not improve starting from epoch 7. This makes sense as it seems like the model achieves close to 100% accuracy on the validation data after 5 epochs. A good indication that the model is not overfitting is that the validation error is lower than the training error, and the validation accuracy is higher than the training accuracy. This is because it indicates that the model is performing well on new unseen data. Due to these findings, we decided to go for 7 epochs for our final model.

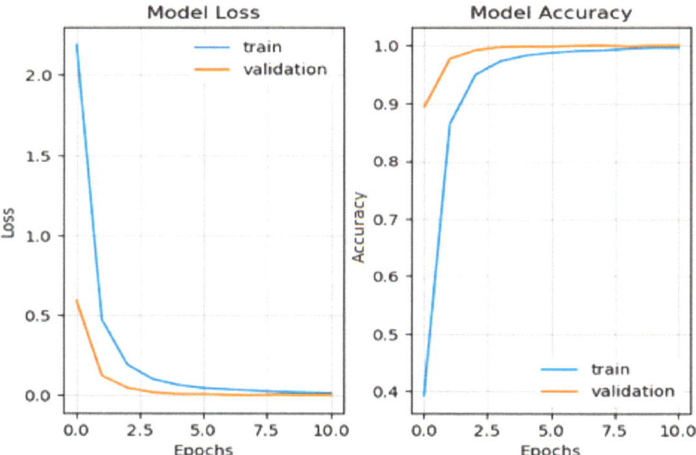

Figure 13. Displaying model Accuracy vs. Validation accuracy and model loss vs. Training loss.

3.3.4. Results Analysis

After we had found all the best parameters for our model, we no longer had a use for validation data and divided the data into a new 70% (train) and 30% (test) split for optimal testing. After training the model, we tested the model on the unseen test data. Table 5 provides a comprehensive performance report of the model for each sign. The precision, recall, and F1-score metrics, all at 99.9%, illustrate the effectiveness of the model. However, a few signs such as 'V', 'N', and 'W' have shown minor errors.

Precision is at 1.00 for almost all signs, with a slight dip to 0.99 for 'V'. This suggests a marginally lower accuracy when predicting this sign. Recall also maintains a high score of 1.00 for most signs, with a minor decrease to 0.99 for 'N' and 'W', indicating slightly lower completeness when identifying these signs. The F1-score is at 1.00 for almost all signs, with a slight decrease to 0.99 for 'V', 'N', and 'W', indicating a minor imbalance between precision and recall for these signs. The model had a training time of 22 s, and predictions on the unseen test data took 22 s.

Table 5. Classification report of the CNN model.

Sign	Precision	Recall	F1-Score	Support	Number of Errors
A	1.00	1.00	1.00	271	0
B	1.00	1.00	1.00	281	1
C	1.00	1.00	1.00	247	0
D	1.00	1.00	1.00	292	1
E	1.00	1.00	1.00	287	0
F	1.00	1.00	1.00	256	0
G	1.00	1.00	1.00	252	0
I	1.00	1.00	1.00	276	0
J	1.00	1.00	1.00	270	0
K	1.00	1.00	1.00	282	0
L	0.99	1.00	0.99	257	0
M	1.00	1.00	1.00	283	0
N	1.00	0.99	1.00	262	2
O	0.99	1.00	0.99	295	0
P	1.00	1.00	1.00	271	0
Q	1.00	1.00	1.00	260	0
R	1.00	1.00	1.00	263	0
S	1.00	1.00	1.00	275	0
T	1.00	1.00	1.00	290	0
U	1.00	1.00	1.00	270	0
V	1.00	0.99	0.99	275	3
W	1.00	1.00	1.00	269	1
X	1.00	1.00	1.00	256	0
Y	1.00	1.00	1.00	281	0
Z	1.00	1.00	1.00	250	0
Æ	1.00	1.00	1.00	259	0
Ø	1.00	1.00	1.00	246	0
Accuracy	1.00	1.00	1.00	7276	8

To gain further insight into the wrongly predicted signs, we decided to display the incorrectly predicted images. In Figure 14, we loop through the indices of the incorrectly predicted images and display each image along with its actual and predicted label. It also displays the top three predicted probabilities for the image as a bar chart, with the color of each bar indicating the confidence of the model for different labels, the actual label (green), wrongly predicted label (red), or another label (gray). Looking at the images, we can see that in the two occurrences where the model predicted L but it was N, the model almost achieved the correct answer. While two errors are a very small sample size, and it is hard to say anything conclusive, it could indicate that L and N share some features. The same situation is with V and O, where the model twice predicted that V was O. Again, the model is very close to labeling it correctly, which indicates that although the model predicted wrong, it was still very close to the correct label. Again, because the sample size of wrongly predicted images is so small, we found it hard to conclude anything and decided that these errors were acceptable.

3.4. SVM-Model

The final model we decided to create was an SVM model. This is because it is a powerful and versatile machine learning model that is commonly used in sign language recognition. Because SVM is great at high-dimensional spaces where the number of samples is greater than the number of dimensions, this also makes it a good fit for implementing it in our research towards NSL recognition because the images contain a lot of features.

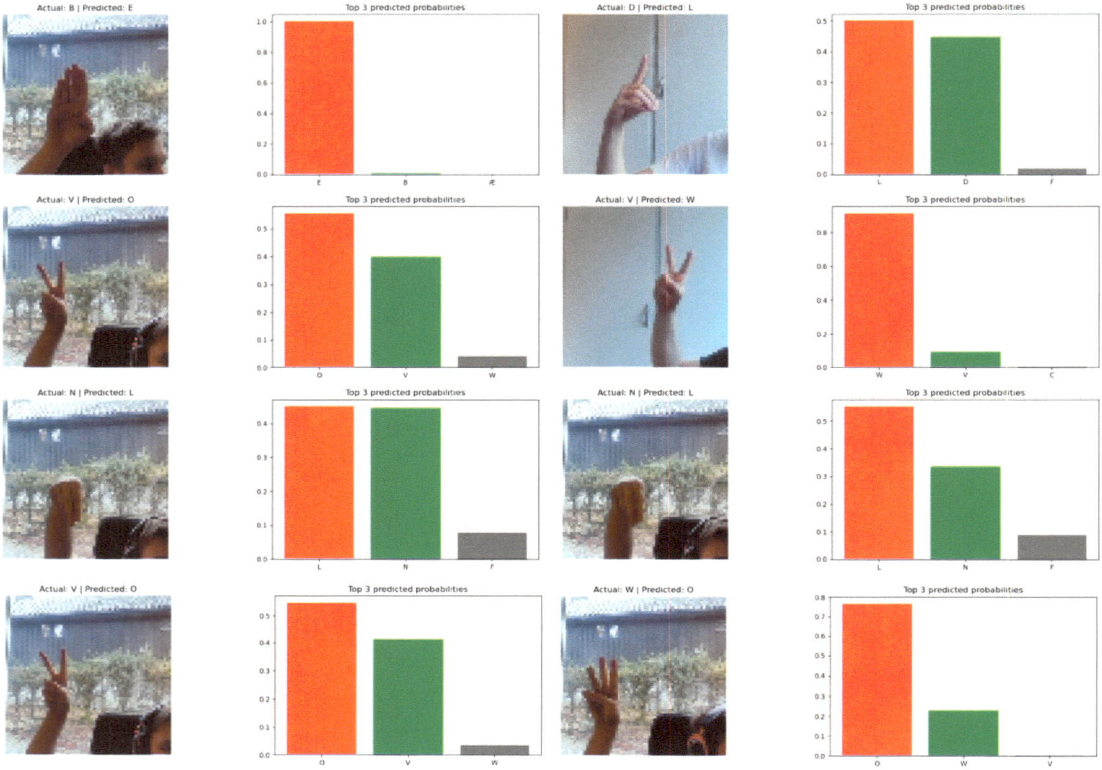

Figure 14. Wrongly predicted images for our CNN model.

3.4.1. Data Preparation

The data preparation steps for our SVM model are very similar to that of the KNN model. The reason for this is twofold, it is not a deep learning model such as CNN, and we, therefore, want to train the model on the most important features. Most of the literature we came across, such as [25,26], used feature extraction, segmentation, or both as part of their data preparation. Such as the other models we have created, the pre-processing of the images was to resize them to a fixed size of 64×64 for consistency and computational efficiency. Figure 15 illustrates the data preparation steps for our SVM model.

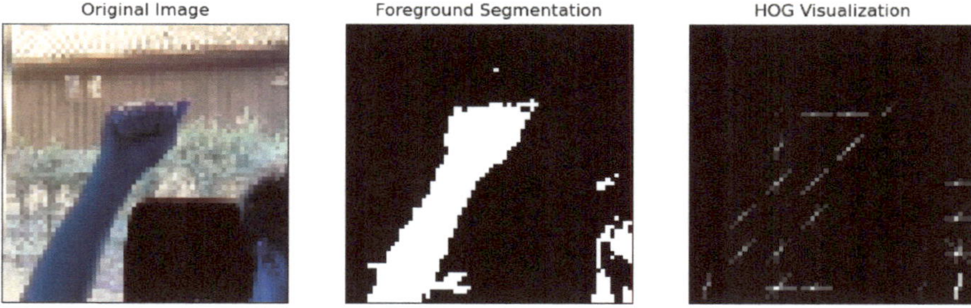

Figure 15. Data preparation steps for the SVM model.

3.4.2. Segmentation

A lot of consideration was given when it came to the image segmentation approach for our SVM model. Our initial thought was to use the same approach we had performed in our KNN model because the LAB color spacing and skin color segmentation we performed for our KNN model were very effective. We did, however, conclude that choosing a different approach for our SVM model would serve greater research value in the field of NSL recognition and decided to choose a different approach.

Such as the segmentation performed in our KNN model, we used LAB color spacing, but instead of segmenting the image based on skin color, we decided to segment the image by separating the foreground from the background. While there are similarities in the approaches, the intent and execution are different.

3.4.3. Feature Extraction

For feature extraction, we decided to go with HOG features as this is a commonly used feature extraction for SVM models. We also found earlier that we had success with HOG for our KNN model. We, therefore, wanted to replicate it for our SVM model without edge detection. In combination with our new approach to segmentation, we concluded that this would lead to valuable research in NSL recognition research.

3.4.4. Model Training and Hyperparameter Tuning

After preparing the dataset and extracting the HOG features, the next step of our process was to proceed with training the SVM model and fine-tuning its hyperparameters. We started by dividing the dataset into training and testing sets, using a 30–70% split to ensure that the model's performance could be evaluated on unseen data. Following this, we conducted a randomized search combined with cross-validation to find the optimal hyperparameters for the model.

To carry out this process, we first created separate parameter grids for each kernel type: linear, radial basis function (RBF), polynomial, and sigmoid. The grids contained various combinations of hyperparameters specific to each kernel, such as regularization parameters 'C', gamma, degree, and coef0. Next, we instantiated RandomizedSearchCV objects for each kernel type to search for the best hyperparameters, utilizing cross-validation. To enhance the search efficiency, we set the number of iterations to 25 for each of the four kernels (total of 100) and employed 5-fold cross-validation. The search process was further expedited by running it in parallel, utilizing all available CPU cores.

Once the randomized search was complete, we extracted the top three models for each kernel type based on their mean cross-validation accuracy. This gave us a total of 12 models to compare. To evaluate the performance of these models on previously unseen data, we trained each of them using the optimal hyperparameters found during the randomized search and tested their accuracy on the validation set.

To visualize and compare the performance of these models, we created a bar plot displaying the accuracy of each model compared with their training time. This allowed us to identify the best-performing and most effective models across different kernel types.

Such as with our KNN model, we decided to conduct two experiments for our SVM model, one with images resized to 64×64 and one SVM model resized to 32×32.

3.4.5. Experiment 1: 64×64 Images

The result of the performance from the top three models for each kernel is displayed in Figure 16. All the models managed to obtain above 99% accuracy. Because there was very little difference in accuracy for the different models, we decided to calculate which model had the best accuracy compared to training time. This was calculated by dividing the accuracy of each model and then their training time to find the optimal model. By considering both accuracy and training time, we found that the kernel that strikes the best balance between performance and computational efficiency is the linear kernel with a C value of 1. This kernel, along with the corresponding hyperparameters, was chosen as the final SVM model for our application.

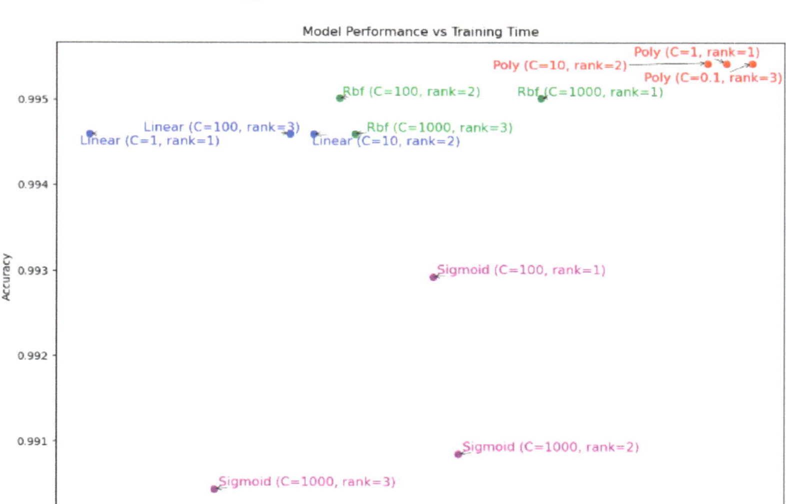

Figure 16. Model performance versus training time for the top three models for each kernel on the 64 × 64 images.

Using these hyperparameters for our SVM model, we split the data into a 70/30 split and trained the model with a linear kernel with a C value of 1. The result of the model on the unseen data is an incredible accuracy of 99.9%.

Table 6 provides a comprehensive performance report of the Support Vector Machine (SVM) model for each sign. The precision, recall, and F1-score metrics, all at 1.00, offer a detailed understanding of the model's performance. However, a few signs such as 'D', 'L', and 'W' have shown minor errors.

Precision is at 1.00 for all signs, with a slight dip to 0.99 for 'V'. This suggests a marginally lower accuracy when predicting this sign. Recall also maintains a high score of 1.00 for most signs, with a minor decrease to 0.99 for 'V', indicating slightly lower completeness when identifying this sign. The F1-score is at 1.00 for all signs, indicating a balance between precision and recall for all signs. The training time for this model was 43 s, and the time it took to make predictions on the unseen data was 53 s, indicating a relatively efficient model.

3.4.6. Experiment 2: 32 × 32 Images

Because the data preparation steps in this experiment are the same as in the 64 × 64 one, we will simply provide the results and discuss them later. Figure 17 displays the RandomizedSearchCV for our 32 × 32 model, and the optimal parameters based on training time and accuracy for this dataset is an RBF kernel with a gamma of 0.1 and C value of 100.

Table 7 presents the performance metrics of the 32 × 32 SVM model for each sign. Despite the overall precision, recall, and F1-score metrics being at 1.00, minor errors were observed for signs 'E', 'I', 'N', 'O', 'Q', and 'W'. The model's precision slightly dipped to 0.99 for 'O', and recall marginally decreased to 0.99 for 'V'. The model's training time was 5.9 s, and the prediction time was 14.9 s, demonstrating its efficiency. With an impressive accuracy of 99.9%, the model's effectiveness in NSL recognition is evident.

Table 6. Classification report of the 64 × 64 SVM mode.

Sign	Precision	Recall	f1-Score	Support	Number of Errors
A	1.00	1.00	1.00	280	0
B	1.00	1.00	1.00	270	0
C	1.00	1.00	1.00	284	0
D	1.00	1.00	1.00	274	1
E	1.00	1.00	1.00	275	0
F	1.00	1.00	1.00	272	0
G	1.00	1.00	1.00	263	0
I	1.00	1.00	1.00	276	0
J	1.00	1.00	1.00	277	0
K	1.00	1.00	1.00	252	0
L	1.00	1.00	1.00	281	1
M	1.00	1.00	1.00	253	0
N	1.00	1.00	1.00	270	0
O	1.00	1.00	1.00	280	0
P	1.00	1.00	1.00	261	0
Q	1.00	1.00	1.00	276	0
R	1.00	1.00	1.00	271	0
S	1.00	1.00	1.00	274	0
T	1.00	1.00	1.00	254	0
U	1.00	1.00	1.00	260	0
V	1.00	0.99	1.00	270	0
W	1.00	1.00	1.00	239	1
X	1.00	1.00	1.00	288	0
Y	1.00	1.00	1.00	260	0
Z	1.00	1.00	1.00	259	0
Æ	1.00	1.00	1.00	271	0
Ø	1.00	1.00	1.00	300	0
Accuracy	1.00	1.00	1.00	7290	3

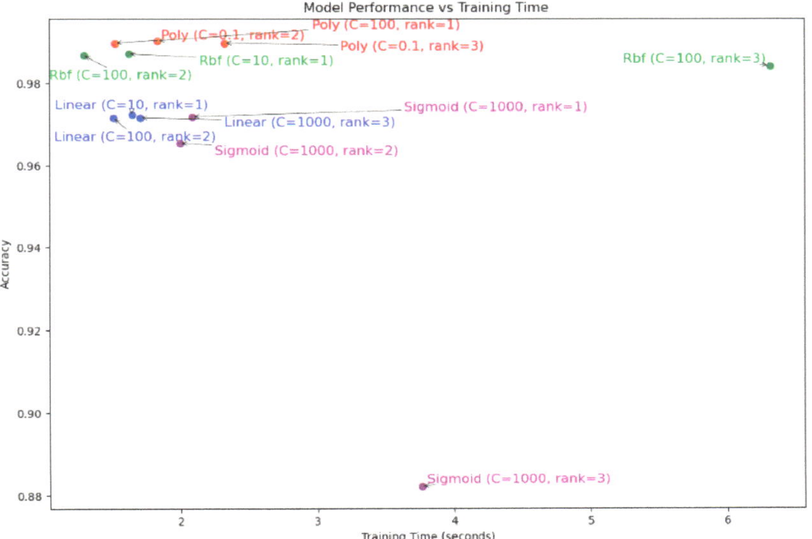

Figure 17. Model performance versus training time for the top three models for each kernel on the 32 × 32 images.

Table 7. Classification report of the 32 × 32 SVM mode.

Sign	Precision	Recall	f1-Score	Support	Number of Errors
A	1.00	1.00	1.00	280	0
B	1.00	1.00	1.00	270	0
C	1.00	1.00	1.00	284	0
D	1.00	1.00	1.00	274	0
E	1.00	1.00	1.00	275	1
F	1.00	1.00	1.00	272	0
G	1.00	1.00	1.00	263	0
I	1.00	1.00	1.00	276	1
J	1.00	1.00	1.00	277	0
K	1.00	1.00	1.00	252	0
L	1.00	1.00	1.00	281	0
M	1.00	1.00	1.00	253	0
N	1.00	1.00	1.00	270	1
O	0.99	1.00	1.00	280	1
P	1.00	1.00	1.00	261	0
Q	1.00	1.00	1.00	276	1
R	1.00	1.00	1.00	271	0
S	1.00	1.00	1.00	274	0
T	1.00	1.00	1.00	254	0
U	1.00	1.00	1.00	260	0
V	1.00	0.99	0.99	270	0
W	1.00	1.00	1.00	239	1
X	1.00	1.00	1.00	288	0
Y	1.00	1.00	1.00	260	0
Z	1.00	1.00	1.00	259	0
Æ	1.00	1.00	1.00	271	0
Ø	1.00	1.00	1.00	300	0
Accuracy	1.00	1.00	1.00	7290	6

4. Results and Discussion

In this section, we will look at the results of our research, discussing the findings from our comparative analysis to find the most effective machine learning models for NSL recognition. All the models are evaluated based on accuracy, training time, and prediction time shown in Table 8. We will also take the time to discuss the different use cases and limitations for each model to offer insight into their suitability for NSL recognition. For consistency in our comparative analysis, all the models were evaluated on the same dataset size with a training and test split of 70/30%.

Table 8. Model comparison of all the created models.

Model	Pre-Processing	Segmentation	Feature Extraction	Accuracy	Training Time	Prediction Time
KNN	64 × 64, Normalization	LAB Color Space & CLAHE, Skin Color detection	HOG, Edge Detection	99.9%	0	188 s
KNN	32 × 32, Normalization	LAB Color Space & CLAHE, Skin Color detection	HOG, Edge Detection	97.2%	0	18 s
SVM	64 × 64, Normalization	Foreground/Background detection using LAB Color space	HOG	99.9%	43 s	53 s
SVM	32 × 32, Normalization	Foreground/Background detection using LAB Color space	HOG	99.9%	6 s	18 s
CNN	64 × 64, Normalization	None	None	99.9%	23 s	21 s

4.1. KNN Model

Accuracy-wise, the KNN model performs well on our self-made dataset achieving results of 99.9% on the 64 × 64 images and 97.2% on the 32 × 32 images. The results also show the effectiveness of the pre-processing, segmentation, and feature extraction performed on the dataset. However, the results in the table illustrate some of the problems of the KNN classification model, as the model that used the 64 × 64 images took 188 s to make predictions on the unseen test set. When we resized the images to 32 × 32, the prediction time significantly decreased to only 18 s, 1/10th of the one trained on 64 × 64 images. These findings align with the general literature of sign language recognition, as KNN quickly loses its efficiency when the dataset becomes too big or has many features.

4.2. SVM Model

The SVM classification model also achieves an impressive accuracy of 99.9% for 64 × 64 and 32 × 32 image sizes, unlike the KNN model, which has different accuracies for the two sizes. Looking beyond accuracy, the model using the 64 × 64 resized images has a training time of 43 s and a prediction time of 53 s, whereas the model using the 32 × 32 resized images has a training time of 6 s and a prediction time of 18 s. That means that the 32 × 32 SVM model is significantly faster than the 64 × 64 SVM model, with an 86% faster training time, a 66% faster prediction time, and overall, 75% faster in combined computational time.

4.3. CNN Model

The CNN model we created was the only model we decided against creating a 32 × 32 version. This was because the model performs extremely well on the 64 × 64 resized images both in terms of accuracy and prediction time, matching the time of the 32 × 32 SVM model. A very successful model can scale if the dataset can be expanded further by using more subjects and background scenarios.

4.4. Discussion

Based on the results from all our models, all three models can perform NSL recognition with high accuracy. However, there are various trade-offs regarding computational efficiency for the different models. This impacts how these models could be used in real-life scenarios or for future research.

For the KNN model, it is clearly effective and quick if the dataset is not complex or big. Because it is a very easy-to-implement model, we still believe that it has its use cases in niche situations where scalability is not needed. Another potential issue with the KNN model we have created is that it uses skin color detection for segmentation. This is a potential limitation for this model in real-world scenarios as the dataset currently mainly consists of individuals with white skin color. Therefore, the model could face problems if a person of different skin color would use it or if the lighting was different.

The SVM model struck a great balance between accuracy and computational efficiency, making it a very versatile model for many NSL recognition tasks. We were surprised to see that the model maintained its accuracy even when the images were resized to 32 × 32, displaying its robustness and adaptability. While the foreground/background segmentation we performed on the dataset for the SVM model might be sensitive to lightning and background, our dataset contains images with various backgrounds, and we are therefore confident that this would also work well for real-world scenarios.

The results of the CNN model were really encouraging as it offered a high accuracy while maintaining a low prediction time. Because CNN models only become more effective as they are exposed to more data, this model could be trained further on more images to grow and expand. The fact that CNN does not require any segmentation or feature extraction makes it a very attractive model because we do not have to consider lighting, skin color, or other uncontrollable factors. Due to its robustness against noise would also serve as an optimal model for future researchers/contributors willing to create an even more comprehensive data set.

While all three of the models show great results in terms of accuracy, the SVM and CNN model stands out as the absolute strongest contenders in terms of accuracy and computational efficiency in terms of predicting new unseen data. The findings from our comparative analysis of image classification models for NSL recognition offer valuable insights that contribute to further research in the field, ultimately identifying the most effective classification model for NSL recognition.

An observation regarding the near-perfect accuracy leads to the consideration of making the dataset more challenging. Computational enrichment could be explored in future research for more rigorous testing of the model's robustness and adaptability, such as adding noise or varying image orientations.

5. Conclusions

A significant portion of the global population struggles with hearing impairments, and image classification models can potentially assist the deaf and hard of hearing by translating sign language. During our research, we explored the state-of-the-art for sign language recognition. While our study found that sign language recognition is well-researched, we identified a significant gap in the literature concerning NSL recognition. To address this gap, we conducted a comparative analysis of various machine learning models to provide further research to determine the most effective classification model for NSL recognition. This research aims to assist in communication capabilities for the deaf and hard of hearing in Norway and hopefully take a step closer to implementing machine learning models to help the deaf and hard of hearing, reducing the communication gap between those who are deaf and hard of hearing and those who are not.

Our literature review revealed several popular machine learning models for sign language recognition. Based on the literature, we chose KNN, CNN, and SVM for our comparative analysis as they represent different paradigms in machine learning and have demonstrated promising results in various applications. Our research aimed to effectively compare these models' data preparation techniques, accuracy, and computational efficiency when applied to NSL recognition.

We created a dataset specifically for NSL recognition. The dataset consists of 24,300 images covering 27 of the 29 NSL alphabet signs, with varying lighting conditions, backgrounds, and hand orientations performed by both male and female signers. The dataset's comprehensiveness and diversity make it a valuable resource for future research in NSL recognition. It also provides a strong foundation for developing and testing machine learning models in this field.

The study conducted a detailed comparative analysis of three popular machine learning models: KNN, CNN, and SVM. This analysis provided insights into the strengths and weaknesses of each model for NSL recognition. It also provided valuable insight into the effectiveness of different data preparation techniques, such as pre-processing, segmentation, and feature extraction. This will assist future researchers of NSL recognition, and we see this contribution as a benchmark for future work.

Based on the comparative analysis, the research identified CNN and SVM as the best-performing models for NSL recognition regarding accuracy, efficiency, adaptability, and scalability. This finding provides a foundation for future research to implement, refine, improve, and scale these models for practical applications.

The contribution and findings of this paper serve as an incentive for future research and development of advanced, accessible, and cost-effective solutions for NSL recognition. Our work also emphasizes the need for such recognition systems in Norway to improve communication between the deaf and hard of hearing and the hearing majority, thereby reducing communication barriers and enhancing overall accessibility.

Author Contributions: Conceptualization, B.S. and S.K.; methodology, B.S. and S.K. All authors have read and agreed to the published version of the manuscript.

Funding: This research received no external funding.

Institutional Review Board Statement: Not applicable.

Informed Consent Statement: Not applicable.

Data Availability Statement: Data are available upon request from the corresponding author.

Conflicts of Interest: The authors declare no conflict of interest.

References

1. World Health Organization. Deafness and Hearing Loss. April 2021. Available online: https://www.who.int/news-room/fact-sheets/detail/deafness-and-hearing-loss (accessed on 15 November 2022).
2. World Federation of the Deaf. Our Work. March 2021. Available online: http://wfdeaf.org/our-work/ (accessed on 15 November 2022).
3. Pigou, L.; Dieleman, S.; Kindermans, P.-J.; Schrauwen, B. Sign language recognition using convolutional neural networks. In Proceedings of the European Conference on Computer Vision, Zurich, Switzerland, 6–7 September 2014; pp. 572–578. [CrossRef]
4. De Meulder, M.; Haualand, H. Sign language interpreting services: A quick fix for inclusion? *Transl. Interpret. Studies. J. Am. Transl. Interpret. Stud. Assoc.* **2021**, *16*, 19–40. [CrossRef]
5. Kudrinko, K.; Flavin, E.; Zhu, X.; Li, Q. Wearable sensor-based sign language recognition: A comprehensive review. *IEEE Rev. Biomed. Eng.* **2020**, *14*, 82–97. [CrossRef] [PubMed]
6. Adeyanju, I.; Bello, O.; Adegboye, M. Machine learning methods for sign language recognition: A critical review and analysis. *Intell. Syst. Appl.* **2021**, *12*, 200056. [CrossRef]
7. Cheok, M.J.; Omar, Z.; Jaward, M.H. A review of hand gesture and sign language recognition techniques. *Int. J. Mach. Learn. Cybern.* **2019**, *10*, 131–153. [CrossRef]
8. Winther, F.Ø. "Hørselshemming", Store Medisinske Leksikon. June 2022. Available online: https://sml.snl.no/h\T1\orselshemming (accessed on 5 March 2023).
9. Peterson, L.E. K-nearest neighbor. *Scholarpedia* **2009**, *4*, 1883. [CrossRef]
10. Guo, G.; Wang, H.; Bell, D.; Bi, Y.; Greer, K. KNN model-based approach in classification. In Proceedings of the OTM Confederated International Conferences "On the Move to Meaningful Internet Systems", Sicily, Italy, 3–7 November 2003; Springer: Berlin/Heidelberg, Germany, 2003.
11. Mahmud, M.; Kaiser, M.S.; Hussain, A.; Vassanelli, S. Applications of deep learning and reinforcement learning to biological data. *IEEE Trans. Neural Netw. Learn. Syst.* **2018**, *29*, 2063–2079. [CrossRef] [PubMed]
12. Xing, J.; Yang, X.; Zhou, X.; Yan, Z.; Zhang, Y. Medical image classification using synergic deep learning. *Inf. Sci.* **2019**, *504*, 130–141. [CrossRef]
13. Cho, K.; van Merriënboer, B.; Gulcehre, C.; Bahdanau, D.; Bougares, F.; Schwenk, H.; Bengio, Y. Learning phrase representations using RNN encoder-decoder for statistical machine translation. In Proceedings of the 2014 Conference on Empirical Methods in Natural Language Processing (EMNLP), Doha, Qatar, 25–29 October 2014; pp. 1724–1734. [CrossRef]
14. Pascanu, R.; Mikolov, T.; Bengio, Y. On the difficulty of training recurrent neural networks. In Proceedings of the International Conference on Machine Learning, Atlanta, GA, USA, 16–21 June 2013; pp. 1310–1318.
15. Chung, J.; Gulcehre, C.; Cho, K.; Bengio, Y. Empirical evaluation of gated recurrent neural networks on sequence modeling. *arXiv* **2014**, arXiv:1412.3555.
16. Yamashita, R.; Nishio, M.; Do, R.K.G.; Togashi, K. Convolutional neural networks: An overview and application in radiology. *Insights Imaging* **2018**, *9*, 611–629. [CrossRef] [PubMed]
17. Srivastava, N.; Hinton, G.; Krizhevsky, A.; Sutskever, I.; Salakhutdinov, R. Dropout: A simple way to prevent neural networks from overfitting. *J. Mach. Learn. Res.* **2014**, *15*, 1929–1958.
18. Glorot, X.; Bengio, Y. Understanding the difficulty of training deep feedforward neural networks. In Proceedings of the Thirteenth International Conference on Artificial Intelligence and Statistics, Sardinia, Italy, 13–15 May 2010; pp. 249–256.
19. He, K.; Zhang, X.; Ren, S.; Sun, J. Delving deep into rectifiers: Surpassing human-level performance on imagenet classification. In Proceedings of the IEEE International Conference on Computer Vision, Las Condes, Chile, 11–18 December 2015; pp. 1026–1034. [CrossRef]
20. Abadi, M.; Barham, P.; Chen, J.; Chen, Z.; Davis, A.; Dean, J.; Devin, M.; Ghemawat, S.; Irving, G.; Isard, M.; et al. TensorFlow: A system for large-scale machine learning. In Proceedings of the 12th USENIX Symposium on Operating Systems Design and Implementation (OSDI 16), Savannah, GA, USA, 2–4 November 2016; pp. 265–283.
21. Paszke, A.; Gross, S.; Chintala, S.; Chanan, G.; Yang, E.; DeVito, Z.; Lin, Z.; Desmaison, A.; Antiga, L.; Lerer, A. Automatic differentiation in PyTorch. In Proceedings of the NIPS 2017 Workshop Autodiff, Long Beach, CA, USA, 9 December 2017.
22. Mannan, M.A. Hypertuned Deep Learning Models for Cybersecurity: An Empirical Study. *IEEE Access* **2022**, *10*.
23. Halder, A. Real-Time Machine Learning-Based Detection of High-Impact Weather Events. *J. Atmos. Ocean. Technol.* **2021**, *38*, 925–944. [CrossRef]

24. Nalepa, G.J.; Awad, A.A. Selecting reliable recommendations for Trust-based Recommender Systems. *Expert Syst. Appl.* **2019**, *125*, 259–271. [CrossRef]
25. Shah, S. Sign Language Recognition Using Deep Learning. *Adv. Intell. Syst. Comput.* **2021**, *1162*, 681–690.
26. Joshi, K. Taguchi method: A pragmatic approach to parameter tuning for machine learning models. *Expert Syst. Appl.* **2020**, *159*, 113576. [CrossRef]
27. Statped. Norsk Tegnordbok. 2022. Available online: https://www.statped.no/tegnordbok (accessed on 8 March 2023).
28. López, V.; Fernández, A.; Herrera, F. On the importance of the validation technique for classification with imbalanced datasets: Addressing covariate shift when data is skewed. *Inf. Sci.* **2014**, *257*, 1–13. [CrossRef]
29. Abadi, M.; Agarwal, A.; Barham, P.; Brevdo, E.; Chen, Z.; Citro, C.; Corrado, G.S.; Davis, A.; Dean, J.; Devin, M.; et al. TensorFlow: Large-Scale Machine Learning on Heterogeneous Distributed Systems. *arXiv* **2016**, arXiv:1603.04467.
30. Kingma, D.P.; Ba, J. Adam: A Method for Stochastic Optimization. In Proceedings of the 3rd International Conference for Learning Representations, San Diego, CA, USA, 7–9 May 2015.

technologies

Communication

Identifying Growth Patterns in Arid-Zone Onion Crops (Allium Cepa) Using Digital Image Processing

David Duarte-Correa [1,†] , Juvenal Rodríguez-Reséndiz [1,*,†] , Germán Díaz-Flórez [2,†] , Carlos Alberto Olvera-Olvera [2,*,†] and José M. Álvarez-Alvarado [1,†]

[1] Facultad de Ingeniería, Universidad Autónoma de Querétaro, Querétaro 76010, Mexico; d.duarte@licore.org (D.D.-C.); jmalvarez@uaq.edu.mx (J.M.Á.-A.)
[2] Unidad Académica de Ingeniería Eléctrica, Universidad Autónoma de Zacatecas "Francisco García Salinas", Jardín Juárez 147, Zacatecas 98000, Mexico; germandiazflorez@gmail.com
* Correspondence: juvenal@uaq.edu.mx (J.R.-R.); colvera@uaz.edu.mx (C.A.O.-O.)
† These authors contributed equally to this work.

Abstract: The agricultural sector is undergoing a revolution that requires sustainable solutions to the challenges that arise from traditional farming methods. To address these challenges, technical and sustainable support is needed to develop projects that improve crop performance. This study focuses on onion crops and the challenges presented throughout its phenological cycle. Unmanned aerial vehicles (UAVs) and digital image processing were used to monitor the crop and identify patterns such as humid areas, weed growth, vegetation deficits, and decreased harvest performance. An algorithm was developed to identify the patterns that most affected crop growth, as the average local production reported was 40.166 tons/ha. However, only 25.00 tons/ha were reached due to blight caused by constant humidity and limited sunlight. This resulted in the death of leaves and poor development of bulbs, with 50% of the production being medium-sized. Approximately 20% of the production was lost due to blight and unfavorable weather conditions.

Keywords: aerial photography; agricultural crop; digital image processing; pattern identification

Citation: Duarte-Correa, D.; Rodríguez-Reséndiz, J.; Díaz-Flórez, G.; Olvera-Olvera, C.A.; Álvarez-Alvarado, J.M. Identifying Growth Patterns in Arid-Zone Onion Crops (Allium Cepa) Using Digital Image Processing. *Technologies* **2023**, *11*, 67. https://doi.org/10.3390/technologies11030067

Academic Editors: Gwanggil Jeon and Imran Ahmed

Received: 10 March 2023
Revised: 27 April 2023
Accepted: 7 May 2023
Published: 10 May 2023

1. Introduction

Currently, agri-food producers require technological tools to increase crop production levels to satisfy the food demand of the population [1–3]. Food security must be provided, since the world's population has a growth rate of approximately 1.09% per year, which demands greater requirements for the agricultural industry to provide increasingly higher yields [4]. However, agricultural production costs ignore the immediate negative impacts on other users [5]. In this context, traditional agricultural practices are causing environmental damage, leading to the degradation of natural resources such as soils due to the excessive use of fertilizers and pesticides that the farmer scatters homogeneously to avoid delays in product development [6,7]. Precision agriculture (PA), a combination of technology and agronomy, offers a viable solution to the challenges faced by traditional farming practices. PA focuses on the integrated management of plots by using technology tools such as the Internet of Things (IoT), machine learning, and drones to gather data on crops' growth and health.

PA seeks to identify the variations in the conditions of the plot to carry out differential spatial management. UAVs (unmanned aerial vehicles) have generated tremendous interest in crop monitoring due to the great potential they can offer in the agricultural sector to verify the parameters that affect the development of plants [8,9]. Aerial images identify growth scans and areas of crop deterioration [10]. This technology helps soil conservation and sustainable agriculture, specifies the causes of soil damage, and identifies damage caused in the early stages of cultivation [11].

PA enables decision-making to improve production through an adequate understanding of the means of agricultural production [12]. Additionally, the current ability to process images with increasingly faster computers aids in analyzing some features of an object. The use of digital image processing (DPI) helps to solve the problems of image segmentation, object area, and characteristics on image-invariant scales, among others. According to [13], information can be obtained with the support of UAVs, which can later be used with different IPR techniques to obtain different behavior patterns in a given area. Various techniques, such as thresholding, different transformations of color spaces, normalized difference vegetation indices (NDVIs), and Excess Green Vegetation (ExG), among others, have been implemented to extract characteristics that define the state of crops such as grains and vegetables [13].

DPI has become an essential tool in precision agriculture. An example of its application is the use of the red–green–blue model (RGB) to carry out Lab color space transformations and the k-means algorithm to extract weeds from onion crops. For example, in [14], the authors mentioned that RGB space colors offer a very high spatial and temporal resolution. In this work, they performed a biomass monitoring system by processing RGB images obtained by a camera mounted on an unmanned aerial vehicle. They reported a 95% accuracy by implementing image analysis instead of invasive sensors. This application highlights the potential for using advanced technologies in PA. DPI enables farmers to detect crop problems early, allowing for targeted interventions and reducing costs while minimizing the environmental impacts. The application of DPI in precision agriculture has the potential to revolutionize the agricultural sector and ensure sustainable agriculture practices [15,16]. Notably, image processing and computer vision applications in the agriculture sector have grown due to reduced equipment costs, increased computational power, and growing interest in non-destructive food-assessment methods [17].

In recent years, there has been increasing interest in using data analysis and machine learning techniques to identify the patterns affecting crop growth and productivity. The authors of [18] presented a Mutual Subspace Method as a classifier in different farm fields and orchards, achieving 75.1% accuracy in high-altitude image-acquisition systems. By analyzing large amounts of data, such as weather patterns, soil quality, and crop yields, farmers and researchers can gain insight into the factors that influence crop growth and use this information to improve agricultural practices. Alibabaei et al. estimated tomato yield based on climate data, irrigation amount, and water content in the soil profile as input variables for Recurrent Neural Networks. The results showed that the Bidirectional Long Short-Term Memory model achieved an R^2 of 0.97 [19].

The remarkable work provides alternative solutions for crop monitoring and preventing disease factors. However, the conditions of an area with an arid climate can affect the quality and quantity of the food harvest, which makes it a challenge to propose alternatives that can improve and make agricultural processes more efficient without compromising land resources [20].

In this context, this study focuses on the state of Zacatecas, Mexico, a significant agricultural region known for producing corn, beans, and chili peppers. This study proposes a solution to increase the production percentage and minimize losses by analyzing the patterns that most impact crop growth during its phenological stage.

Crops suffer losses during their development cycle, and it is necessary to increase production each time. Therefore, this work aims to analyze which patterns most affect crop growth during its phenological stage and identify these patterns at an early stage. In this way, a proposed solution to increase the production percentage of the crop is offered, and a determination is made of which patterns most influence crop growth throughout its phenological stage in the state of Zacatecas, Mexico. It was concluded that humidity, cloudy weather, and leaf blight greatly impacted crop development and overall agricultural production. This fungal disease can damage the foliage and ultimately lead to crop loss.

2. Materials and Methods

2.1. Development of Experimentation

Crop development for this work occurs in the municipality of Fresnillo, Zacatecas, Mexico, with Cartesian coordinates 23°06′30.6″ N102°38′26.3″ W, as shown on the maps in Figure 1. A UAV Phantom 4 Pro with an RGB camera, a GPS/GLONASS system, and a 20-megapixel camera 4K resolution flew over onion crops (Allium Cepa) to capture photographs throughout the crop cycle. It presented a flight every week from March to July 2019 for 16 weeks, obtaining an average of 758 images in a land area of approximately three hectares at 20 m in height and a series of 5m. These series verified the optimal height, starting at 2 m and continuing at 5, 10, 15, 20, and successively up to 120 m.

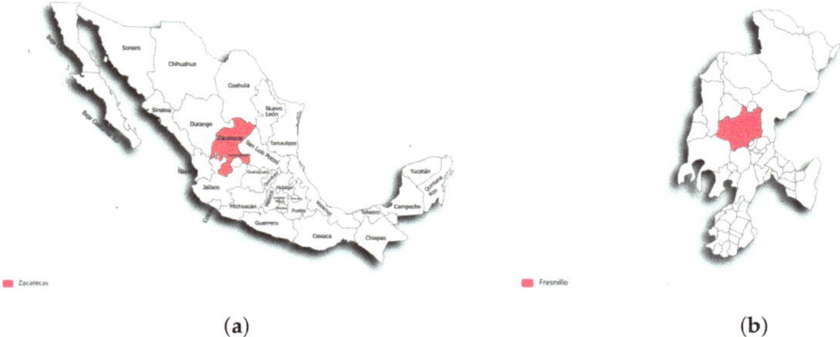

(**a**)　　　　　　　　　(**b**)

Figure 1. (**a**) Location of Zacatecas state in Mexico. (**b**) Location of the municipality of Fresnillo in the state of Zacatecas.

2.2. Data Processing

Image processing allows the detection of the vegetal area. Using grayscales allowed spotting damage due to blight, and a transformation to Jet coloration helped recognize wet areas. For weed detection, a color transformation of the RGB-CIELAB plane was made using the algorithm of near neighbors. See Figure 2.

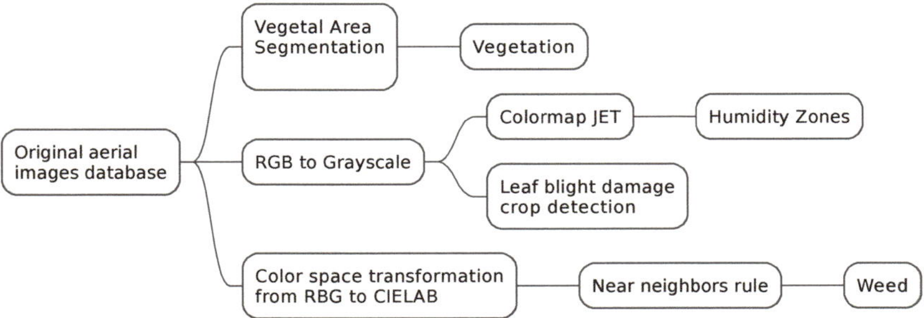

Figure 2. General diagram from data processing.

2.2.1. Vegetal Area Segmentation

In another work, ref. [21] used a color segmentation technique to obtain and separate green areas in an image by calculating the excess green value ExG, as shown in Equation (1).

Due to variations in illumination in photographs, an experimental modification without significant luminosity variation was made, described in Equation (2), as suggested by Tang and Liu [22], in order to trace a green sphere in a video, looking for a better way to segment the plant area.

The following images compare the segmentation performed using the ExG index and the one proposed under the name *justGreen*, as determined by Equations (1) and (2).

$$ExG = 2g - r - b \tag{1}$$

$$justGreen = g - \frac{r}{2} - \frac{b}{2} \tag{2}$$

where r, g, and b are red, green, and blue color channels in the RGB space, respectively.

2.2.2. Blight Damage Detection

The grayscale conversion model, described in Mathworks™ [23] and indicated by Equation (3), was applied to RGB images, as mentioned by [22]. Only the luminescence values were implemented, omitting saturation and hue, obtaining a linear combination of RGB values in a grayscale.

$$grayscale = 0.299R + 0.587G + 0.114B \tag{3}$$

where R, G, and B are the color channels in the RGB space. Grayscale images have 8-bit values from 0 to 255, allowing threshold values for image segmentation.

2.2.3. Humidity Detection

Each pixel in the image grayscale is mapped to the colormap jet (200). Transforming the color to the jet plane generates an RGB image with uninterrupted areas of pure color from a binary image, generating a multicolored image.

2.2.4. Weed Detection

To perform weed detection in the CIELAB, color-space-based segmentation (CIE 1976 $L\,a\,b$) was implemented as a model used to describe all colors the human eye can perceive. It is described in Equation (4), an international standard for color measurement specified by CIE in 1976 [24]. This color space consists of an L brightness layer with a black–white ratio from 0 to 100, with the chromaticity layer a indicating where the color falls along the green–red axis, with negative values for green and positive for red. The chromaticity layer b indicates the blue–yellow axis with negative values for blue and positive values for yellow.

$$
\begin{aligned}
L &= 116\, f\left(\frac{Y}{Y_0}\right) - 16 \\
a &= 500\left[f\left(\frac{X}{X_0}\right) - f\left(\frac{Y}{Y_0}\right)\right] \\
b &= 200\left[f\left(\frac{Y}{Y_0}\right) - f\left(\frac{Z}{Z_0}\right)\right]
\end{aligned}
\tag{4}
$$

The transformation of primary colors is performed and integrated into the XYZ color model, where X, Y, and Z represent vectors in a three-dimensional space of the color described in Equation (5).

$$
\begin{aligned}
X &= 0.4303R + 0.3416G + 0.1784B \\
Y &= 0.2219R + 0.7068G + 0.0713B \\
Z &= 0.0202R + 0.1296G + 0.9393B
\end{aligned}
\tag{5}
$$

where R, G, and B refer to the red, green, and blue colors of the RGB color model, while X, Y, and Z represent the CIE color set.

f is described by Equation (6):

$$f(q) = \sqrt[3]{q}; \qquad q > 0.008856$$
$$f(q) = 7.787q + \tfrac{16}{116} \quad q \leq 0.008856 \tag{6}$$

and the CIE XYZ are the tristimulus values of the reference white point. For this study, illuminant D65 was used, where $[X0 \quad Y0 \quad Z0] = [0.9504, \quad 1.0, \quad 1.0888]$, simulating midday light with a correlated color temperature of 6504 K.

In the color space in this representation, L represents the luminosity of an object, a represents the variation from green to red, and b represents the variation from blue to yellow. With this color transformation, a sample region was selected for each color. The average color of each sample region was calculated in space ab, using Equation (4). These samples were used as color markers to classify each pixel in the image. Once the color label was found, the near-neighbors technique was implemented, which is a machine-learning method that classifies an unknown sample according to its neighbors [25]. Each pixel is classified based on calculating the Euclidean distance between that pixel and each color marker, as indicated by Equation (7).

$$d(x,y) = \sqrt{(a - marker1)^2 + (b - marker2)^2} \tag{7}$$

where a and b are the CIELAB color channels a and b while $marker1$ and $marker2$ are the values of the markers found for the pixels. The smallest distance denotes the match with a particular color label. The label corresponding to the green color was selected, and the erosion and dilation operations were applied to extract the weed areas.

3. Results

The temperature and rain conditions obtained throughout the phenological cycle are shown in Figure 3. The minimum temperatures varied between 7 °C and 15 °C. The maximum temperatures fluctuated between 20 °C and 26 °C, remaining stable.

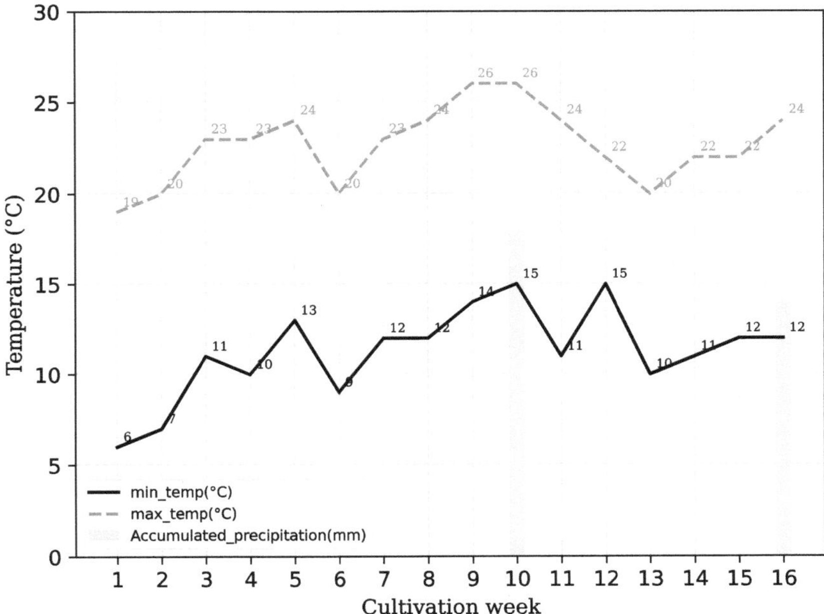

Figure 3. Temperature and precipitation.

Accumulated precipitation was nil for most of the period, except for weeks 10 (18 mm) and 16 (14 mm). Due to low sun conditions, an ideal environment for fungus propagation was generated, causing significant damage. Both graphs were created with data extracted from AccuWeather© [26].

Figure 4 compares the crops' temporal development and the different processes applied to the selected representative images obtained from the plot, where the Original Image (OI), Vegetation Image (VI), Leaf Blight Image (LBI), Humidity Zones Image (HZI), Weed Images (WI), and Error Identification Images (EII) were analyzed. These terms represent the subtraction of OI and WI when analyzing the error between both.

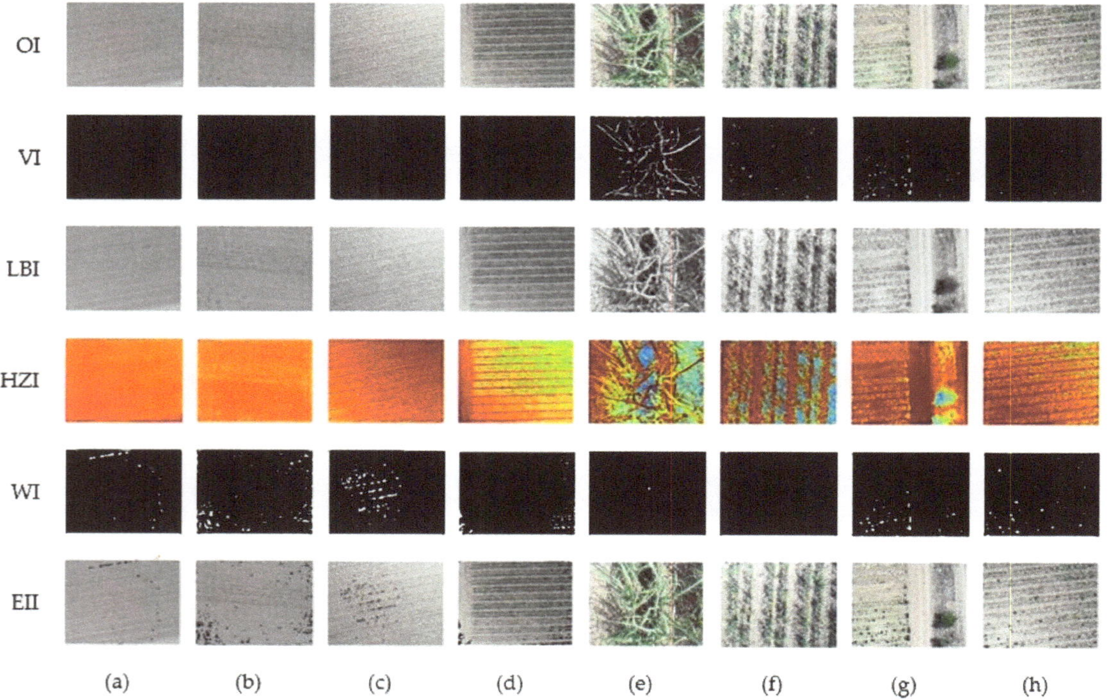

Figure 4. General processing stages; images of crop development at 20 m in different weeks: (**a–c**) in week 4, (**d**) in week 9 (**g,h**) in week 16; (**e**) image at 2 m in week 13, (**f**) image at 5 m in week 15.

In this selection, the segmentation of the areas was applied in VI green using Equation (1), representing vegetation segmentation, including cultivation and weeds.

In LBI, the RGB image is taken from the grayscale image to highlight the damage caused by white leaf blight, while the rest of the image contains grayer tones, making detection at low altitudes easy. At heights greater than 5 m, the damage caused by fungi is not visually detectable enough. HZI is the result of making a Jet-type mapping of the grayscale images, in which the dry areas are highlighted in reddish color while the humid areas are presented lighter colors, in blue passing through yellow.

The RGB to CIELAB transformation was applied to detect weeds using the nearby neighbors' decision rule and Euclidean distance metric with erosion and dilation. Results were affected by image brightness, with correct results only obtained in certain lighting conditions [18], such as a cloudy day between 9:30 am to 11:00 am during week 16.

In the first weeks, abrupt changes affecting onion development were undetected. At the same time, it was not possible to detect the crop at heights greater than 20 m. Therefore, there were no significant results in the first four weeks.

On the right side of the images, a sample of the analysis is presented at different heights (2, 5, and 20 m). The ideal height to detect vegetation areas, even with visible damage, was determined to be 5 m. However, the humid areas are visible at 20 m since they do not depend on the height of the plant but on water leaks or excessive irrigation.

3.1. Segmentation of Plant Area

The plant area was segmented using two different indices. First, the ExG index was used. However, this index produced a poor response in photographs with excessive or poor lighting. Therefore, the plant area was also segmented using JustGreen, an experimental index described in Equation (2). JustGreen provided a more robust response to the amount of illumination, allowing a more accurate estimation of the plant area than the ExG index. Figure 5 compares the ExG and JustGreen indices.

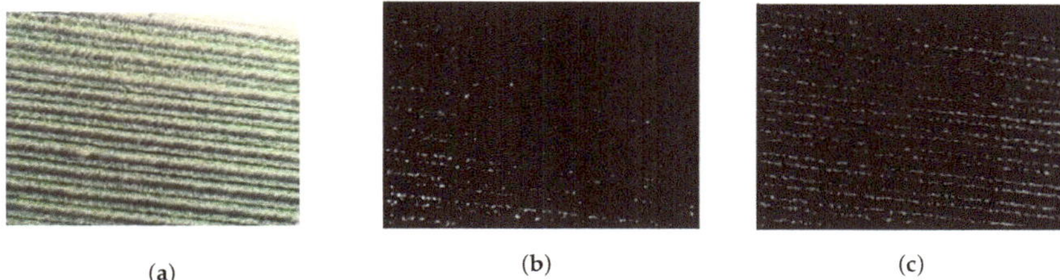

(a)   (b)   (c)

Figure 5. Comparison between the ExG and JustGreen indices: (**a**) original image, (**b**) segmentation by ExG, (**c**) segmentation by JustGreen.

As shown in Figure 5b, the plant area is lost in photographs with different lighting. In Figure 5c, the result of the JustGreen concerning the ExG is more reliable than the original image in Figure 5a The JustGreen implemented transformation, allowing the vegetation only to be seen, as shown in Figure 4 VI. This figure presents a sample of the area obtained in Equation (1) to estimate the vegetation and obtain its segmentation.

3.2. Leaf Blight Damage

The damage caused by leaf blight can stop the growth of the plant. Hence a gray-scale transformation was applied to highlight the blight damage of the objects in the background to properly present the damage by the fungus.

As shown in Figure 6a, an adequate inspection is more difficult when objects are near the plant.

(a)   (b)

Figure 6. Detection of leaf blight: (**a**) original image at a height of 2 m; (**b**) grayscale image showing leaf blight damage.

Figure 6b further separates the white area generated by the fungus from the rest of the image in darker grays.

The damage is not clearly distinguished at a height greater than 5 m.

3.3. Humidity Detection

When using the gray-scale images towards the jet200 colormap, it was possible to distinguish dry from wet soil, which allowed for detecting areas with excessive humidity and/or waterlogging. The row HZI in Figure 4 shows the performance of the jet plane in detecting wet areas. The humidity was visible at different heights. The dry soil appeared reddish, while the humid area was shown in light colors close to blue. This mapping is affected by the plant's shade due to humidity.

3.4. Detection of Weed

Figure 4 shows that the EII parameter indicates poor results for the decision rule of nearby neighbors, as lighting variations in each set of images affect the algorithm's ability to detect weeds in most of the data. For the early weeks, it presented many errors, but in the last week, it obtained hits in detection Figure 5c, with image Figure 6b showing the error obtained in the algorithm. Figure 5a is the original image.

Figure 7 shows the area with weeds detected within the crop using the CIELAB transformation, the rule of close neighborhoods, and dilation. The original image at 20 m high is shown in Figure 7a, as well as the overgrown area detected in Figure 7b. The detection error is visualized by subtracting the two images in Figure 7c (overlapping them).

The final production of the crop was 25.00 $\frac{tons}{ha}$. In comparison, the agri-food and fisheries information service (SIAP) [27] reports a general production in Zacatecas of 40.166 $\frac{tons}{ha}$ in 2019 throughout the agricultural year, in which higher irrigation is presented as temporary in July . This is due to weather conditions and environmental damage.

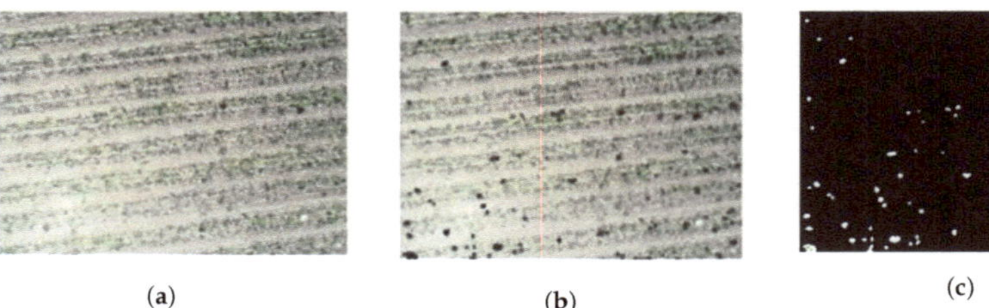

(a)   (b)   (c)

Figure 7. Weed detection: (**a**) original image, (**b**) subtract from the original minus the detected area, (**c**) weed positions.

4. Discussion

In [28], a threshold must be applied to each plant color, whether healthy or diseased, for example, between the healthy crop and the blight damage that causes different coloration in the plant.

The segmentation using Equation (1) allowed the plant area to be extracted without using a segmentation threshold.

Using the algebraic transformation of Equation (1) has the advantage of segmenting the entire plant area with gray colors.

This study [29] required median filters and Otsu thresholding to detect disease spots, with blight as the cause of discoloration in the crop.

The use of gray scales in the identification of diseased areas results in a low computational cost and, thus, shorter processing time, compared to the processing applied by them, as an alternative to the plant-to-plant detection of diseased areas. To detect humid areas, laboratory analysis of soil samples or the use of instruments that indirectly help to measure humidity or/and temperature is commonly required. However, it is only feasible to place a few sensors since it becomes costly. The proposed method is a non-destructive alternative that estimates the wet area in the entire crop. This allows farmers to focus on areas with prolonged humidity as a possible risk factor for fungal damage or diseases caused by blight. Farmers can take appropriate measures to prevent fungal damage by identifying these areas. The weed detection technique presented the need to manually define the size of the weed, making it a deficiency in non-uniform areas of weed development. However, as seen in Figure 6, it has good detection at 20 m high, regardless of the lighting in the photograph.

The blight appeared because the temperatures remained stable, averaging 10 to 25 °C. This stable temperature range facilitated its development, resulting in the deterioration of the plant's growth. As a result of the presence of blight, the growth of the plant is deteriorated.

5. Conclusions

Identifying diseased areas using grayscale images is an effective and low-cost approach to the detection of plant disease. Moreover, the results demonstrated a simplified representation of the plant tissue, which can help to reduce the computational cost of image-processing techniques. This method also provides a non-destructive alternative for estimating wet areas in the crop, allowing farmers to focus on areas with prolonged humidity and prevent fungal damage or diseases by blight. However, the weed-detection technique presented some limitations in non-uniform areas of weed development. The blight appeared due to stable temperatures that facilitated its development, resulting in the deterioration of the plant's growth. Overall, this study highlights the potential of digital image-processing techniques in improving crop management and identifying areas for targeted intervention, ultimately contributing to more sustainable and productive agricultural practices.

Monitoring the onion crop allowed us to identify parameters that affect its development, such as weeds, water leaks, excess humidity, vegetation deficit, and a crucial parameter known as blight.

Blight developed due to the lack of constant irrigation in adequate amounts, resulting in approximately 16 mm of precipitation received in the tenth weekend. During this period, the presence of a fungus due to the cloudy climate and low sunlight significantly affected the last weeks of crop development.

In addition, the irrigation deficit resulted in a lower yield than in the state of Zacatecas, where only 50% of medium-sized onions were produced. Approximately 20% of the production was lost due to the blight and on-site irrigation conditions.

To control the blight problem in this variety, it is recommended to irrigate the crops adequately and not to expose them to water stress. However, over-watering should also be avoided since excessive humidity in the soil and high temperatures create ideal conditions for fungi and weed development.

Author Contributions: Conceptualization, D.D.-C. and J.R.-R.; Methodology, G.D.-F. and J.M.Á.-A.; Software, D.D.-C.; Validation, D.D.-C. and G.D.-F.; Formal analysis, J.R-R.; Investigation, D.D.-C.; Resources, J.R.-R. and G.D.-F.; Data curation, D.D.-C. and J.R.-R.; Writing original draft preparation, review, and editing, C.A.O.-O., J.R.-R. and G.D.-F. All authors have read and agreed to the published version of the manuscript.

Funding: This research received no external funding.

Institutional Review Board Statement: Not applicable.

Informed Consent Statement: Not applicable.

Data Availability Statement: Not applicable.

Acknowledgments: The authors thank CONACYT for funding this work and Ruben García for the edition of the manuscript.

Conflicts of Interest: The authors declare no conflict of interest.

References

1. Bakthavatchalam, K.; Karthik, B.; Thiruvengadam, V.; Muthal, S.; Jose, D.; Kotecha, K.; Varadarajan, V. IoT Framework for Measurement and Precision Agriculture: Predicting the Crop Using Machine Learning Algorithms. *Technologies* **2022**, *10*, 13. [CrossRef]
2. Tatas, K.; Al-Zoubi, A.; Christofides, N.; Zannettis, C.; Chrysostomou, M.; Panteli, S.; Antoniou, A. Reliable IoT-Based Monitoring and Control of Hydroponic Systems. *Technologies* **2022**, *10*, 26. [CrossRef]
3. Sanida, M.V.; Sanida, T.; Sideris, A.; Dasygenis, M. An Efficient Hybrid CNN Classification Model for Tomato Crop Disease. *Technologies* **2023**, *11*, 10. [CrossRef]
4. Mirás-Avalos, J.M.; Rubio-Asensio, J.S.; Ramírez-Cuesta, J.M.; Maestre-Valero, J.F.; Intrigliolo, D.S. Irrigation-Advisor—A Decision Support System for Irrigation of Vegetable Crops. *Water* **2019**, *11*, 2245. [CrossRef]
5. Padilla-Bernal, L.E.; Reyes-Rivas, E.; Pérez-Veyna, Ó. Evaluación de un cluster bajo agricultura protegida en México. *Contaduría Adm.* **2012**, *57*, 219–237. [CrossRef]
6. Ferentinos, K.P. Deep learning models for plant disease detection and diagnosis. *Comput. Electron. Agric.* **2018**, *145*, 311–318. [CrossRef]
7. Liu, J.; Wang, X. Tomato diseases and pests detection based on improved Yolo V3 convolutional neural network. *Front. Plant Sci.* **2020**, *11*, 898. [CrossRef]
8. Callejero, C.P.; Salas, P.; Mercadal, M.; Seral, M.A.C. Experiencias en la adquisición de imágenes para agricultura a empresas de drones españolas. In *Nuevas Plataformas y Sensores de Teledetección*; Editorial Politécnica de Valencia: Zaragoza, Spain, 2017.
9. Cunha, J.P.A.R.d.; Sirqueira, M.A.; Hurtado, S.M.C. Estimating vegetation volume of coffee crops using images from unmanned aerial vehicles. *Eng. Agrícola* **2019**, *39*, 41–47. [CrossRef]
10. González, A.; Amarillo, G.; Amarillo, M.; Sarmiento, F. Drones Aplicados a la Agricultura de Precisión. *Publ. Investig.* **2016**, *10*, 23–37. [CrossRef]
11. Lottes, P.; Khanna, R.; Pfeifer, J.; Siegwart, R.; Stachniss, C. UAV-based crop and weed classification for smart farming. In Proceedings of the 2017 IEEE International Conference on Robotics and Automation (ICRA), Singapore, 29 May–3 June 2017; pp. 3024–3031. [CrossRef]
12. Rocha De Moraes Rego, C.A.; Penha Costa, B.; Valero Ubierna, C. Agricultura de Precisión en Brasil. In Proceedings of the VII Congreso de Estudiantes Universitarios de Ciencia, Tecnología e Ingeniería Agronómica, Madrid, Spain, 5–6 May 2015; p. 3.
13. Poojith, A.; Reddy, B.V.A.; Kumar, G.V. Image processing in agriculture. In *Image Processing for the Food Industry*; World Scientific Publishing Co Pte Ltd'.: Singapore, 2000; pp. 207–230. [CrossRef]
14. Ballesteros, R.; Ortega, J.F.; Hernandez, D.; Moreno, M.A. Onion biomass monitoring using UAV-based RGB imaging. *Precis. Agric.* **2018**, *19*, 840–857. [CrossRef]
15. Ballesteros, R.; Ortega, J.F.; Hernández, D.; Moreno, M.A. Applications of georeferenced high-resolution images obtained with unmanned aerial vehicles. Part I: Description of image acquisition and processing. *Precis. Agric.* **2014**, *15*, 579–592. [CrossRef]
16. Timsina, J. Can Organic Sources of Nutrients Increase Crop Yields to Meet Global Food Demand? *Agronomy* **2018**, *8*, 214. [CrossRef]
17. Patrício, D.I.; Rieder, R. Computer vision and artificial intelligence in precision agriculture for grain crops: A systematic review. *Comput. Electron. Agric.* **2018**, *153*, 69–81. [CrossRef]
18. Din, N.U.; Naz, B.; Zai, S.; Ahmed, W. Onion Crop Monitoring with Multispectral Imagery Using Deep Neural Network. *Int. J. Adv. Comput. Sci. Appl.* **2021**, *12*, 303–309. [CrossRef]
19. Alibabaei, K.; Gaspar, P.D.; Lima, T.M. Crop yield estimation using deep learning based on climate big data and irrigation scheduling. *Energies* **2021**, *14*, 3004. [CrossRef]
20. Shorabeh, S.N.; Kakroodi, A.; Firozjaei, M.K.; Minaei, F.; Homaee, M. Impact assessment modeling of climatic conditions on spatial-temporal changes in surface biophysical properties driven by urban physical expansion using satellite images. *Sustain. Cities Soc.* **2022**, *80*, 103757. [CrossRef]
21. Meyer, G.E.; Neto, J.C. Verification of color vegetation indices for automated crop imaging applications. *COmputers Electron. Agric.* **2008**, *63*, 282–293. [CrossRef]
22. Liu, Q.; Xiong, J.; Zhu, L.; Zhang, M.; Wang, Y. Extended RGB2Gray conversion model for efficient contrast preserving decolorization. *Multimed. Tools Appl.* **2017**, *76*, 14055–14074. [CrossRef]
23. rgb2gray. Available online: https://www.mathworks.com/help/matlab/ref/rgb2gray.html (accessed 25 September 2019).
24. Connolly, C.; Fleiss, T. A study of efficiency and accuracy in the transformation from RGB to CIELAB color space. *IEEE Trans. Image Process.* **1997**, *6*, 1046–1048. [CrossRef]
25. Donaldson, R. Approximate formulas for the information transmitted by a discrete communication channel (Corresp.). *IEEE Trans. Inf. Theory* **1967**, *13*, 118–119. [CrossRef]
26. AccuWeather. El Tiempo en México. Available online: https://www.accuweather.com/es/mx/fresnillo/236598/may-weather/236598 (accessed on 19 October 2019).
27. SIAP. *Avance de Siembras y Cosechas*; SIAP: Chiba, Japan, 2019.

28. Netto, A.F.A.; Martins, R.N.; de Souza, G.S.A.; de Moura Araújo, G.; de Almeida, S.L.H.; Capelini, V.A. Segmentation of rgb images using different vegetation indices and thresholding methods. *Nativa* **2018**, *6*, 389–394. [CrossRef]

29. AS, M.; Abdullah, H.; Syahputra, H.; Benaissa, B.; Harahap, F. An Image Processing Techniques Used for Soil Moisture Inspection and Classification. In Proceedings of the 4th International Conference on Innovation in Education, Science and Culture, ICIESC 2022, Medan, Indonesia, 11 October 2022. [CrossRef]

Article

Image-Based Quantification of Color and Its Machine Vision and Offline Applications

Woo Sik Yoo [1,2,*], **Kitaek Kang** [1], **Jung Gon Kim** [1] and **Yeongsik Yoo** [3,*]

[1] WaferMasters, Inc., Dublin, CA 94568, USA
[2] Institute of Humanities Studies, Kyungpook National University, Daegu 41566, Republic of Korea
[3] College of Liberal Arts, Dankook University, Yongin 16890, Republic of Korea
* Correspondence: woosik.yoo@wafermasters.com (W.S.Y.); ysyoophd@dankook.ac.kr (Y.Y.)

Abstract: Image-based colorimetry has been gaining relevance due to the wide availability of smart phones with image sensors and increasing computational power. The low cost and portable designs with user-friendly interfaces, and their compatibility with data acquisition and processing, are very attractive for interdisciplinary applications from art, the fashion industry, food science, medical science, oriental medicine, agriculture, geology, chemistry, biology, material science, environmental engineering, and many other applications. This work describes the image-based quantification of color and its machine vision and offline applications in interdisciplinary fields using specifically developed image analysis software. Examples of color information extraction from a single pixel to predetermined sizes/shapes of areas, including customized regions of interest (ROIs) from various digital images of dyed T-shirts, tongues, and assays, are demonstrated. Corresponding RGB, HSV, CIELAB, Munsell color, and hexadecimal color codes, from a single pixel to ROIs, are extracted for machine vision and offline applications in various fields. Histograms and statistical analyses of colors from a single pixel to ROIs are successfully demonstrated. Reliable image-based quantification of color, in a wide range of potential applications, is proposed and the validity is verified using color quantification examples in various fields of applications. The objectivity of color-based diagnosis, judgment and control can be significantly improved by the image-based quantification of color proposed in this study.

Keywords: color sensing; colorimetry; image processing; image analysis; machine vision; offline analysis

Citation: Yoo, W.S.; Kang, K.; Kim, J.G.; Yoo, Y. Image-Based Quantification of Color and Its Machine Vision and Offline Applications. *Technologies* **2023**, *11*, 49. https://doi.org/10.3390/technologies11020049

Academic Editors: Gwanggil Jeon, Imran Ahmed, Alessandro Tognetti and Manoj Gupta

Received: 7 March 2023
Revised: 22 March 2023
Accepted: 24 March 2023
Published: 29 March 2023

1. Introduction

Color is an important factor in our perception of substances. While variations in luminance can be caused by both changes in substances and changes in illumination, variations in color are highly diagnostic for changes in the physical and/or chemical properties of substances [1–3]. The color of a substance depends on the wavelengths of light absorbed. The perceived color is the complementary or subtractive light spectra, as light reaching the eye lacks the absorbed light spectra. What color (more specifically light spectra) is absorbed depends on the chemical nature of the substance. Color matching is the most basic and important task in color vision, and it forms the basis of colorimetry for a wide range of applications [1–3].

In the early experiments on color matching, homogeneously colored stimuli were used [1]. These experiments were intended to assess the sensitivities of the human visual system to different colors to help us better understand the perception of color in the vision of human beings at the level of the photoreceptors. However, the objects within our visual environment contain very few surfaces that are precisely uniform or identical in color. Moreover, these surfaces are made of a wide range of substances and geometries [4]. Different substances with different surface properties affect the perception of color in the human eye due to variations in reflection and scattering properties. As a result, we perceive variations in reflected light spectra (i.e., color).

Colorimetric information, including brightness, contrast, and uniformity, are frequently used in characterizing and describing properties of objects and substances of interest, both in daily life and in specific circumstances. Its applications are very broad from art, the fashion industry, food science, medical science, oriental medicine, agriculture, geology, chemistry, biology, material science, environmental engineering, etc. [5–20]. For ease of access and portable solutions, smartphone-based spectrometers and colorimeters have been studied worldwide [21–30]. The post-processing capability of acquiring images for colorimetric applications is very important to developing optimized algorithms. Once an optimized algorithm is developed, it can be integrated with image-capturing devices for machine vision applications.

We have reported newly developed image analysis software (PicMan, WaferMasters, Inc., Dublin, CA, USA) with image-capturing and video-recording functions with multiple application examples in various fields, including engineering, cultural heritage evaluation/conservation, archaeology, chemical, biological and medical applications [6,31–38]. The software has also been used for various research activities by third-party research groups and proved its capability of quantitative color analysis for various objectives in color pigment, painting cultural property, material science, and cell biology studies [7,8,39–41]. For machine vision applications, the video clip analysis technique and utilization of commercially available general image sensors, such as digital cameras, USB cameras, smart phones and camcorders using the newly developed software, have been demonstrated [7,31,38].

In this paper, the image-based quantification of color and its machine vision and offline applications in interdisciplinary fields using specifically developed image analysis software (PicMan) are introduced with examples. The colorimetric information of individual pixels and regions of interest (ROIs) in acquired images were extracted in RGB (red, green and blue) intensity, HSV (hue, saturation and value) and CIE L*a*b* values. The conversion of RGB intensity values into equivalent Munsell color indices and hexadecimal color codes has also been demonstrated for the potential expansion of machine vision and offline colorimetric applications.

2. Experimental

To demonstrate the image-based quantification of color, three types of images were used. Color information from a single pixel to predetermined sizes/shapes of areas, including customized regions of interest (ROIs) from various digital images of dyed T-shirts, assays, color charts, and tongues from various sources, including the internet, have been extracted. Average color and histograms of color components such as RGB, HSV, CIE L*a*b*, Munsell color and hexadecimal color of ROIs were extracted. Colorimetric analysis was conducted using newly developed image analysis software (PicMan from WaferMasters, Inc., Dublin, CA, USA) [6–8,31–41]. For machine vision applications, PicMan can be connected with image sensor devices, including USB cameras and digital microscopes, to acquire snapshot photographs and video images.

For the first set of images, twenty-nine dyed T-shirt images, downloaded from an internet shopping site [42], were selected as the first test image for color analysis demonstration. For the second set of images, an image of methylene blue (MB) solutions, with different concentrations ranging from 0.1 to 10 ppm, were selected from a publicly available literature source [23]. For the third set of images, a commercially available AquaChek Color Chart, which indicates the chemical levels in hot tub water was selected as a quantifiable set of colors from a chart [43]. Total hardness, total chlorine/total bromine, free chlorine, pH, total alkalinity, and cyanuric acid levels can be read from the AquaChek test strips color chart. As the last example of color images, a tongue diagnosis chart [44] used in traditional oriental medicine (traditional Chinese medicine (TCM, 中医, Zhōngyī), traditional Korean medicine (한방, 韓方, Hanbang) and traditional Chinese medicine in Japan (漢方, Kampo)) was selected. A block diagram from image acquisition, color space conversion, image analysis, and results output is illustrated in Figure 1. A customized system can be configured with any imaging device and a PC with the image analysis software, PicMan.

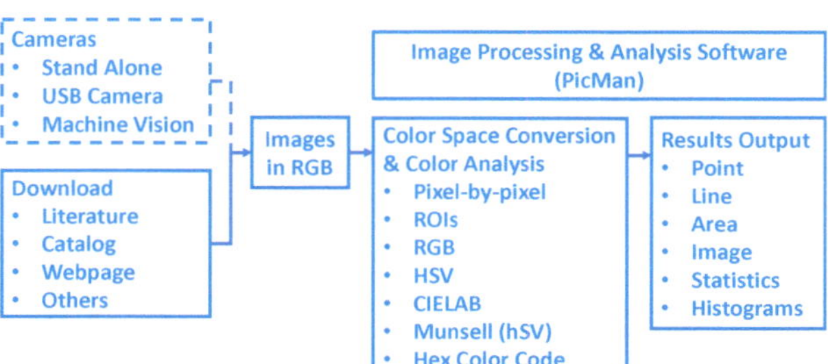

Figure 1. Block diagram from image acquisition, color space conversion, image analysis, and results output.

Quantitative colorimetric information on individual pixels, lines and regions of interest (ROIs) in digital images from various sources has been extracted to demonstrate the feasibility of offline colorimetric analysis with traceability for further statistical analysis. Machine vision application examples of the colorimetric analysis, monitoring and quality control capabilities are provided towards the end of this paper. Once this technique is proved to be effective, careful consideration and optimization should be applied in hardware configurations, illumination, calibration and optical properties of target objects for reliable results.

3. Results

Color of areas of interest, single pixel to predetermined sizes/shapes of areas from four different types of objective images were quantified using newly developed image analysis software, PicMan. The analyzed color information of single pixel to predetermined sizes/shapes of areas was exported in commonly used color formats in RGB, HSV, CIE L*a*b*, Munsell color and hexadecimal codes to assist us in image analysis and quality monitoring/control. The RGB color system is one of many color-defining systems [45–47]. For optical image display in monitoring screen or projection devices, an RGB color system is used and all colors are uniquely identified as RGB-based hexadecimal color codes. A 24-bit color signal (8-bit per channel for red, green and blue channels) can display 2^{24} (=16,777,216) colors [6–8]. The histogram and statistical analysis of extracted color information assists us in finding correlations between color and input variables/conditions of objects of interest at the time of image acquisition. Detailed examples using four types of images from different disciplines are described in the following subsections.

3.1. Dyed T-Shirts

Twenty-nine tie-dye T-shirt images, selected for color analysis demonstration, are shown in Figure 2. The term tie-dye is used to describe a number of resist dyeing techniques. The process of "tie-dye" typically consists of a few 'resist' preparation steps before tying (binding) with string or rubber bands, as in folding, twisting, pleating, or crumpling fabric or clothes. The application of dye or dyes takes place to the tied fabrics. The preparation steps of the fabric before the application of dye are called 'resists,' as they partially or completely prevent the coloring the fabric from the applied dye(s). Tie-dye can be used to create a wide variety of designs on fabric. Standard and popular patterns of tie-dye are the spiral, peace sign, diamond. A "marble effect" can be added to beautiful works in this manner. It is a perfect example of images being enhanced for color analysis demonstration.

Figure 3 shows the twenty-nine tie-dye T-shirt images opened using the image processing software, PicMan, for color analysis. The colorful line intensity graph shows RGB (red, green and blue) channel brightness as a dotted red vertical line in 256 levels (0–255). The grey-colored line intensity graph shows the weighted average brightness of RGB channels of individual pixels on the red dotted vertical line. The weighted average brightness was calculated using (R + 2G + B)/4, considering the luminosity curve and the color filter array pattern used in typical CMOS image sensors [48]. In the Bayer color filter mosaic, each two-by-two sub-mosaic contains one red, two green, and one blue filter, each filter covering one pixel sensor [45] for considering the human luminosity factor. Color information of various ROIs from a single pixels to squares and circles of different sizes were quantified in various color-describing formats, such as RGB, HSV, L*a*b*, Munsell color and hexadecimal codes. The color information of any shape and size, for example, an individual tie-dye T-shirt or an entire image of twenty-nine T-shirts, can be analyzed and exported either pixel by pixel or as average color in various color formats of choice for further analysis, color monitoring and quality control.

Examples of color information extraction and RGB intensity line graph across the line A-A' on six dyed T-shirt images (far left column of Figure 1) are shown in Figure 4. The RGB intensity line graph in Figure 2 was plotted using exported pixel-by-pixel RGB intensity values across the line A-A'.

Table 1 shows the summary of color analysis results on individual T-shirts. The measured area of individual T-shirts in the image was counted in the number of pixels. The weighted averages of RGB intensity and their standard deviations of individual T-shirts were calculated. The average intensity and standard deviation of red, green and blue channels of pixels within twenty-nine individual T-shirts were calculated. The average HSV values, average CIELAB L*a*b* values and average Munsell color values were calculated based on RGB values of all pixels within the individual T-shirt images. In addition, the average color using the hexadecimal code was determined for easy understanding. The image analysis software is capable of calculating width, height, circumference, area/circumference ration, circularity, and many more important parameters to characterize the individual T-shirts (or any shape and size of ROIs specified). All numbers are unique characteristics of individual T-shirts regarding the size, shape, and color. These numbers can be used for finding correlation with statistics of human reaction, impression, preference, etc., towards the human interface algorithm development, machine learning (ML), artificial intelligence (AI), psycho-visual modeling, human visual system modeling, product design, and so on.

Figure 2. Twenty-nine dyed T-shirt images (**a–ac**) are shown for color analysis demonstration.

Figure 3. Examples of color information extraction of individual pixels and regions of interest (ROIs) with different sizes and shapes from twenty-nine dyed T-shirt images.

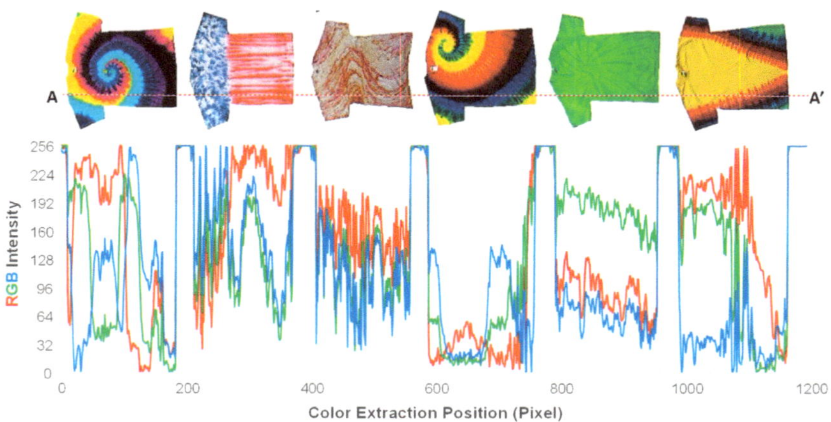

Figure 4. Examples of color information extraction of individual pixels across the line A-A' on six dyed T-shirt images in the left column of Figure 1.

Table 1. Statistical summary of color information on twenty-eight dyed T-shirts. (Mean and standard deviation of brightness and RGB values, as well as average HSV (for hue, saturation, value) values, CIE L*a*b* values, Munsell color values, hex color codes for average color of individual dyed T-shirts). Intensity means and standard deviations of red, green and blue channels were shown in representing colors.

ID	Area (Pixel)	Brightness Intensity Mean	Brightness Intensity StdDev	Red Mean	Red StdDev	Green Mean	Green StdDev	Blue Mean	Blue StdDev	H	S	V	L*	a*	b*	h	V	C	Hex Color Code	Color
a	24,742	97.1	61.0	100.2	88.8	92.4	81.3	105.0	72.5	280.0	0.12	0.41	40.2	5.7	−5.7	5 P	3	4	#645C68	
b	21,366	115.9	34.5	87.2	31.7	121.3	37.0	135.2	35.4	197.5	0.36	0.53	48.8	−8.4	−11.5	7.5 B	5	2	#577987	
c	25,275	136.3	32.3	205.5	25.5	89.0	36.9	163.1	35.5	321.5	0.57	0.80	54.7	54.8	−17.9	2.5 RP	5	12	#CD58A3	
d	25,365	100.7	55.9	158.0	44.9	73.2	58.5	99.8	64.1	341.6	0.54	0.62	42.7	38.1	1.9	10 RP	3	8	#9E4963	
e	23,411	109.5	40.7	178.5	51.6	95.3	54.5	70.4	46.9	13.9	0.61	0.70	49.6	31.3	29.2	10 R	5	8	#B25F46	
f	21,958	155.8	46.4	179.8	58.2	140.0	53.5	164.9	54.8	323.1	0.22	0.70	62.5	18.8	−7.0	10 P	8	2	#B38CA4	
g	19,368	189.5	25.5	227.8	20.3	214.2	23.2	103.3	42.1	53.7	0.55	0.89	84.7	−9.5	55.6	7.5 Y	8	8	#E3D667	
h	23,574	102.6	56.1	81.2	61.6	108.4	67.1	113.7	52.8	189.4	0.28	0.44	43.8	−8.6	−6.0	10 BG	3	2	#516C71	
i	24,971	92.7	40.1	102.4	74.1	76.4	46.2	117.3	55.6	278.0	0.35	0.46	36.5	19.3	−19.3	5 P	5	6	#664C75	
j	22,275	123.7	31.8	35.9	37.9	139.4	33.4	181.7	28.8	197.3	0.81	0.71	54.2	−14.0	−30.2	10 B	7	10	#238BB5	
k	19,739	137.4	36.6	165.0	30.6	131.6	39.4	122.7	40.4	12.6	0.26	0.65	57.7	11.6	9.8	10 R	5	4	#A5837A	
l	25,247	92.9	58.7	65.3	56.0	105.0	58.7	97.9	62.3	168.0	0.38	0.41	41.4	−16.1	0.0	2.5 BG	4	2	#416961	
m	23,587	122.2	46.5	114.5	45.4	112.7	47.6	150.3	45.8	243.2	0.25	0.59	48.7	9.4	−20.3	10 PB	6	2	#727096	
n	24,079	89.6	62.3	118.5	76.1	83.7	77.5	74.1	65.2	12.3	0.37	0.46	38.8	13.3	11.1	10 R	5	4	#76534A	
o	21,736	77.0	43.9	121.3	75.5	58.1	48.5	71.9	55.9	347.6	0.52	0.47	33.0	28.9	4.9	7.5 RP	5	4	#793A47	
p	24,273	79.3	56.0	117.1	92.3	73.7	72.5	54.1	49.8	18.1	0.54	0.46	35.6	16.5	19.1	2.5 YR	4	6	#754936	
q	25,645	102.1	51.5	84.5	62.3	95.0	52.2	135.7	49.5	228.2	0.38	0.53	40.6	6.9	−24.1	7.5 PB	5	4	#545E87	
r	23,373	116.8	34.9	100.3	80.0	102.8	45.2	162.8	42.9	238.1	0.38	0.64	45.4	14.5	−32.7	7.5 PB	2	8	#6466A2	
s	24,780	132.8	28.7	81.2	32.5	166.1	27.0	119.3	31.0	146.8	0.51	0.65	62.0	−36.9	16.6	5 G	7	4	#51A677	
t	23,795	144.7	24.1	143.8	24.2	144.1	24.2	148.0	23.8	228.0	0.03	0.58	59.8	0.4	−2.3	7.5 PB	6	4	#8F9094	
u	22,178	132.8	26.1	97.0	31.7	182.0	22.7	71.8	32.3	106.4	0.61	0.71	66.5	−46.4	47.1	10 GY	8	8	#60B547	
v	20,812	164.5	28.9	153.6	63.8	171.6	36.7	162.7	58.5	150.0	0.11	0.67	68.4	−8.1	2.5	7.5 G	7	2	#99ABA2	
w	22,275	119.0	46.7	88.0	54.9	109.3	54.5	171.0	46.6	224.8	0.49	0.67	46.9	9.7	−35.6	7.5 PB	7	6	#586DAB	
x	24,872	143.8	52.0	196.0	68.9	144.5	86.4	91.7	56.5	30.6	0.53	0.76	63.6	13.1	35.5	7.5 YR	7	4	#C3905B	
y	25,236	125.0	40.3	164.7	30.6	106.0	45.5	124.6	42.1	340.7	0.36	0.64	51.2	26.3	−0.7	5 RP	4	2	#A4697C	
z	25,609	109.0	53.3	161.4	68.9	115.9	73.4	44.3	39.6	36.4	0.73	0.63	51.9	11.1	44.8	10YR	7	8	#A1732C	
aa	26,092	183.5	42.9	206.9	26.6	170.2	52.1	188.3	46.0	330.0	0.17	0.81	73.2	16.1	−4.3	2.5 RP	7	4	#CEAABC	
ab	23,683	112.4	43.1	193.5	39.0	97.9	60.8	61.9	41.6	16.4	0.68	0.76	52.1	35.6	37.7	10 R	7	8	#C1613D	
ac	23,160	140.8	36.6	88.4	54.9	167.4	41.4	141.4	51.8	160.3	0.47	0.65	63.0	−30.7	6.0	10 G	7	4	#58A78D	

The RGB color model is an additive color model in which the red, green and blue primary colors of light are added together to reproduce a broad range of colors. For 24-bit colors, 8-bit brightness (or intensity) information in the range of 0 to 255 is used per color channel. The CIELAB color space (L*a*b*) is a color space defined by the International Commission on Illumination (abbreviated CIE) (1976). It expresses color as three values in a cartesian coordinate. L* is for perceptual lightness and a* and b* are for the four unique colors of human vision: red, green, blue and yellow. For Munsell color, The H (hue) number is converted by mapping the hue rings into numbers between 0 and 100. The C and V are the same as the normal chroma and value. A hexadecimal color is specified with the rrggbb format, where rr (red), gg (green) and bb (blue) are hexadecimal integers between 00 and ff, specifying the intensity of the color.

Figure 5 shows RGB intensity histograms of the twenty-nine T-shirt images. The brightness distribution of RGB values of entire pixels within individual T-shirt images were extracted and binned from 0 to 255 intensity category. The histogram contains all information on the color of individual T-shirts. The areas of each RGB histogram curve should be identical to the number of pixels. The histogram data ignore the x, y coordinate information of individual pixels of the image. The order of pixels in the image is ignored. It treats the group of pixels as a totally mixed 'soup' of RGB pixels with brightness ranging from 0 to 255. The histogram provides an attribute of RGB brightness population, but it is very difficult to imagine the resulting color from the histogram. It is useful to detect saturation and range of brightness used in individual channels.

Figure 5. RGB intensity histograms of twenty-nine T-shirt images (**a–ac**).

For easy recognition, the twenty-nine dyed T-shirt images and their corresponding average color are shown in Figure 6. When multiple, highly contrasting colors are used, the corresponding average color does not seem to match our impression (a, f, h, i, n, o, p, v, w, x, z of Figure 6). They are calculated from the arithmetic mean of RGB brightness values of entire pixels. It represents the average color of totally mixed 'RGB pixel solutions.' In many cases, simple statistics cannot represent reality, but it can make sense and has its own value and usage as one of many quantifiable characterization factors.

Figure 6. Twenty-nine dyed T-shirt images and their corresponding averaged color (**a–ac**).

3.2. Methylene Blue (MB) Solutions

For testing color analysis sensitivity of liquid samples, an image of methylene blue (MB: Methylthioninium chloride ($C_{16}H_{18}ClN_3S$)) solutions was adapted from the published literature [23]. The MB solutions with different concentrations are shown in Figure 5. The concentration of the MB solution varies from 0 ppm (100% distilled water (DW)) to 10 ppm (most right) in eight steps. RGB values of 15×15 pixel square areas of MB solutions from 0.1 to 10 ppm were extracted from the image for quantification of color change. The color change of MB solutions with respect to MB concentration is easily seen. As the MB concentration is increased, the red channel intensity is rapidly decreased. The green channel intensity started to decrease above MB concentration of 5.0 ppm. It coincided with the red intensity reduction with the increase in MB concentration reported from the smart phone spectrometer study [23].

Figure 7 shows the corresponding colors of the eight MB solutions with 0 to 10.0 ppm concentrations. Their RGB intensity and weighted average brightness (intensity) are shown as line intensity graphs (a partial screen capture image of PicMan) in Figure 8. Trends of RGB channel intensity as a function of MB concentration can be used to formulate the relationship between apparent color and MB concentration. Unlike red channel brightness, both green and blue channel intensities remained meaningful and showed clear trends above noise level in the entire MB concentration range. The response surface method (RSM) can be used to find the relationships between MB concentration and RGB intensities of tested MB solutions.

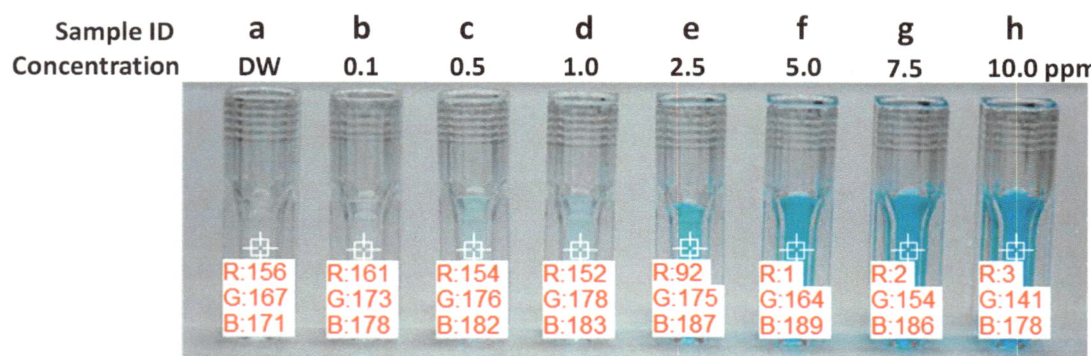

Figure 7. RGB values of MB solutions from 0.1 to 10 ppm (adapted from [23]. (15 × 15 pixel square areas were chosen for color analysis).

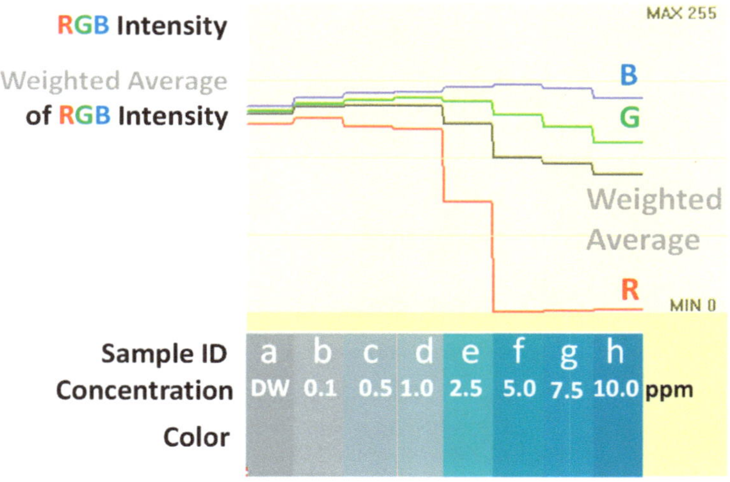

Figure 8. Representation of MB solutions from 0.1 to 10 ppm (adapted from [23] color and RGB values and weighted RGB values extracted using PicMan). The weighted average of RGB values were calculated by using a formula (R + 2G + B)/4.

Table 2 summarizes the color of MB solutions with different concentrations in various color expression formats. All values are averages of 15×15 pixel square areas of MB solutions from 0.1 to 10 ppm, as shown in Figure 7. The color was quantified for objective analysis. As noticed from the RGB values of MB solutions with different concentrations, the red channel intensity starts from 161 for 0.1 ppm and decreases with the increment of concentration. At the concentration between 1.0 and 2.5 ppm, the red component of longer wavelength photons were absorbed. As a result, almost no red color signal is detected. The green color intensity remained constant around 175 up to the concentration of 1.0 ppm and started to decrease its intensity as the concentration is increased. The blue color intensity remained very strong above 178 for all concentrations and showed the maximum intensity at around 2.5 ppm. By using this color change with the concentration, the concentration of MB solutions can be estimated from color analysis results.

Table 2. Summary of color analysis results of MB solutions from 0.1 to 10 ppm in various color expression methods (RGB, weighted RGB, HSV, CIE L*a*b*, Munsell and Hexadecimal codes). Representing colors for sample a–h are shown for easy recognition. Intensity of red, green and blue channels were shown in representing colors.

Sample ID		a	b	c	d	e	f	g	h
Concentration (ppm)		DW	0.1	0.2	0.5	1.0	2.5	5.0	10.0
Representing Color									
RGB Color	R	156	161	154	152	92	1	2	3
	G	167	173	176	178	175	164	154	141
	B	171	178	182	183	187	189	186	178
	Average **	165	171	172	172	157	129	124	115
HSV Color	H	196	197.6	192.9	189.7	187.6	188	190.4	192.7
	S	0.09	0.1	0.15	0.17	0.51	0.99	0.99	0.98
	V	0.67	0.7	0.71	0.72	0.73	0.74	0.73	0.7
CIELAB Color	L	67.8	70	70.4	70.8	66.9	61.9	58.7	54.3
	a	−3.06	−3.18	−6.29	−7.8	−21.91	−26.39	−22.26	−18.15
	b	−3.4	−4.01	−5.63	−5.54	−13.92	−22.89	−26.1	−28.26
Munsell Color	Munsell-h	7.5B	7.5B	5B	5B	2.5B	2.5B	2.5B	7.5B
	V	8	6	8	8	9	5	7	6
	C	2	2	2	2	4	6	8	8
Hexadecimal Color	Color Code	#9CA7AB	#A1ADB2	#9AB0B6	#98B2B7	#5CAFBB	#01A4BD	#029ABA	#038DB2

** Weighted average calculated using $(R + 2G + B)/4$.

3.3. Water Quality Inspection Color Chart

Pools and hot tubs require periodic water quality tests to maintain the quality of water for the health and safety of users. Paper-based test strips are commonly used for water quality checks. Color change, after dipping a test strip into the testing water, is used as an indicator for various testing categories, such as the total hardness, total chlorine/total bromine, free chlorine, pH, total alkalinity, and cyanuric acid levels. A commercially available AquaChek Color Chart [43], which indicates the chemical levels in hot tub water, was selected as the color analysis example. The color of the test strip after dipping is compared with a color chart to interpret the test results and find a remedy for any issue.

Figure 9 shows a commercially available AquaChek Color Chart, which indicates the chemical levels in hot tub water. A printable AquaChek 7-in-1 Color Chart for a specific test strip product can be digitally downloaded and referenced for interpretation of proper test results. The downloaded AquaChek 7-in-1 Color Chart can be printed out on white filter papers with colored dots to prevent a color shift between a digital file and its printed chart.

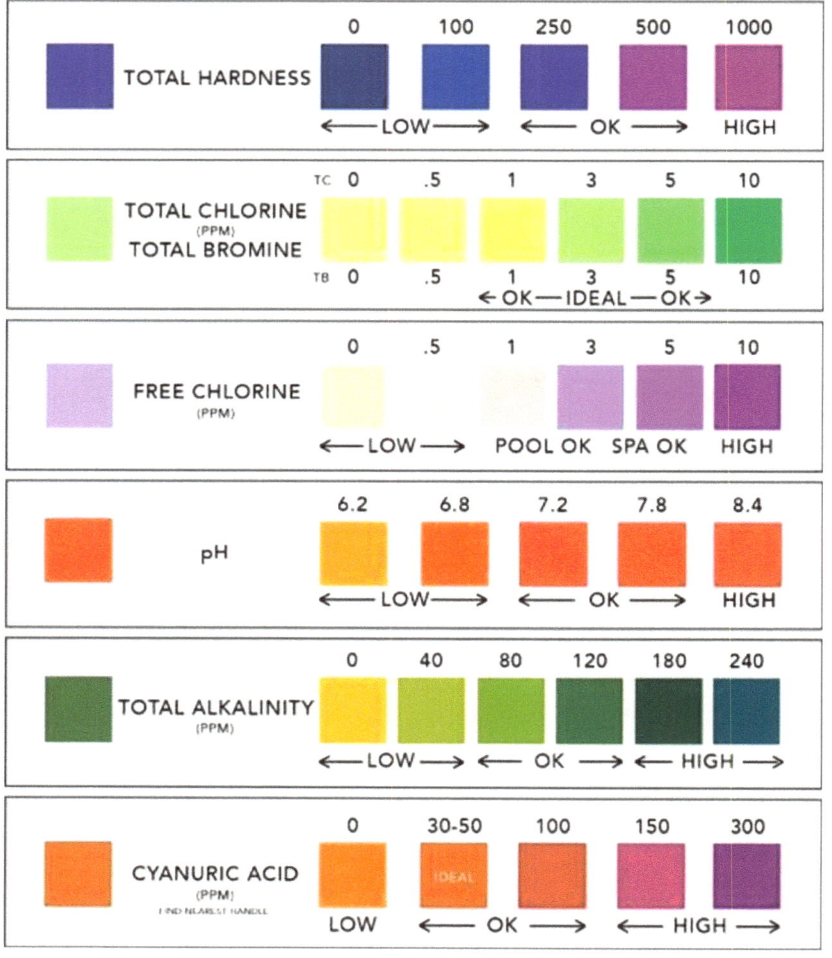

Figure 9. A commercially available AquaChek Color Chart, which indicates the chemical levels in hot tub water [43].

There are seven elements to the AquaChek Test Strips. From top to bottom, Total Hardness, Total Chlorine, Total Bromine, Free Chlorine, pH, Total Alkalinity, and Cyanuric Acid levels are tested and read from the chart. It is based on visual inspection and interpretation and is subjective to an individual's color vision.

Table 3 shows the color quantification summary of the AquaChek Color Chart, which indicates the chemical levels in hot tub water.

Table 3. Summary of color analysis results of a commercially available AquaChek Color Chart, which indicates the chemical levels in hot tub water [43]. Intensity of red, green and blue channels were shown in representing colors.

		R	G	B	Average **	H	S	V	HexCode	Color	Remarks
Total Hardness	Pre Test	64	43	160	77.5	250.8	0.73	0.63	#402BA0		Pre Test
	0	4	25	146	50.0	231.1	0.97	0.57	#041992		Low
	100	39	56	171	80.5	232.3	0.77	0.67	#2738AB		Low
	250	66	43	158	77.5	252.0	0.73	0.62	#422B9E		OK
	500	135	59	158	102.8	286.1	0.63	0.62	#873B9E		OK
	1000	145	32	139	87.0	303.2	0.78	0.57	#91208B		High
Total Chlorine (ppm) Total Bromine	Pre Test	188	220	150	194.5	87.4	0.32	0.86	#BCDC96		Pre test
	0	254	255	168	233.0	60.7	0.34	1.00	#FEFFA8		
	0.5	241	254	171	230.0	69.4	0.33	1.00	#F1FEAB		
	1	232	246	159	220.8	69.7	0.35	0.96	#E8F69F		OK
	3	185	216	139	189.0	84.2	0.36	0.85	#B9D88B		Ideal
	5	146	197	121	165.3	100.3	0.39	0.77	#92C579		OK
	10	77	162	97	124.5	134.1	0.52	0.64	#4DA261		
Free Chroline (ppm)	Pre Test	184	160	226	182.5	261.8	0.29	0.89	#B8A0E2		Pre Test
	0	254	253	206	241.5	58.8	0.19	1.00	#FEFDCE		Low
	0.5	246	249	228	243.0	68.6	0.08	0.98	#F6F9E4		Low
	1	231	223	217	223.5	25.7	0.06	0.91	#E7DFD9		Poor
	3	173	139	208	164.8	269.6	0.33	0.82	#AD8BD0		Spa OK
	5	159	106	189	140.0	278.3	0.44	0.74	#9F6ABD		Spa OK
	10	128	29	153	84.8	287.9	0.81	0.60	#801D99		High
pH	Pre Test	225	79	37	105.0	13.4	0.84	0.88	#E14F25		Pre Test
	6.2	242	174	63	163.3	37.2	0.74	0.95	#F2AE3F		Low
	6.8	235	106	45	123.0	19.3	0.81	0.92	#EB6A2D		Low
	7.2	227	54	37	93.0	5.4	0.84	0.89	#E33625		OK
	7.8	224	45	33	86.8	3.8	0.85	0.88	#E02D21		OK
	8.4	214	45	34	84.5	3.7	0.84	0.84	#D62D22		High
Total Alkalinity (ppm)	Pre Test	91	116	71	98.5	93.3	0.39	0.45	#5B7447		Pre Test
	0	229	190	66	168.8	45.6	0.71	0.90	#E5BE42		Low
	40	163	168	53	138.0	62.6	0.68	0.66	#A3A835		Low
	80	138	158	60	128.5	72.2	0.62	0.62	#8A9E3C		OK
	120	73	110	57	87.5	101.9	0.48	0.43	#496E39		OK
	180	35	82	48	61.8	136.6	0.57	0.32	#235230		High
	240	35	88	98	77.3	189.5	0.64	0.38	#235862		High
Cyanuric Acid (ppm)	Pre Test	205	99	37	110.0	22.1	0.82	0.80	#CD6325		Pre Test
	0	229	136	49	137.5	29.0	0.79	0.90	#E58831		Low
	30–50	207	93	38	107.8	19.5	0.82	0.81	#CF5D26		OK
	100	190	38	32	74.5	2.3	0.83	0.75	#BE2620		OK
	150	177	37	124	93.8	322.7	0.79	0.69	#B1257C		High
	300	121	27	127	75.5	296.4	0.79	0.50	#791B7F		High

** Weighted average calculated using (R + 2G + B)/4.

3.4. Tongue Color

As an effective alternative medicine, oriental medicine (called by different names in different countries: traditional Chinese medicine (TCM, 中医, Zhōngyī), traditional Korean medicine (한방, 韓方, Hanbang) and traditional Chinese medicine in Japan (漢方, Kampo)) utilizes tongue diagnosis as a major method to assess the patient's health conditions. Oriental medicine doctors examine the color, shape, and texture of a patient's tongue [12–20,44]. Tongue color, shape and texture can give the pre-disease indications without any significant health problems or disease-related symptoms. It provides a basis for preventive medicine and useful advice to a patient for lifestyle adjustment. However, traditional tongue diagnosis by visual inspection has limitations, as the inspection process is subjective and inconsistent between doctors. To provide more consistent and objective health assessments through tongue diagnosis, quantification of tongue color and texture from digital images seem to be a very attractive approach. Many researchers have reported their ideas and approaches, and they seen very encouraging results for years [12–20,44].

Figure 10 shows a sample tongue image from an acupuncturist's site [44] and color analysis results using the color analysis of ROIs. Figure 10a,b are sample tongue photos of a patient. The tongue photos are covered with a tongue chart used in oriental medicine.

Figure 10c,d are areas of each section with a number of pixels and a tongue repainted by representing average colors with corresponding hexadecimal color codes of the individual sections. The color difference between (a) and (d) can be used to quantify color variation and texture of a real tongue from image-based color quantification. The average tongue color can also be identified after image processing and used for tongue color-based diagnosis of a patient's health conditions. Since the authors are not medical doctors, they cannot diagnose health conditions of the patient. However, the quantified color information can be very useful for medical doctors to standardize their experience-based knowledge into verifiable datasets. Table 4 summarized the color analysis results of a sample tongue photo of a patient [44] by section in various color expression formats.

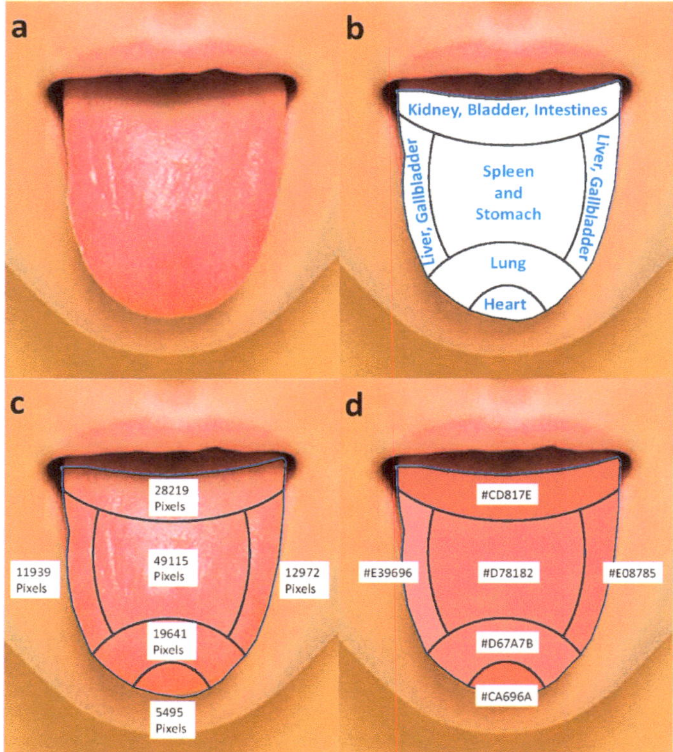

Figure 10. (**a**) A sample tongue photo of a patient, (**b**) a tongue chart used in oriental medicine, (**c**) areas of each section, and (**d**) a tongue repainted by representing average colors with corresponding hexadecimal color codes of individual sections of a tongue.

Table 4. Summary of color analysis results of a sample tongue photo of a patient [44].

No.	Region of Interest	Area (Pixel)	Red					Green					Green					Hexadecimal Color		HSV			CIELAB			Munsell Color		
			Mean	Min	Max	Range	Stdev	Mean	Min	Max	Range	Stdev	Mean	Min	Max	Range	Stdev	Hex Code	Color	H	S	V	L*	a*	b*	h	V	C
1	Heart	5495	202.6	47	227	180	30.9	105.7	81	217	136	12.7	106.1	67	230	163	18.0	#CA696A		359.4	0.48	0.79	55.8	38.5	17.1	2.5 R	7	6
2	Lung	19,641	214.5	47	244	197	24.1	122.3	82	217	135	12.0	123.3	82	230	148	13.0	#D67A7B		359.3	0.43	0.84	61.3	35.8	15.2	2.5 R	7	6
3	Spleen, Stomach	11,939	215.8	47	252	205	39.7	129.7	82	220	138	16.5	130.9	88	230	142	15.8	#D78182		359.3	0.4	0.84	63.1	33.3	13.8	2.5 R	7	6
4	Liver, Gallbladder (Left)	49,115	227.7	47	255	208	21.0	150.3	82	225	143	17.7	150.1	113	230	117	17.4	#E39696		0	0.34	0.89	69.7	29.0	12.0	5 R	6	2
5	Liver, Gallbladder (Right)	12,972	224.4	47	255	208	21.8	135.4	82	217	135	15.7	133.5	87	230	143	16.0	#E08785		1.3	0.41	0.88	65.6	33.8	15.7	5 R	6	10
6	Kidney, Bladder, Intestines	28,219	205.0	47	255	208	32.5	129.5	26	217	191	28.9	126.9	29	230	201	29.9	#CD817E		2.3	0.39	0.8	61.8	29.0	14.0	2.5 R	8	4

Zhang et al. proposed an in-depth systematic tongue color analysis system including a tongue image capture device for medical applications in 2013 [18]. A tongue color gamut was used in the study. Tongue foreground pixels are first extracted and assigned to one of 12 colors representing this gamut. The ratio of each color for the entire image is calculated and forms a tongue color feature vector to suppress subjectivity and significantly improve the objectivity of tongue color determination. A relationship between the health condition of the human body and its tongue color has been demonstrated using more than 1054 tongue images of 143 healthy patients and 902 patients with health issues. The patients with health issues were classified into 13 disease groups of more than 10 tongue images and one miscellaneous group. This study has reported that a given tongue sample can be classified into one of two classes (healthy and disease) with an average accuracy of 91.99%. Moreover, disease tongue images can be classified into three clusters, and within each cluster, most of the illnesses are distinguished from one another. In total, 11 illnesses had a classification rate greater than 70%. These results in colorimetry were very encouraging for oriental medicine.

A Precise and Fast Tongue Segmentation System Using U-Net with a Morphological Processing Layer called TongueNet has been reported by a research group of the University of Macau [19]. Quantification of tongue color using ML has been reported by Japanese and Malaysia-Japan international research groups [17].

4. Discussion

The characteristics of a color are determined by three different elements in the HSV color space: hue, chroma (saturation) and value (luminance). Hue is the most recognizable characteristic of a color and most people refer to hue as color. There are an infinite number of possible hues and would be difficult to precisely describe with words in any languages. Chroma refers to the purity and intensity of a color. High-chroma colors look rich and full, while low-chroma colors look dull or pale. Value is the lightness or darkness characteristic of a color. Light colors are often called tints, and dark colors are often referred to as shades. Personal impression is very subjective, and verbal expression of personal impression is not suited for precise description of color. For objective characterization of color, quantitative characterization of color is required.

The quantification capability of color information can be very useful for color monitoring, color control, machine vision, ML, AI-based algorithm development and smart phone app development for daily life and sophisticated applications. For accurate color quantification, illumination and image acquisition methods have to be optimized in addition to the image processing algorithm development. Medical image processing and analysis became a very active research field [11–20,38,41]. Image classification, deep convolution (DL), deep learning and ML examples and their approaches have been reported [49–52]. A DL neural network used for image classification is an important part of DL and has great significance in the field of computer vision towards medical image analysis and diagnosis applications. It would help us to navigate realistic directions for the development of ML, DL and AI-based algorithms for medical applications.

The quantitative colorimetric analysis technique has been extremely useful for digital forensic studies of cultural heritage identification processes, in identifying particular printing techniques through ink tone analyses and character image comparisons [53–56]. The world's oldest metal-type printed book (The song of enlightenment (南明泉和尚頌證道歌) in Korea in 1239) has been identified by comparing six nearly identical books from Korea from the 13th to 16th centuries, the Jikji (直指) printed in 1377 and the Gutenberg 42-line Bible printed in 1455 using this image analysis technique. Diagnoses of the conservation status of painting cultural heritage and color fading characteristics of pigments were also very promising [7,8]. Many other applications based on color and shape extraction techniques using newly developed image processing/analysis software (PicMan) were introduced [57]. Image processing software applications combined with imaging devices have been reported in previous papers [6,31,38,56].

5. Conclusions

The concept of an image-based colorimetry technique has been introduced. Due to the wide availability of smart phones with image sensors and increasing computational power, a smart-phone-based, low-cost colorimeter has become a very attractive tool. The low cost and portable designs with user-friendly interfaces, and their compatibility with data acquisition and processing, make a perfect fit for interdisciplinary applications in a wide range of fields. Some examples are from art, the fashion industry, food science, medical science, oriental medicine, agriculture, geology, chemistry, biology, material science, environmental engineering, etc.

In this study, the image-based quantification examples of color using specifically developed image processing/analysis software were described. The feasibility of machine vision and offline applications in interdisciplinary fields using software was demonstrated. Examples of color information extraction from a single pixel to predetermined sizes/shapes of areas, including customized ROIs from various digital images of dyed T-shirts, color chart, assays, and tongues, are demonstrated. Conversion between RGB, HSV, CIELAB, Munsell color, and hexadecimal color codes for extracted colors from a single pixel to ROIs were demonstrated. The capability of color extraction and color conversion between different color spaces is very useful for machine vision and offline applications in various fields. Histograms and statistical analyses of colors from a single pixel to ROIs are successfully demonstrated using four different fields of applications.

Flat colors, colors with textures, segregated colors, and totally mixed colors appear very different to the human eye depending on the individual's perception. Color-based judgments by humans are very subjective. The judgments are strongly dependent on the examiner's vision, experience and knowledge. We were able to demonstrate image-based color quantification techniques using newly developed image processing software for various fields of applications.

Reliable image-based quantification of color, in a wide range of potential applications, including ML, AI, algorithm development and new smart phone app development, are proposed as realistic tasks to be worked on. The successful development and demonstration of image-based quantification of color of any size and shape of interest, using specifically developed software, makes customization easier for new applications that require quantitative colorimetry.

Author Contributions: All authors equally contributed in this study. Conceptualization, W.S.Y., K.K. and Y.Y.; material preparation, Y.Y., J.G.K. and W.S.Y.; methodology, Y.Y and J.G.K..; software, W.S.Y. and K.K.; data acquisition and analysis, Y.Y., J.G.K. and W.S.Y.; writing—original draft preparation, review and editing, W.S.Y. and Y.Y. All authors have read and agreed to the published version of the manuscript.

Funding: This research received no external funding.

Conflicts of Interest: Woo Sik Yoo, Kitaek Kina and Jung Gon Kim were employed by WaferMasters, Inc. The remaining author declares that the research was conducted in the absence of any commercial or financial relationships that could be construed as a potential conflict of interest. All authors declare no conflict of interest.

References

1. Giesel, M.; Gegenfurtner, K.R. Color appearance of real objects varying in material, hue, and shape. *J. Vis.* **2010**, *10*, 1–21. [CrossRef] [PubMed]
2. Dische, Z. Qualitative and quantitative colorimetric determination of heptoses. *J. Biol. Chem* **1953**, *204*, 983–998. [CrossRef] [PubMed]
3. Nimeroff, I. Colorimetry, National Bureau of Standards Monograph 104. 1968. Available online: https://nvlpubs.nist.gov/nistpubs/Legacy/MONO/nbsmonograph104.pdf (accessed on 28 February 2023).
4. Bäuml, K.H. Simultaneous color constancy: How surface color perception varies with the illuminant. *Vis. Res.* **1999**, *39*, 1531–1550. [CrossRef] [PubMed]

5. Mohtasebi, A.; Broomfield, A.D.; Chowdhury, T.; Ravi Selvaganapathy, P.; Peter Kruse, P. Reagent-Free Quantification of Aqueous Free Chlorine via Electrical Readout of Colorimetrically Functionalized Pencil Lines. *ACS Appl. Mater. Interfaces* **2017**, *9*, 20748–20761. [CrossRef] [PubMed]

6. Yoo, W.S.; Kim, J.G.; Kang, K.; Yoo, Y. Development of Static and Dynamic Colorimetric Analysis Techniques Using Image Sensors and Novel Image Processing Software for Chemical, Biological and Medical Applications. *Technologies* **2023**, *11*, 23. [CrossRef]

7. Chua, L.; Quan, S.Z.; Yan, G.; Yoo, W.S. Investigating the Colour Difference of Old and New Blue Japanese Glass Pigments for Artistic Use. *J. Conserv. Sci.* **2022**, *38*, 1–13. [CrossRef]

8. Eom, T.H.; Lee, H.S. A Study on the Diagnosis Technology for Conservation Status of Painting Cultural Heritage Using Digital Image Analysis Program. *Heritage* **2023**, *6*, 1839–1855. [CrossRef]

9. Lehnert, M.S.; Balaban, M.O.; Emmel, T.C. A new method for quantifying color of insects. *Fla. Entomol.* **2011**, *94*, 201–207. [CrossRef]

10. Taweekarn, T.; Wongniramaikul, W.; Limsakul, W.; Sriprom, W.; Phawachalotorn, C.; Choodum, A. A novel colorimetric sensor based on modified mesoporous silica nanoparticles for rapid on-site detection of nitrite. *Microchim. Acta* **2020**, *187*, 643. [CrossRef]

11. Kim, J.; Wu, Y.; Luan, H.; Yang, D.S.; Cho, D.; Kwak, S.S.; Liu, S.; Ryu, H.; Ghaffari, R.; Rogers, J.A. A Skin-Interfaced, Miniaturized Microfluidic Analysis and Delivery System for Colorimetric Measurements of Nutrients in Sweat and Supply of Vitamins Through the Skin. *Adv. Sci.* **2022**, *9*, 2103331. [CrossRef]

12. Horzov, L.; Goncharuk-Khomyn, M.; Hema-Bahyna, N.; Yurzhenko, A.; Melnyk, V. Analysis of tongue colorassociated features among patients with PCR-confirmed Covid-19 infection in Ukraine. *Pesqui Bras. Odontopediatria Clín. Integr.* **2021**, *21*, e0011. [CrossRef]

13. Chen, H.S.; Chen, S.M.; Jiang, C.Y.; Zhang, Y.C.; Lin, C.Y.; Lin, C.E.; Lee, J.A. Computational tongue color simulation in tongue diagnosis. *Color Res. Appl.* **2021**, *9*, 2103331. [CrossRef]

14. Segawa, M.; Iizuka, N.; Ogihara, H.; Tanaka, K.; Nakae, H.; Usuku, K.; Hamamoto, Y. Construction of a Standardized Tongue Image Database for Diagnostic Education: Development of a Tongue Diagnosis e-Learning System. *Front. Med. Technol.* **2021**, *3*, 760542. [CrossRef] [PubMed]

15. Xie, J.; Congcong Jing, C.; Zhang, Z.; Xu, J.; Ye Duan, Y.; Xu, D. Digital tongue image analyses for health assessment. *Med. Rev.* **2021**, *1*, 172–198. [CrossRef]

16. Sun, Z.M.; Zhao, J.; Qian, P.; Wang, Y.Q.; Zhang, W.F.; Guo, C.R.; Pang, X.Y.; Wang, S.C.; Li, F.F.; Li, Q. Metabolic markers and microecological characteristics of tongue coating in patients with chronic gastritis. *BMC Complement. Altern. Med.* **2013**, *13*, 227. Available online: http://www.biomedcentral.com/1472-6882/13/227 (accessed on 28 February 2023). [CrossRef] [PubMed]

17. Kawanabe, T.; Kamarudin, N.D.; Ooi, C.Y.; Kobayashi, F.; Mi, X.; Sekine, M.; Wakasugi, A.; Odaguchi, H.; Hanawa, T. Quantification of tongue colour using machine learning in Kampo medicine. *Eur. J. Integr. Med.* **2016**, *8*, 932–941. [CrossRef]

18. Zhang, B.; Wang, X.; You, J.; Zhang, D. Tongue Color Analysis for Medical Application. *Evid.-Based Complement. Altern. Med.* **2013**, *2013*, 264742. [CrossRef]

19. Zhou, J.; Zhang, Q.; Zhang, B.; Chen, X. TongueNet: A Precise and Fast Tongue Segmentation System Using U-Net with a Morphological Processing Layer. *Appl. Sci.* **2019**, *9*, 3128. [CrossRef]

20. Takahoko, K.; Iwasaki, H.; Sasakawa, T.; Suzuki, A.; Matsumoto, H.; Iwasaki, H. Unilateral Hypoglossal Nerve Palsy after Use of the Laryngeal Mask Airway Supreme. *Case Rep. Anesthesiol.* **2014**, *2014*, 369563. [CrossRef]

21. Kılıç, V.; Alankus, G.; Horzum, N.; Mutlu, A.Y.; Bayram, A.; Solmaz, M.E. Single-Image-Referenced Colorimetric Water Quality Detection Using a Smartphone. *ACS Omega* **2018**, *3*, 5531–5536. [CrossRef]

22. Alberti, G.; Zanoni, C.; Magnaghi, L.R.; Biesuz, R. Disposable and Low-Cost Colorimetric Sensors for Environmental Analysis. *Int. J. Environ. Res. Public Health* **2020**, *17*, 8331. [CrossRef] [PubMed]

23. Kılıç, V.; Horzum, N.; Solmaz, M.E. From Sophisticated Analysis to Colorimetric Determination: Smartphone Spectrometers and Colorimetry. In *Color Detection*; IntechOpen: London, UK, 2018. [CrossRef]

24. O'Donoghue, J. Simplified Low-Cost Colorimetry for Education and Public Engagement. *J. Chem. Educ.* **2019**, *96*, 1136–1142. [CrossRef]

25. Hermida, I.D.P.; Prabowo, B.A.; Kurniawan, D.; Manurung, R.V.; Sulaeman, Y.; Ritadi, M.A.; Wahono, M.D. Use of Smartphone Based on Android as a Color Sensor. In Proceedings of the 2018 Electrical Power, Electronics, Communications, Controls and Informatics Seminar (EECCIS), Batu, Indonesia, 9–11 October 2018; pp. 424–429. [CrossRef]

26. Han, P.; Dong, D.; Zhao, X.; Jiao, L.; Lang, Y. A smartphone-based soil color sensor: For soil type classification. *Comput. Electron. Agric.* **2016**, *123*, 232–241. [CrossRef]

27. Alawsi, T.; Mattia, G.P.; Al-Bawa, Z.; Beraldi, R. Smartphone-based colorimetric sensor application for measuring biochemical material concentration. *Sens. Bio-Sens. Res.* **2021**, *32*, 100404. [CrossRef]

28. Chellasamy, G.; Ankireddy, S.R.; Lee, K.N.; Govindaraju, S.; Yun, K. Smartphone-integrated colorimetric sensor array-based reader system and fluorometric detection of dopamine in male and female geriatric plasma by bluish-green fluorescent carbon quantum dots. *Mater. Today Bio* **2021**, *12*, 100168. [CrossRef]

29. Böck, F.C.; Helfer, G.A.; da Costa, A.B.; Dessuy, M.B.; Flôres Ferrão, M. PhotoMetrix and colorimetric image analysis using smartphones. *J. Chemom.* **2020**, *34*, e3251. [CrossRef]
30. Wongniramaikul, W.; Kleangklao, B.; Boonkanon, C.; Taweekarn, T.; Phatthanawiwat, K.; Sriprom, W.; Limsakul, W.; Towanlong, W.; Tipmanee, D.; Choodum, A. Portable Colorimetric Hydrogel Test Kits and On-Mobile Digital Image Colorimetry for On-Site Determination of Nutrients in Water. *Molecules* **2022**, *27*, 7287. [CrossRef]
31. Yoo, Y.; Yoo, W.S. Turning Image Sensors into Position and Time Sensitive Quantitative Colorimetric Data Sources with the Aid of Novel Image Processing/Analysis Software. *Sensors* **2020**, *20*, 6418. [CrossRef]
32. Kim, G.; Kim, J.G.; Kang, K.; Yoo, W.S. Image-Based Quantitative Analysis of Foxing Stains on Old Printed Paper Documents. *Heritage* **2019**, *2*, 2665–2677. [CrossRef]
33. Yoo, W.S.; Kim, J.G.; Kang, K.; Yoo, Y. Extraction of Colour Information from Digital Images Towards Cultural Heritage Characterisation Applications. *SPAFA J.* **2021**, *5*, 1–14. [CrossRef]
34. Yoo, W.S.; Kang, K.; Kim, J.G.; Yoo, Y. Extraction of Color Information and Visualization of Color Differences between Digital Images through Pixel-by-Pixel Color-Difference Mapping. *Heritage* **2022**, *5*, 3923–3945. [CrossRef]
35. Yoo, Y.; Yoo, W.S. Digital Image Comparisons for Investigating Aging Effects and Artificial Modifications Using Image Analysis Software. *J. Conserv. Sci.* **2021**, *37*, 1–12. [CrossRef]
36. Kim, J.G.; Yoo, W.S.; Jang, Y.S.; Lee, W.J.; Yeo, I.G. Identification of Polytype and Estimation of Carrier Concentration of Silicon Carbide Wafers by Analysis of Apparent Color using Image Processing Software. *ECS J. Solid State Sci. Technol.* **2022**, *11*, 064003. [CrossRef]
37. Yoo, W.S.; Kang, K.; Kim, J.G.; Jung, Y.-H. Development of Image Analysis Software for Archaeological Applications. *Adv. Southeast Asian Archaeol.* **2019**, *2*, 402–411.
38. Yoo, W.S.; Han, H.S.; Kim, J.G.; Kang, K.; Jeon, H.-S.; Moon, J.-Y.; Park, H. Development of a tablet PC-based portable device for colorimetric determination of assays including COVID-19 and other pathogenic microorganisms. *RSC Adv.* **2020**, *10*, 32946–32952. [CrossRef] [PubMed]
39. Wakamoto, K.; Otsuka, T.; Nakahara, K.; Namazu, T. Degradation Mechanism of Pressure-Assisted Sintered Silver by Thermal Shock Test. *Energies* **2021**, *14*, 5532. [CrossRef]
40. Jo, S.-I.; Jeong, G.-H. Single-Walled Carbon Nanotube Synthesis Yield Variation in a Horizontal Chemical Vapor Deposition Reactor. *Nanomaterials* **2021**, *11*, 3293. [CrossRef]
41. Zhu, C.; Espulgar, W.V.; Yoo, W.S.; Koyama, S.; Dou, X.; Kumanogoh, A.; Tamiya, E.; Takamatsu, H.; Saito, M. Single Cell Receptor Analysis Aided by a Centrifugal Microfluidic Device for Immune Cells Profiling. *Bull. Chem. Soc. Jpn.* **2019**, *92*, 1834–1839. [CrossRef]
42. Men's T-Shirts. Available online: https://www.pinterest.com/pin/624733779558363886/ (accessed on 28 February 2023).
43. AquaChek Color Chart. Available online: https://www.masterspaparts.com/aquachek-color-chart/ (accessed on 28 February 2023).
44. What Is Chinese Tongue Diagnosis? Available online: https://www.lucyclarkeacupuncture.co.uk/what-is-chinese-tongue-diagnosis/ (accessed on 28 February 2023).
45. Color Model. Available online: https://en.wikipedia.org/wiki/Color_model (accessed on 28 February 2023).
46. Color Space. Available online: https://en.wikipedia.org/wiki/Color_space (accessed on 28 February 2023).
47. Color Conversion. Available online: https://en.wikipedia.org/wiki/HSL_and_HSV (accessed on 28 February 2023).
48. Palum, R. Image Sampling with the Bayer Color Filter Array. In *PICS 2001: Image Processing, Image Quality, Image Capture, Systems Conference, Montréal, QC, Canada, 22–25 April 2001*; The Society for Imaging Science and Technology: Cambridge, MA, USA; pp. 239–245. ISBN 0-89208-232-1.
49. Zhao, M.; Liu, Q.; Jha, A.; Deng, R.; Yao, T.; Mahadevan-Jansen, A.; Tyska, M.J.; Millis, B.A.; Huo, Y. VoxelEmbed: 3D Instance Segmentation and Tracking with Voxel Embedding based Deep Learning. In *Machine Learning in Medical Imaging*; Lian, C., Cao, X., Rekik, I., Xu, X., Yan, P., Eds.; Lecture Notes in Computer Science; Springer: Cham, Switzerland, 2021; Volume 12966, pp. 437–446. [CrossRef]
50. Yao, T.; Qu, C.; Liu, Q.; Deng, R.; Tian, Y.; Xu, J.; Jha, A.; Bao, S.; Zhao, M.; Fogo, A.B.; et al. Compound Figure Separation of Biomedical Images with Side Loss. In *Deep Generative Models, and Data Augmentation, Labelling, and Imperfections*; Lecture Notes in Computer Science; Springer: Cham, Switzerland, 2021; Volume 13003, p. 183. [CrossRef]
51. Zheng, Q.; Yang, M.; Zhang, Q.; Zhang, X. Fine-grained image classification based on the combination of artificial features and deep convolutional activation features. In Proceedings of the 2017 IEEE/CIC International Conference on Communications in China (ICCC), Qingdao, China, 22–24 October 2017; pp. 1–6. [CrossRef]
52. Zheng, Q.; Yang, M.; Tian, X.; Wang, X.; Wang, D. Rethinking the Role of Activation Functions in Deep Convolutional Neural Networks for Image Classification. *Eng. Lett.* **2020**, *28*, EL_28_1_11.
53. Yoo, W.S. The World's Oldest Book Printed by Movable Metal Type in Korea in 1239: The Song of Enlightenment. *Heritage* **2022**, *5*, 1089–1119. [CrossRef]
54. Yoo, W.S. How Was the World's Oldest Metal-Type-Printed Book (The Song of Enlightenment, Korea, 1239) Misidentified for Nearly 50 Years? *Heritage* **2022**, *5*, 1779–1804. [CrossRef]

55. Yoo, W.S. Direct Evidence of Metal Type Printing in The Song of Enlightenment, Korea, 1239. *Heritage* **2022**, *5*, 3329–3358. [CrossRef]
56. Yoo, W.S. Ink Tone Analysis of Printed Character Images towards Identification of Medieval Korean Printing Technique: The Song of Enlightenment (1239), the Jikji (1377) and the Gutenberg Bible (~1455). *Heritage* **2023**, *6*, 2559–2581. [CrossRef]
57. PicManTV. Available online: https://www.youtube.com/@picman-TV (accessed on 28 February 2023).

technologies

Article

Mobilenetv2_CA Lightweight Object Detection Network in Autonomous Driving

Peicheng Shi [1,*], Long Li [1], Heng Qi [1] and Aixi Yang [2]

[1] School of Mechanical Engineering, Anhui Polytechnic University, Wuhu 241000, China
[2] Polytechnic Institute, Zhejiang University, Hangzhou 310027, China
* Correspondence: shipeicheng@126.com

Abstract: A lightweight network target detection algorithm was proposed, based on MobileNetv2_CA, focusing on the problem of high complexity, a large number of parameters, and the missed detection of small targets in the target detection network based on candidate regions and regression methods in autonomous driving scenarios. First, Mosaic image enhancement technology is used in the data pre-processing stage to enhance the feature extraction of small target scenes and complex scenes; second, the Coordinate Attention (CA) mechanism is embedded into the Mobilenetv2 backbone feature extraction network, combined with the PANet and Yolo detection heads for multi-scale feature fusion; finally, a Lightweight Object Detection Network is built. The experimental test results show that the designed network obtained the highest average detection accuracy of 81.43% on the Voc2007 + 2012 dataset, and obtained the highest average detection accuracy of 85.07% and a detection speed of 31.84 FPS on the KITTI dataset. The total amount of network parameters is only 39.5 M. This is beneficial to the engineering application of MobileNetv2 network in automatic driving.

Keywords: object detection; attention mechanism; lightweight network; autonomous driving

Citation: Shi, P.; Li, L.; Qi, H.; Yang, A. Mobilenetv2_CA Lightweight Object Detection Network in Autonomous Driving. *Technologies* **2023**, *11*, 47. https://doi.org/10.3390/technologies11020047

Academic Editors: Gwanggil Jeon and Imran Ahmed

Received: 5 February 2023
Revised: 1 March 2023
Accepted: 21 March 2023
Published: 23 March 2023

1. Introduction

With the development of autonomous driving technology in recent years, object detection has become a research focus in the field of autonomous driving, and it is crucial to achieve the fast and accurate detection of common objects in real traffic scenes. Traditional object detection algorithms usually rely on specific application scenarios and manually designed features with poor generalization abilities [1,2]. With the wide application of Convolutional Neural Networks (CNN) in the field of computer vision in recent years, object detection based on CNN [3–5] has achieved an excellent performance. Ge Z et al. [6] balanced the detection accuracy and speed using the decoupled head decoupling head, anchor free method and SimOTA sample selection method. Xu S et al. [7] used the efficient task alignment head, advanced label assignment strategy and fine-grained objective loss function, which implements an efficient object detector. However, traditional CNN usually requires a large quantity of parameters to achieve satisfactory precision. Liu Z et al. [8] created a convolution-free model for image classification by stacking multiple Transformer blocks to process sequences of non-overlapping image patches. Compared with the CNN-based model, the Transformer-based model has a larger receptive field but requires a system with a large amount of training data and model parameters to achieve superior performance. The redundant attention (Attention) in the Transformer will also lead to a huge computational cost. This is extremely unhelpful to deploying the model on an autonomous driving platform. In real embedded autonomous driving platforms, the design trend in the neural network is to explore lightweight and efficient network architectures with an acceptable performance for mobile devices to improve network detection efficiency.

For the lightweight design of neural networks, it is common practice to compress the existing convolutional neural networks or design a new lightweight network architecture.

Han S et al. [9] proposed the use of network pruning to prune unimportant weights in neural networks, which reduces a large quantity of parameters in the fully connected layer by weight pruning, but may not sufficiently reduce the computational cost of the convolutional layer due to the irregular sparsity in the pruned network. Li H et al. [10] pruned filters to obtain effective CNNs using l_1 parametric regularization; these filters are considered to have little impact on the output precision, which reduces the calculation cost by removing the overall filters in the network and its connected feature maps. This approach does not lead to sparse connectivity patterns compared to the pruning weights shown in the literature [9]. Rastegari M et al. [11] quantified the weights and activations as 1-bit data, while the filter and convolution layers, improved to binary, use binary operations to approximate convolution, achieving a certain degree of compression and acceleration ratio. However, the overall parametric reduction in the network is not significant. Hinton G et al. [12] introduced the knowledge distillation method. This transfers the knowledge from the larger model to the smaller model, which leads to a poor generalization ability within the whole network. Han K et al. [13] used a new Ghost module, which first controls the total number of ordinary convolutions and then generates more feature maps using linear operations to reduce the total number of parameters without changing the size of the feature maps. These methods of compressing the existing convolutional neural networks usually mean that the convolutional neural networks tend to be shallow. The detection efficiency of shallow neural networks is generally far lower than that of deep neural networks, and will be limited by the pre-trained deep neural networks used as the baseline. Therefore, to reduce the complexity of the network while achieving both good detection precision and speed, a future development trend is to redesign a lightweight network architecture with high efficiency.

The main idea of designing a lightweight neural network architecture is to design a more efficient network structure by optimizing the computational method of convolution to effectively reduce the computational effort during convolutional computation. Iandola F N et al. [14] proposed the SqueezeNet network, which is constructed by the basic module of Fire module, in which the Squeeze layer uses 1×1 convolutional kernels to reduce the number of parameters, and the Expand layer uses 1×1 and 3×3 convolutional kernels to obtain the corresponding feature maps and then stitch them together to obtain the output of the Fire module. However, this has a weak efficiency for complex problems due to the low number of parameters. Zhang X et al. [15] improved the conventional residual structure by replacing the first 1×1 convolution layer that is connected to the input feature map with grouped convolution and using the Channel shuffle operation to interact information on the output results of each group of grouped convolutions, achieving a network performance with reduced computation. However, this produces a boundary effect; that is, an output channel comes from a small part of the input channel. Howard A.G [16] proposed replacing the general convolution layer with the depth separable convolution, which greatly reduces the network parameters; however, feature degradation occurs during the training process. Sandler M et al. [17] proposed resolving problem of feature degradation in MobileNetv1 using the inverted residual structure "dimension raising-convolution-dimension reduction", and the depth separable convolution is used to reduce the calculation amount in the intermediate convolution operation and improve the efficiency of the convolution calculation. However, this convolution may not extract enough information about a small number of deep-level features.

Figure 1 clearly demonstrates the superiority of our model in terms of the number of network parameters and the average detection degree. Our algorithm achieved the highest detection accuracy with the lowest amount of network parameters, and has a good application prospect.

In this paper, based on previous research, we further explore the design of lightweight and high-precision object detection networks. Our contributions are as follows: (1) embed coordinate attention (CA) into the lightweight backbone feature extraction network MobileNetv2, which improves the backbone feature extraction network's ability to initially extract features; (2) accurately integrate the improved combination of the backbone feature extraction network and the multi-scale feature fusion network (PANet), which further enriches the network's ability to extract location information and semantic information; (3) fast and accurate target detection is realized on the autonomous driving dataset KITTI, which is beneficial to the algorithm during automatic driving applications.

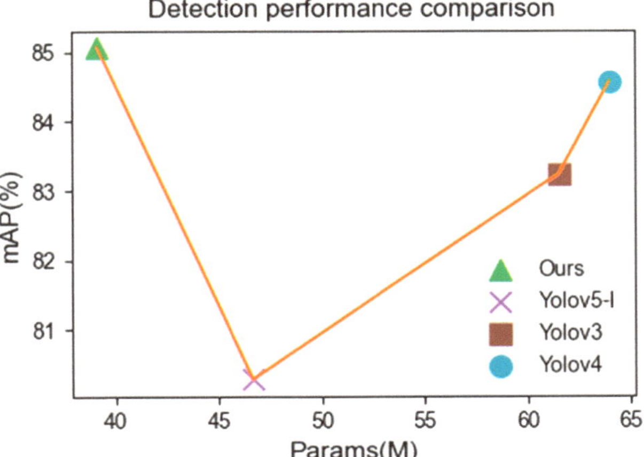

Figure 1. Advanced model performance comparison chart.

2. Algorithm Design

To achieve a lightweight object detection network with high precision and speed for object detection, a lightweight network model based on MobileNetv2_CA was proposed. The overall structure of the network model is shown in Figure 2. First, the improved lightweight Mobilenetv2 [17] network was used as the backbone feature extraction network for preliminary feature extraction, and the coordinated attention mechanism (CA) [18] was embedded to adaptively extract the channel relationships and location information to enrich feature extraction. Then, features at different scales were used as inputs for maximum pooling at different scales in the SPP [19] module to isolate significant contextual features, and the features were fed into the PANet [20] network for further enhanced feature extraction and multi-scale fusion. Finally, the three enhanced feature layers that were processed by, and input into, several convolutions were processed by 3×3 convolutional feature integration and 1×1 convolutional adjustment of the number of channels using the Yolo [5] detection head to obtain the prediction results. After decoding, score ranking and a non-maximum suppression screening of the prediction results, the detection results were obtained. The entire network combines the advantages of lightweight convolutional neural networks, a coordinate attention (CA) mechanism and the PANet multi-scale feature fusion network to construct an efficient and lightweight object detection network.

2.1. Pretreatment Stage

To reduce the network's demand for device memory during feature extraction, and to enhance the feature extraction of small object scenes and multiple complex scenes, a novel Mosaic image enhancement method inspired by Yolov4 [21] was used in the data pre-processing stage. We experimentally determined the optimal scaling factor, original image warping ratio and channel warping for KITTI images. Scale represents the scaling ratio of the original image, which was set between 0.25 and 2 in this article. The jitter represents the distortion ratio of the width and height of the original image. In this article, we set jitter = 0.3, which means the distortion was between 0.7/1.3 (0.538) and 1.3/0.7 (1.857). Hue = 0.1, sat = 0.5 and val = 0.5, which represent the distortion of the three channels in the hsv color gamut, namely, hue (H), saturation (S) and lightness (V), respectively. These hyperparameters are related to image size, object class and density. The distribution of small object scenes and complex scenes in datasets was generally uneven, which meant that the learning of small objects and complex scenes in the training process was always less than adequate. By using the Mosaic image enhancement method, the possibility of obtaining multiple complex scenes and small object scenes after traversing four images greatly increases, which helps the network to accurately detect small objects and multiple complex scene targets by learning object features during the training process. The test results are shown in Figure 2.

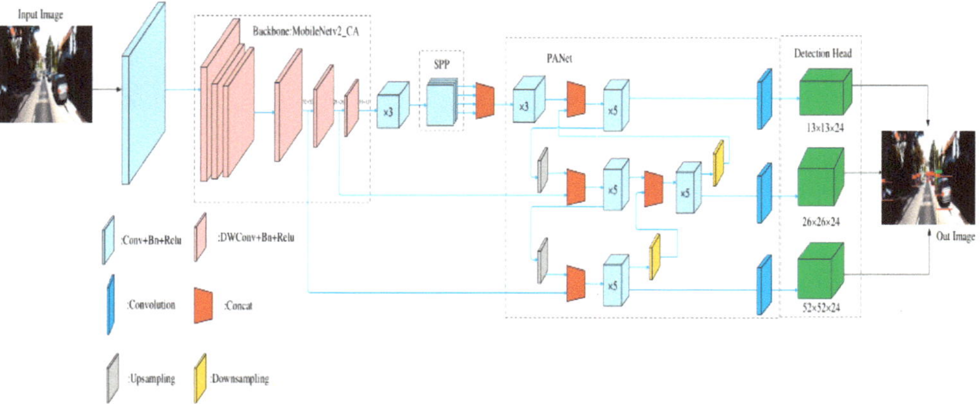

Figure 2. The input image is passed through the backbone feature extraction network to the multi-scale pooling layer, multi-scale fusion is performed, and, finally, the detection result is output.

Figure 3a shows the effect of data enhancement for small object scenes, and the small object in the distance was successfully captured. Figure 3b shows the effect of data enhancement for obscured objects and objects in low-light-state scenes, and the object in an obscured and low-light state was successfully captured. Figure 3c is the detection effect diagram of our model with Mosaic data enhancement turned off, and Figure 3d is the detection effect diagram of our model with Mosaic data enhancement turned on, which confirms that, if there is no data enhancement of the images, the detector will suffer from missed detections.

2.2. Improvement in Mobilenetv2 Backbone Network

To achieve the goal of a lightweight network, the backbone feature extraction network uses and improves MobileNetv2 [17]. The MobileNetV2 network model has a total of 17 Bottleneck layers (each Bottleneck contains two point-by-point convolutional layers and one depth convolutional layer), one standard convolutional layer (conv) and two point-by-point convolutional layers (pw conv), with a total of 54 trainable parameter layers. This is a series of Bottleneck layers stacked with different operation step stride(s). For repeated Bottleneck layers, generally only the first layer of s is 2, and the rest of the layers are 1.

When the Stride is 2, MobileNetv2 Bottleneck optimizes the network with a Linear Bottleneck Block. The fewer the number of channels in the feature map, the less the computation of the convolutional layer. However, it is difficult to extract enough feature information using only feature maps with a low number of channels. MobileNetV2 establishes a trade-off between these two, taking the 1×1 pointwise convolution to raise the dimension and using the Relu6 activation function instead of the Relu activation function. This is followed by 3×3 depth convolution and the use of the Relu6 activation function; then, 1×1 pointwise convolution is used to down-dimension and, after down-dimensioning, the linear activation function is used, the structure of which is shown in Figure 4. This convolution operation is beneficial to reduce the number of parameters.

Figure 3. Test effect of Mosaic image enhancement method in different scenarios. (**a**) Small object scenes; (**b**) obscured objects and objects in low-light state scenes; (**c**) with the Mosaic data enhancement detection effect diagram closed; (**d**) with the Mosaic data enhancement detection effect diagram open.

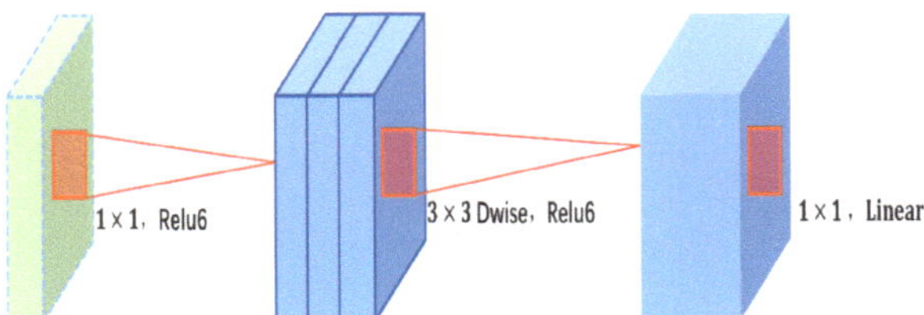

Figure 4. Linear Bottleneck Block.

When the Stride is 1, MobileNetv2 Bottleneck optimizes the network using the Inverted Residuals Block to enhance the feature extraction ability of MobileNetv2, a lightweight backbone feature extraction network, and causes the network to adaptively adjust the inter-channel relationship and location information to improve feature extraction. In this paper, the coordinate attention (CA) [18] mechanism proved to be an efficient attention mechanism. To demonstrate its advantages, different attention mechanisms are embedded in the same backbone network and trained and tested on the same dataset. The Pascal VOC [22] test set was used, and the object detection results are shown in Figure 5. When the same lightweight object detector—Yolov4Tiny—was used, the network adding our coordinated attention mechanism CA achieved the highest mAP of 77.54%, which is higher than the results achieved by SE [23], at 77.32% mAP, and CBAM [24], at 77.27% mAP.

However, the parameter amount in the CA network was only 5.983 M, which is lower than the 6.004 M achieved by CBAM. Compared with SE, CA only increases the parameter amount by 0.02 M, but increases mAP by 0.22%. Compared with the other two attention mechanisms, SE and CBAM, the CA attention mechanism achieves the best detection at the cost of a small increase in the amount of parameters and is highly competitive. Inspired by this, we inserted this mechanism into the backbone network Mobilenetv2.

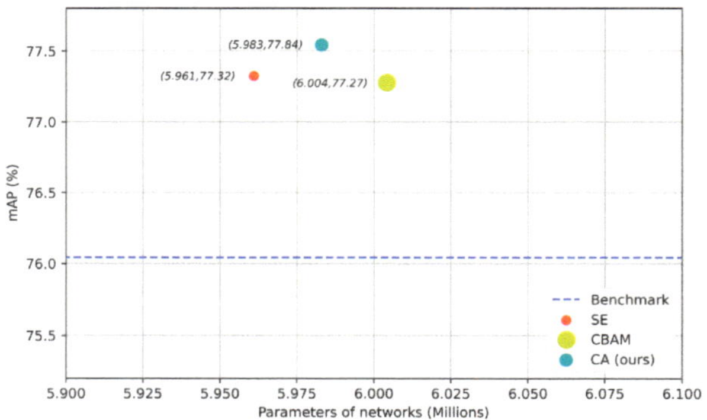

Figure 5. Performance comparison of different attention mechanisms.

We embedded the mechanism in an Inverted Residuals Block, which is located between 3×3 deep convolution and 1×1 reduced-dimensional point convolution, forming the Inverted Residuals_CA Block. This was embedded into the Inverted Residuals Block using the following steps:

(1) In the channel attention, input X is given. Each channel was encoded along the horizontal and vertical coordinates using two spatial range pool kernels, (H,1) and (1, W), respectively. The output of the C-th channel at height H can be determined using Formula (1):

$$z_c^h(h) = \frac{1}{W} \sum_{0 \leq i < W} x_c(h, i) \tag{1}$$

Similarly, the output of the C-th channel with width W can be determined using Formula (2):

$$z_c^w(w) = \frac{1}{H} \sum_{0 \leq j < H} x_c(j, w) \tag{2}$$

The above two transformations aggregate features along two spatial directions, which helps the network to more accurately locate the objects of.

(2) After enabling the global receive field and encoding the exact location information using the aggregated feature mappings generated by Formulas (1) and (2), these were concatenated and input to the 1×1 convolutional transformation function f to generate Formula (3):

$$f = \delta\left(F_1\left(\left[z^h, z^w\right]\right)\right) \tag{3}$$

wherein [] denotes the cascade operation along the spatial dimension, δ is the nonlinear activation function, $f \in R^{C/r \times (H+W)}$ is the intermediate feature mapping that horizontally and vertically encodes spatial information and r is the reduction rate of the control block size. f was divided into two independent tensors, $f^h \in R^{C/r \times H}$ and $f^w \in R^{C/r \times W}$, along the spatial dimension. The other two 1×1 convolution transformations, F_h and F_w, were used to transform f^h and f^w into tensors with the same number of channels, which can be used to obtain Formulas (4) and (5), respectively:

$$g^h = \sigma\left(F_h\left(f^h\right)\right) \tag{4}$$

$$g^w = \sigma(F_w(f^w)) \tag{5}$$

wherein σ is sigmoid function. Outputs g^h and g^w were expanded. The output of the coordinate attention block Y can be formulated as Equation (6):

$$y_c(i,j) = x_c(i,j) \times g_c^h(i) \times g_c^w(j) \tag{6}$$

Each element in the two attention mappings reflects the presence or absence of the object of interest in the corresponding row and column.

The structure of the Inverted Residuals_CA Block after embedding the Coordinate Attention (CA) mechanism is shown in Figure 6. This can first aggregate features along two spatial directions, and then encode the generated feature maps as a pair of direction-aware and position-sensitive attention maps. These attention maps can be complementarily applied to the input feature map to enhance the representation of the objects of interest and fully extract the feature information of the image.

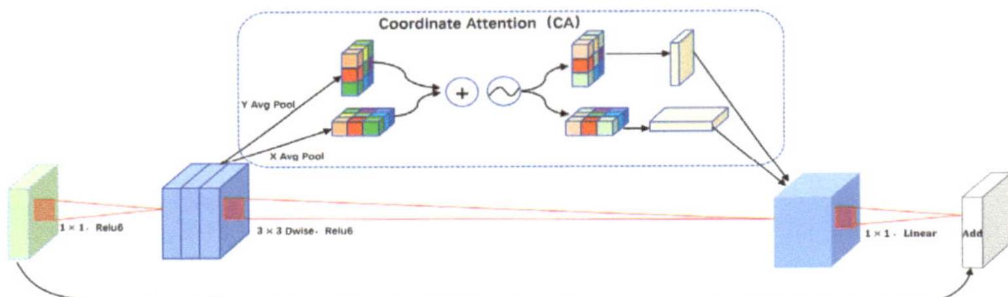

Figure 6. The overall structure diagram after embedding the coordinated attention mechanism, called the Inverted Residuals_CA Block.

After embedding the CA mechanism, the improved MobileNetv2 was renamed the MobileNetv2_CA network. A structural table is shown in Table 1. Input indicates the size of each module input in the network; operator denotes the modules experienced by each feature layer; t indicates the expansion factor, which is always applied to the input size; c indicates the number of channels output after passing through each feature layer; n indicates the number of bottlenecks; and s indicates the first-layer stride of each operation.

Table 1. Overall structure of Mobilenetv2_CA.

Input	Operator	t	c	n	s
$416^2 \times 3$	Conv2d	-	32	1	2
$208^2 \times 32$	Bottleneck	1	16	1	1
$208^2 \times 16$	Bottleneck	6	24	2	2
$104^2 \times 24$	Bottleneck	6	32	3	2
$52^2 \times 32$	Bottleneck	6	64	4	2
$26^2 \times 64$	Bottleneck	6	96	3	1
$26^2 \times 96$	Bottleneck	6	160	3	2
$13^2 \times 160$	Bottleneck	6	320	1	1
$13^2 \times 320$	Conv2d1 $\times$ 1	-	1280	1	1
$7^2 \times 1280$	Avgpool7 $\times$ 7	-	-	1	-
$1 \times 1 \times 1280$	Conv2d1 $\times$ 1	-	k	-	

The network adaptive adjustment channel relationship and location information was extracted by MobileNetv2_CA backbone features, and the effective feature layer with the c = 32 output channels was extracted using the Inverted Residuals_CA Block at the seventh layer of the network when s = 1. The effective feature layer with c = 96 output channels was extracted using the Inverted Residuals_CA Block at the fourteenth layer of the network when s = 1. The effective feature layer with c = 320 output channels was extracted using the Inverted Residuals_CA Block at the eighteenth layer of the network when s = 1. The interaction between different channels' information at different scale feature layers, and the aggregation of different directional features, was achieved. Three feature layers for different scales with different feature directivities could be preliminarily extracted and used as the basic features of the next multi-scale feature fusion.

2.3. Multi-Scale Feature Fusion and Post-Processing

The three different scales of feature layers extracted by the backbone feature extraction network were located in the middle layer, the lower middle layer and the bottom layer. In the feature extraction process, the feature information extracted by the shallow layer network and the deep layer network was different: the shallow layer feature learning was more focused on the location information of the object, while the deep layer feature learning was more focused on the semantic information of the object. To make full use of the feature information contained in the feature maps at different scales, the feature information at different scales, extracted from the seventh, fourteenth and eighteenth layers of the network, was input into PANet network [20] for feature fusion to obtain rich feature information. As shown in Figure 7, in this paper, the PANet network first spliced the features extracted from the 13 × 13 scale of the eighteenth layer of the MobileNetv2_CA backbone network with the Concat tensor, convolved the pooled feature layers several times for upsampling, and spliced them with the features extracted from the 26 × 26 scale of the fourteenth layer of the backbone feature network using the Concat tensor after convolution. After further convolutional processing, another upsampling processing was performed. The features extracted from the 52 × 52 scale of the seventh layer of the backbone network after convolutional processing were spliced using the Concat tensor to create a path incorporating rich location information. Then, the features extracted from the 52 × 52 scale of the seventh layer of the backbone network were downsampled and Concat tensor spliced with the features extracted from the 26 × 26 scale of the fourteenth layer of the backbone network. After convolutional processing, another downsampling process was performed to splice the features extracted from the 13 × 13 scale of the eighteenth layer of the network using the Concat tensor, thus creating a path that incorporates rich semantic information. Finally, the feature fusion network fused deep and shallow feature information to achieve feature enhancement, avoiding the information loss caused by the use of single features.

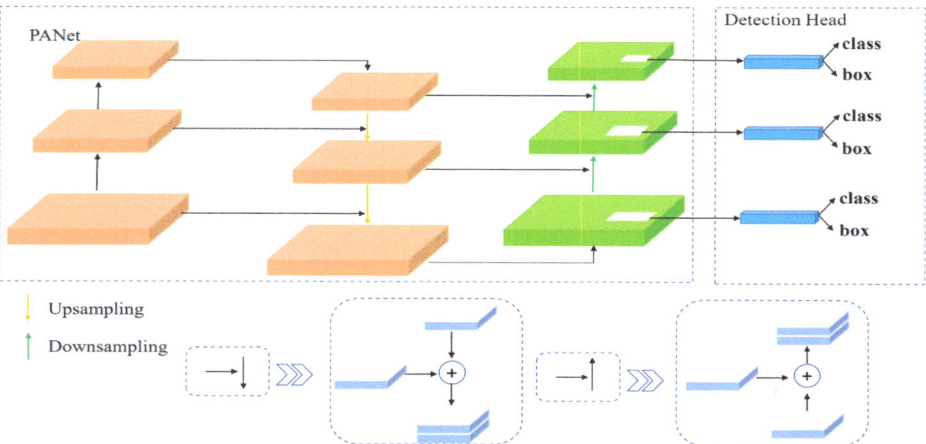

Figure 7. Multi-scale feature fusion and prediction were used to fuse location and semantic information at different scales.

The fully fused feature information was input into Yolo Head [5] for 3 × 3 convolutional feature integration and 1 × 1 convolutional adjustment of the number of channels. Each effective feature layer divided the whole image into grids corresponding to length and width. Multiple a priori boxes set by the network were built at the center of the grid, and the network judged whether these boxes contain objects and the type of object to output prediction results. The prediction results were decoded to obtain the center and size of the prediction frame, which enabled object localization regression. The object detection results were obtained by sorting the obtained prediction results into scores, removing detection results with scores greater than a set threshold in each category and filtering them according to NMS non-maximum suppression using the confidence level.

3. Experiments

3.1. Dataset and Experimental Environment

The experiments in this paper used the VOC [22] dataset and KITTI [25] dataset for validation. The KITTI dataset contains 6000 images for training and validation, and 1481 images for testing. The KITTI dataset has three classes: car, pedestrian and cyclist. This dataset has a total of 7481 pictures, which were divided according to the ratio of the training set + validation set: test set = 9:1. 5984 pictures were taken as the training set, 748 pictures were used as the verification set and 749 pictures were used as the test set. For the VOC dataset, this paper mainly used four of these categories: car, bus, motorbike and person. This dataset contained a total of 9963 images, which were divided according to the ratio of training set + validation set: test set = 9:1. 7970 images were taken as the training set, 997 images were used as the validation set and 996 images were used as the test set. In Table 2, the upper part shows the division of the Kitti dataset, and the lower part shows the division of the VOC.

Table 2. Dataset partition details.

Class	Number	Car	Cyclist	Pedestrian	
train&val	6732	30,005	1480	4277	
test	749	3256	147	432	
Class	Number	Car	Bus	Motorbike	Person
train&val	8967	11,167	1329	1522	41,146
test	996	1201	213	325	4528

Our computation environment for inference included a 4 cores Silver 4110 CPU, 16 GB of DDR4 memory and a GTX 2080 Ti GPU, the software environment includes Python 3.8, cuda11.0 and Pytorch1.7.0.

3.2. Optimizer and Loss Function

Our network was pre-trained on the ImageNet [26] dataset using Mobilenetv2 after embedding coordinate attention (CA) as the backbone of the entire network. In this paper, experiments were conducted on the KITTI dataset. The concrete process is as follows: the Adam [27] optimizer was used to train the network on the training set. During the first 50 epochs, the training of the backbone network was frozen, and only the model was slightly adjusted. The feature extraction network did not change. The initial learning rate was set to 10^{-3} and the batch size was set to 16. After 50 epochs, the network backbone training was unfrozen. After the feature extraction network changed, all the parameters of the network would then change. The initial learning rate was 10^{-4} and the batch size was 8. The multiplication factor and weight attenuation of the updated learning rate were 0.94 and 0.0005, respectively. The IoU threshold was set to 0.5. The mean average precision (mAP) of each class was evaluated on the validation set.

The loss function is an important index to evaluate the training effect of the network, and the corresponding loss function in this paper consisted of three parts: the location loss function ($Loss_{cIou}$), the confidence loss function ($Loss_{con}$) and the class loss function ($Loss_{cls}$). The total loss function was $Loss = r_1 Loss_{loc} + r_2 Loss_{con} + r_3 Loss_{cls}$, where r_1 represents the localization loss function coefficient, r_2 represents the confidence loss function coefficient and r_3 represents the category loss function coefficient. $r_1 = 0.05, r_2 = \frac{input_{[0]} \times input_{[1]}}{416^2}$,

$r_3 = 1 \times \frac{numclasses}{80}$. In different datasets, the values of r_1 and r_2 will change according to the specific situation. Additionally, the loss function of confidence and class adopted the cross-entropy loss function, and the loss function of location adopted CIoU [28] loss. CIoU loss function makes the object box regression more stable, and it does not face problems such as scattering during IoU training, is sensitive to scale transformation and converges faster than IoU loss function. Then, for the bounding box localization loss $Loss_{loc}$, the IoU loss function needs to be replaced by the CIoU loss function, which can be expressed as:

$$\begin{cases} \alpha = \frac{v}{1-IoU+v} \\ v = \frac{4}{\pi^2}\left(\arctan\frac{\omega^{gt}}{h^{gt}} - \arctan\frac{\omega}{h}\right)^2 \end{cases} \tag{7}$$

where ω and ω^{gt} and h and h^{gt}, respectively, represent the width and height of the prediction frame and the true frame. The loss function $Loss_{loc}$ of the corresponding CIoU can be expressed as:

$$Loss_{loc} = 1 - CIoU = 1 - IoU + \frac{\rho^2\left(b, b^{gt}\right)}{c^2} + \alpha v \tag{8}$$

The prediction confidence loss function $Loss_{con}$ is measured by the cross-entropy loss function, which can be expressed as:

$$\begin{aligned} Loss_{con} = &- \sum_{i=0}^{s^2} \sum_{j=0}^{B} I_{i,j}^{obj}\left[\hat{C}_i^j \log\left(C_i^j\right) + \left(1 - \hat{C}_i^j\right)\log\left(1 - C_i^j\right)\right] \\ &- \lambda_{noobj} \sum_{i=0}^{s^2} \sum_{j=0}^{B} I_{i,j}^{noobj}\left[\hat{C}_i^j \log\left(C_i^j\right) + \left(1 - \hat{C}_i^j\right)\log\left(1 - C_i^j\right)\right] \end{aligned} \tag{9}$$

where s^2 is the number of divided grids and B is the number of prior frames contained in each grid. $I_{i,j}^{obj}$ and $I_{i,j}^{noobj}$ indicate whether the j-th prior frame of the i-th grid contains an object and whether it is set to 1 or 0. λ_{noobj} is the confidence error weight of the prior frame without objects; it has an extremely small value because positive samples and negative samples in prior frame are extremely unbalanced (there are very few prior frames with objects).

The class loss $Loss_{cls}$ was used to measure the class error between the prediction frame and the true frame, and was measured by the cross-entropy loss function, which can be expressed as:

$$Loss_{cls} = - \sum_{0}^{s^2} I_{i,j}^{obj} \sum_{c \in classes}\left[\hat{P}_i^j \log\left(P_i^j\right) + \left(1 - \hat{P}_i^j\right)\log\left(1 - P_i^j\right)\right] \tag{10}$$

where P and $\hat{P}$ represent the class probabilities of the prediction frame and the true frame, respectively.

3.3. Evaluation Indexes

General object detection evaluation indexes were adopted: precision (P), recall rate (R), average precision (AP) and mean average precision (mAP), represented by:

$$P = \frac{TP}{TP + FP} \tag{11}$$

$$R = \frac{TP}{TP + FN} \tag{12}$$

$$mAP = \frac{\sum_{i=1}^{k} AP_i}{k} \tag{13}$$

$$AP = \int_{0}^{1} P(R)dR \tag{14}$$

where true positive (TP) represents the true case, false positive (FP) represents the false positive case, false negative (FB) represents the false negative case and k is the category.

In this paper, the recall R was divided into six points (0, 0.2, 0.4, ..., 1.0), and the precision P under each recall was obtained according to the calculation of PASCAL VOC [22]. Then, the P–R curve was obtained, and the area enclosed by the P–R curve and the coordinate axis denotes the AP value. For each class, the AP was calculated in the above manner, and the average of the AP for all classes was the mAP.

3.4. Analysis of Experimental Results

To verify the effectiveness of using the Mosaic image enhancement method and adding the coordinate attention (CA) mechanism to Mobilenetv2—the lightweight backbone feature extraction network—ablation experiments were designed for different modules. Ablation experiments were performed on two different datasets: KITTI and VOC.

Performing ablation experiments of the same module using a variety of different datasets can effectively verify whether each module can adapt to different task requirements in different scene datasets, and verify the actual contribution of each module to the network.

$\sqrt{}$ in Table 3 represents the participation of the module. This module is represented in Table 3. A denotes use of the Mobilenetv2 backbone network, B is with the addition of the coordinate attention (CA) mechanism to the backbone network, and C is with the addition of the Mosaic image enhancement to the data pre-processing stage. To verify that using the Mosaic image enhancement method and a lightweight backbone network incorporating a coordinated attention (CA) mechanism can improve the precision of network detection, a comparative experiment was designed.

Table 3. Ablation experiment.

	Model	A	B	C	mAP (%)	Promotion Rate (%)
	Mobilenetv1-PANet				81.86	/
KITTI	Mobilenetv2-PANet	$\sqrt{}$			84.56	2.70
	Mobilenetv2-CA-PANet	$\sqrt{}$	$\sqrt{}$		84.79	2.93
	Mobilenetv2-CA-M-PANet	$\sqrt{}$	$\sqrt{}$	$\sqrt{}$	**85.07**	**3.21**
	Mobilenetv1-PANet				79.72	/
VOC	Mobilenetv2-PANet	$\sqrt{}$			80.12	0.4
	Mobilenetv2-CA-PANet	$\sqrt{}$	$\sqrt{}$		81.12	1.4
	Mobilenetv2-CA-M-PANet	$\sqrt{}$	$\sqrt{}$	$\sqrt{}$	**81.43**	**1.71**

Networks with the network backbone Mobilenetv1, networks with the network backbone Mobilenetv2, networks with a coordinated attention (CA) mechanism added to backbone Mobilenetv2 and networks with a coordinated attention (CA) mechanism added to backbone Mobilenetv2 using Mosaic image enhancement were trained separately and evaluated on two different datasets: KITTI and VOC. $\sqrt{}$ in Table 3 represents the participation of this module. By using Mobilenetv2 as the backbone network, compared with the Mobilenetv1 backbone network, the accuracy improved by 2.7% on the KITTI dataset and 0.4% on the VOC dataset, but the detection speed was almost unchanged. Mobilenetv2 is shown to be an efficient and lightweight backbone network. After adding the attention mechanism CA to Mobilenetv2, the network considered both the relationship between the channels and the location information, which enhances the network's sensitivity to direction and location information, strengthens the extraction of important features, and improves the network's learning ability. The network accuracy was further improved—by 0.23% on the KITTI dataset and by 1% on the VOC dataset—proving the contribution of CA. After using the Mosaic image enhancement method, the network can detect objects outside the normal context, making the feature extraction of various objects more complete and continuous, allowing the attainment of more abundant image features. The detection accuracy was further improved on the KITTI dataset by 0.28%. On the VOC dataset, the detection accuracy was further improved by 0.31%, proving the contribution of Mosaic image enhancement. Compared with the original network, using

Mobilenetv1 as the backbone, the detection accuracy on the KITTI dataset was increased by 3.21%, and the detection accuracy on the VOC dataset increased by 1.71%, which improved the overall generalization performance of the network.

There are many kinds of current object detection networks. To verify the effectiveness of the networks in this paper, a variety of object detection networks were selected for experimental comparison on the KITTI dataset. Faster R-CNN [3], SSD [4], and YOLOv3 [5], YOLOv4 [21], YOLOv5-l [29] and YOLOv4-tiny [30] algorithms were selected to be trained on the KITTI dataset and compared with this detection network on the test set, focusing on average class accuracy, the number of model parameters, and detection speed.

The parameters in Table 4 refer to the number of parameters of the network, and FPS refers to the number of images processed by the network each second, which can be used to characterize the network's detection speed performance. As can be seen in Table 4, compared with the Faster R-CNN algorithm, the detection speed of this paper was 3.1 times faster, the detection accuracy of the car improved by 15.14%, the detection accuracy of pedestrians improved by 18.71%, the detection accuracy of cyclists improved by 22.52% and the average detection accuracy (mAP) improved by 18.79%. However, the number of participants significantly decreased by 28.8%. Compared with the SSD algorithm, although the detection speed of this paper's algorithm was reduced, the detection accuracy for cars improved by 9.57%, for pedestrians by 28.07%, for cyclists by 23.22% and the average detection accuracy (mAP) improved by 20.23% for the same number of model parameters. Compared with the YOLOv3 algorithm, the detection speed is slightly slower, the detection precision of cars is slightly higher, the detection precision of pedestrians improved by 2.73%, the detection precision of cyclists improved by 2.75%, the average detection precision (mAP) of this paper increased by 1.84% and the number of participants was only 63.8% of the YOLOv3 algorithm's total number of participants. The network performance was better than that of the YOLOv3 algorithm. The network performance of this method was better than that of YOLOv3. Compared with YOLOv4, our algorithm improved by 0.67% in car detection accuracy, with almost the same level of accuracy as when detecting pedestrians. Additionally, an improvement of 1.05% was obtained in the detection accuracy of cyclists, and 0.51% in the average detection accuracy. Compared with YOLOv5, our algorithm improved by 6.23% in the detection accuracy of cars, 2.97% in the detection accuracy of pedestrians, 5.15% in the detection accuracy of cyclists and 4.79% in average detection accuracy. Compared with YOLOv4-tiny, our algorithm improved the average detection accuracy by 24.32%. We focused more on high detection accuracy than on increasing the detection speed. Taken together, the number of network parameters was significantly reduced in this paper, achieving the goal of excellent detection precision with a lightweight network.

Table 4. Performance comparison of different detection networks on KITTI dataset.

Model	AP_{car}	$AP_{pedestrian}$	$AP_{cyclist}$	mAP (%)	Parameters	FPS
Faster R-CNN [3]	80.22	53.15	65.46	66.28	136.7 M	10.33
SSD [4]	85.79	43.79	64.76	64.84	23.9 M	72.36
YOLOv3 [5]	95.34	69.13	85.23	83.23	61.5 M	45.03
YOLOv4 [21]	94.69	**72.05**	86.93	84.56	63.95 M	34.79
YOLOv5-l [29]	89.13	68.89	82.83	80.28	46.64 M	39.51
YOLOv4-tiny [30]	82.11	39.80	61.13	60.75	**8.879 M**	**129.32**
MobileNetv2_CA (ours)	**95.36**	71.86	**87.98**	**85.07**	39.1 M	31.84

In autonomous driving scenarios, cars, pedestrians and cyclists are the most common detection objects, and it is important to achieve high precision detection on the basis of real-time detection. During object detection, the real-time detection requirement is satisfied when the number of images per second that are processed by the network is more than 30 FPS. Therefore, the requirement of real-time detection can be met. Table 4 shows that that the Faster R-CNN network does not meet the requirements of real-time detection, and the SSD network, although it achieved a faster detection speed, has a detection accuracy that is too low to meet the requirements of high-precision detection. Both the YOLOv3 network and the MobileNetv2_CA network in this paper can achieve real-time and high-precision detection. Therefore, the detection precision of the two networks are further compared in Table 5, using the KITTI validation set and three different evaluation criteria. It can be seen that, under the three different criteria, the proposed network has a 2.25%, 1.38% and 0.9% higher detection precision than the YOLOv3 in the car class. The precision of pedestrian detection is 8.8%, 7.65% and 0.8% higher than that of YOLOv3, respectively. The detection precision of cyclists is 0.95%, 0.53% and 0.88% higher than that of YOLOv3, respectively. In conclusion, the detection precision of all the networks in this paper is better than that of the YOLOv3 network, and the number of parameters is only 63.58% of that of the YOLOv3 network. Although the FPS is slightly lower than that of the YOLOv3 network, because the Coordinate Attention (CA) mechanism added in this paper takes time to extract rich features, the detection precision of this network is higher if the real-time detection FPS is greater than 30 FPS. A scheme combining a better detection speed and precision was achieved under the premise of ensuring a lightweight network.

Table 5. Performance comparison using KITTI validation set (where Easy, Moderate and Hard refer to the precision under easy, moderate and hard standards, respectively).

		Average Precision (%)		
Model	Standard	Car	Pedestrian	Cyclist
	Easy	83.72	68.97	89.66
YOLOv3 [5]	Moderate	85.51	60.52	80.57
	Hard	77.16	59.33	80.20
	Easy	**85.97**	**77.77**	**90.61**
MobileNetv2_CA (ours)	Moderate	**86.89**	**68.17**	**81.10**
	Hard	**78.06**	**60.13**	**81.08**

In the relevant traffic scenes of the autonomous driving KITTI dataset, vehicles (car) and pedestrians (pedestrian) are the main detection objects, and the pedestrian (pedestrian) class is a small target compared to the vehicle (car). During the data pre-processing stage, the Mosaic data enhancement method is used, and the network enhances the sensitivity of the network, allowing for it to detect small targets through pre-processing methods such as scaling the target image. When the backbone feature extraction network extracts relevant features, it coordinates the attention mechanism. In this algorithm, according to the uneven distribution of the large and small objects in the image, the information between the channels can be cooperatively processed, and more attention is paid to the adaptive extraction of the position information of objects of different sizes compared with the comparison algorithm without the attention mechanism. This increases the location information extraction of small targets. Therefore, compared with the vehicle (car) class detection accuracy, there were 2.25%, 1.38% and 0.9% gains under the different standards. In the pedestrian (pedestrian) class detection accuracy, higher gains of 8.8%, 7.65% and 0.8% were obtained under different standards. To show the difference in time taken by different layers in specific detection tasks in more detail, this paper split and refined different parts of the model for time series analysis.

Experiments on a single dataset may not be very convincing, and more datasets need to be tested and verified to draw better conclusions. Therefore, we selected several categories that are more common in autonomous driving scenarios for comparison. The results show that, on the Pascal Voc2007 + 2012 dataset, as show in Table 6, our method improved the bus category by 0.29% compared to Faster R-CNN. The results were also 0.76% higher than SSD, 1.66% higher than Yolov3, 2.22% higher than Faster R-CNN in the motorbike category, 1.62% higher than SSD and 2.07% higher than Yolov3. In the person category, compared with Faster R-CNN, accuracy improved by 0.06%—by 5.31% compared with SSD and by 0.1% compared with Yolov3. Compared with Faster R-CNN on mAP, accuracy improved by 1.15%, was 3% higher than SSD and was 2.07% higher than Yolov3.

Compared with YOLOv4, our algorithm showed little difference in terms of detection accuracy and detection speed, but the number of parameters in our model was almost half that of YOLOv4. Compared with YOLOv5, our algorithm improved the average detection accuracy by 6.26%. Compared with YOLOv4-tiny, our algorithm improved the average detection accuracy by 4.89%.

Table 6. Pascal Voc2007 + 2012 dataset detection performance.

Model	AP_{car}	AP_{bus}	$AP_{motorbike}$	AP_{person}	mAP (%)
Faster R-CNN [3]	88.71	86.69	84.81	85.37	80.28
SSD [4]	88.37	86.22	85.41	80.12	78.43
YOLOv3 [5]	**91.54**	85.32	84.96	85.33	79.36
YOLOv4 [21]	90.61	**92.89**	82.76	87.28	81.05
YOLOv5-l [29]	88.63	90.54	76.43	**87.96**	75.17
YOLOv4-tiny [30]	90.76	83.42	84.03	84.34	76.54
MobileNetv2_CA (ours)	91.11	86.98	**87.03**	85.43	**81.43**

As shown in Table 7, we also explored the effect of using different original image scaling ratios and width-to-height distortion ratios on the detection accuracy, and finally determined our parameter range to achieve the highest detection accuracy.

Table 7. Effect of different original image scaling ratios and width-to-height distortion ratios on detection accuracy.

r	mAP (%)
r1	84.29
r2	84.37
Ours	**85.07**

3.5. Network Detection Effect

To intuitively reflect the detection performance of this model, some pictures with a complex image environment that are difficult to distinguish were selected from the KITTI test set for detection. Figure 8 shows the small object scene. Figure 8a shows the Faster R-CNN detection effect. Although it can achieve more accurate object detection, its detection speed is too slow to meet the real-time detection requirements. Figure 8b shows the SSD detection effect, which shows a more serious leakage phenomenon and does not detect any object. Figure 8c shows the detection effect of YOLOv3 with some missed detections. Figure 8d shows the detection effect of the network in this paper, which can more comprehensively and accurately detect the emerged objects. Figure 9 shows low-light scenes. Figure 9a shows the Faster R-CNN detection effect in low-light intensity scenes, while the vehicle located on the right side of the scene in the low-light environment was missed. Figure 9b shows the SSD detection effect, where both the vehicle on the right side in the low-light environment and the vehicle on the left side, which belonged to a smaller target, were missed. Figure 9c shows the detection effect of YOLOv3, which contains a high number of parameters despite achieving real-time detection of the target. Figure 9d shows the detection effect of the network in this paper, which can precisely detect both vehicles in the low-light environment as well as vehicles belonging to smaller objects. Figure 10 shows a complex scene containing small objects, occlusion, high light intensity and low light intensity, etc. Figure 10a shows the Faster R-CNN detection effect. Although it can achieve a more accurate target detection, its detection speed is too slow to meet the real-time detection requirements. Figure 10b shows the effect of SSD detection, and the detection phenomenon was missed for vehicles at a low light intensity and rear-obstructed vehicles. Figure 10c shows the detection effect of YOLOv3, which can better detect the object, and Figure 10d shows the detection effect of this network, which can correctly and simultaneously detect all the objects at a low light intensity, high light intensity and obscured objects. In summary, the network used in this paper can quickly and accurately complete detection tasks in different scenarios; therefore, the network proposed in this paper is a real-time object detection network with a light weight and high precision. To more clearly show the difference in the detection results of different models for targets in different KITTI dataset scenes, we visualized the final detection results of different models in a heat map. The first column in Figure 11 shows the detection results of our work. It is clear that our model can accurately locate and detect hard-to-differentiate persons with the right shade in the first scenario, hard-to-differentiate two-vehicle targets with the left shade in the second scenario, and the smaller vehicle in the opposite lane and the vehicle target in the same direction in the third scenario. These scenarios took place in a poor lighting environment. In the fourth scenario, the smaller vehicle in the opposite lane and the vehicle in the same direction were accurately located and detected. This demonstrates the better robustness of our model compared to YOLOv4, YOLOv5, and YOLOv4-tiny in handling autonomous driving scenarios.

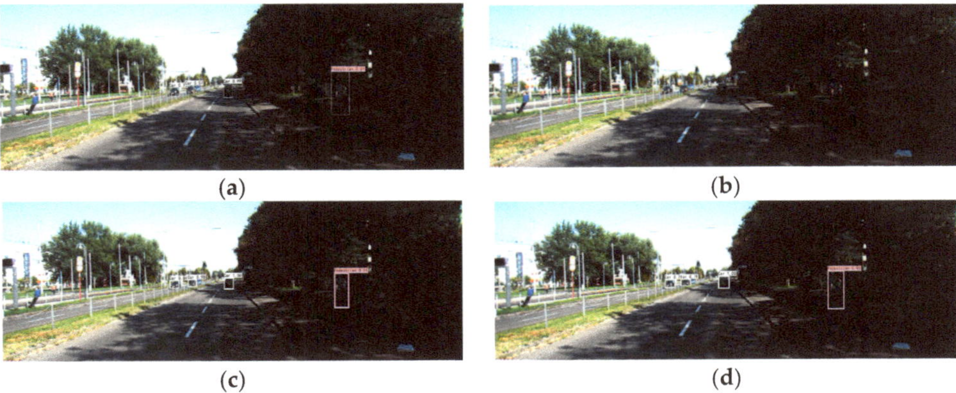

Figure 8. Small target scenes. (**a**) Faster R-CNN detection effect, (**b**) SSD detection effect, (**c**) YOLOv3 detection effect, (**d**) MobileNetv2_CA detection effect.

Figure 9. Low-light scenes. (**a**) Faster R-CNN detection effect, (**b**) SSD detection effect, (**c**) YOLOv3 detection effect, (**d**) MobileNetv2_CA detection effect.

(a)   (b)

(c)   (d)

Figure 10. Complex scenes. (**a**) Faster R-CNN detection effect, (**b**) SSD detection effect, (**c**) YOLOv3 detection effect, (**d**) MobileNetv2_CA.

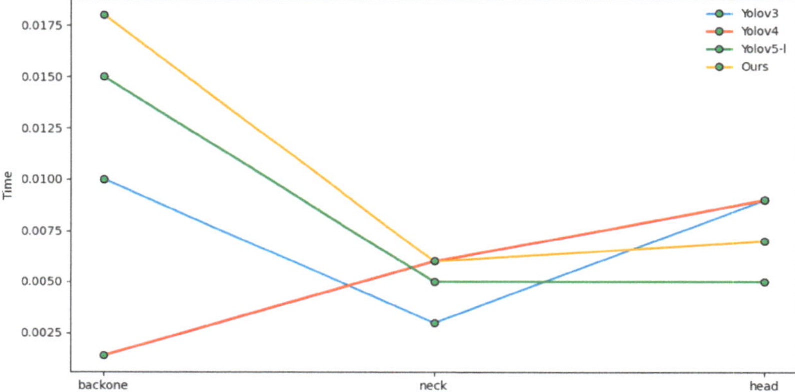

Figure 11. Time-consuming difference diagram of different modules.

After comprehensively weighing the detection accuracy and the number of network parameters, our method could achieve real-time detection. A runtime comparison of each part of various detectors is shown in Figure 11. To show the time-consuming differences between different layers during specific detection tasks in more detail, this paper split and refined the different parts of the model to analyze the runtime. Our model mainly uses the attention mechanism to extract backbone features. This is slightly more time-consuming, but real-time detection is still possible.

At the same time, in order to clearly show the superiority of our algorithm model in capturing small objects or occluded objects, we drew the heat map shown in Figure 12 below, from which it can be seen that our algorithm model uses the attention mechanism in the process of network training and learning, strengthens some signals in the heat map and suppresses other parts, making the network extraction features more accurate.

Figure 12. Heat map visualization results of detection results for different models in different scenarios in the KITTI dataset. The first column shows our results, the second column shows the results for YOLOv4, the third column shows the results for YOLOv5-l and the fourth column shows the results for YOLOv4-tiny.

4. Discussion

This paper studies and explores a lightweight object detection network, which reduces the complexity of the object detection network and can achieve efficient and accurate object detection in autonomous driving embedded platform devices. We propose a method of feature extraction, using an improved lightweight neural network as the backbone. First, the Mosaic image enhancement method was used, and an efficient lightweight backbone was used. These were combined with a coordinated attention mechanism to guide the initial feature extraction of the entire network, and then combined with a multi-scale feature fusion network to fuse image features with location information and semantic information to obtain richer image features. Finally, a multi-scale predictor was used for prediction to obtain the detection results. In this article, our related experiments show that our method has obvious advantages in terms of network parameters over the comparison network, which obtained only 39.1 M, and is even better in terms of robustness. Whether referring to the average precision of 85.07% on the KITTI dataset or the detection accuracy of 81.43% on the Voc2007 + 2012 dataset, our method achieved robustness. Regarding the detection accuracy in the automatic driving scene, which requires high detection accuracy, achieving a robust accuracy performance is of great significance. Our method also met the speed requirements of automatic driving target detection in terms of delay, reaching a 31.84 FPS detection speed performance. In terms of comprehensive computing resources, robustness and real-time performance, our method achieved the best performance coupling. As an efficient target detection algorithm, the lightweight neural network can also be applied to some other fields. For example, in the field of smart cities, our algorithm can be deployed in smart traffic lights. It can also be used to monitor the real-time flow of people and vehicles at an intersection, and intelligently change the duration of traffic lights according to the flow size.

5. Conclusions

The purpose of this paper was to explore a lightweight and efficient target detection algorithm that can be efficiently deployed on smart cars. Target detection is different when not using simple scenes. Our method was used in complex traffic scenes. We used an efficient coordination mechanism for lightweight backbone network feature extraction and fusion, focusing on optimizing pre-processing methods and multi-scale feature fusion networks. In this way, our work achieved a better performance than previous work in autonomous driving scenarios. It realized the fast and accurate detection of targets in various traffic scenarios and has great application potential in actual autonomous driving scenarios. In future work, our focus will be on further optimizing our detection speed. The next step is to continue to optimize the network model to improve the detection speed and accuracy, allowing for the network to adapt to more complex traffic scenes and achieve a robust detection in actual complex traffic scenes.

Author Contributions: P.S.: Conceptualization, Methodology, Investigation, Funding acquisition. L.L.: Software, Data Curation, Writing—Original Draft. H.Q.: Visualization, Validation. A.Y.: Writing—Review and Editing, Supervision. All authors have read and agreed to the published version of the manuscript.

Funding: This work was supported by the financial support of the key research and development projects of Anhui (202104a05020003), Anhui development and reform commission support of the R&D and innovation project ([2020]479).

Institutional Review Board Statement: Not applicable.

Informed Consent Statement: Not applicable.

Data Availability Statement: The data used in this study are available upon request.

Conflicts of Interest: The authors declare no conflict of interest.

References

1. Dalal, N.; Triggs, B. Histograms of oriented gradients for human detection. In Proceedings of the IEEE Conference on Computer Vision and Pattern Recognition, San Diego, CA, USA, 20–26 June 2005; IEEE Computer Society: San Diego, CA, USA, 2005; pp. 886–893.
2. Lowe, D.G. Distinctive image features from scale-invariant key points. *Int. J. Comput. Vision.* **2004**, *60*, 91–110. [CrossRef]
3. Ren, S.; He, K.; Girshick, R.; Sun, J. Faster r-cnn: Towards real-time object detection with region proposal networks. *Adv. Neural Inf. Process. Syst.* **2015**, *28*, 2969239–2969250. [CrossRef] [PubMed]
4. Liu, W.; Anguelov, D.; Erhan, D.; Szegedy, C.; Reed, S.; Fu, C.Y.; Berg, A.C. Ssd: Single shot multibox detector. In Proceedings of the European Conference on Computer Vision; Springer: Cham, Switzerland, 2016; pp. 21–37.
5. Redmon, J.; Farhadi, A. Yolov3: An incremental improvement. *arXiv* **2018**, arXiv:1804.02767.
6. Ge, Z.; Liu, S.; Wang, F.; Li, Z.; Sun, J. Yolox: Exceeding yolo series in 2021. *arXiv* **2021**, arXiv:2107.08430.
7. Xu, S.; Wang, X.; Lv, W.; Chang, Q.; Cui, C.; Deng, K.; Wang, G.; Dang, Q.; Wei, S.; Du, Y.; et al. PP-YOLOE: An evolved version of YOLO. *arXiv* **2022**, arXiv:2203.16250.
8. Liu, Z.; Hu, H.; Lin, Y.; Yao, Z.; Xie, Z.; Wei, Y.; Ning, J.; Cao, Y.; Zhang, Z.; Dong, L.; et al. Swin transformer v2: Scaling up capacity and resolution. In Proceedings of the IEEE/CVF Conference on Computer Vision and Pattern Recognition, New Orleans, LA, USA, 18–24 June 2022; pp. 12009–12019.
9. Han, S.; Mao, H.; Dally, W.J. Deep compression: Compressing deep neural networks with pruning, trained quantization and huffman coding. *arXiv* **2015**, arXiv:1510.00149.
10. Li, H.; Kadav, A.; Durdanovic, I.; Samet, H.; Graf, H.P. Pruning filters for efficient convnets. *arXiv* **2016**, arXiv:1608.08710.
11. Rastegari, M.; Ordonez, V.; Redmon, J.; Farhadi, A. Xnor-net: Imagenet classification using binary convolutional neural networks. In Proceedings of the European Conference on Computer Vision; Springer: Cham, Switzerland, 2016; pp. 525–542.
12. Hinton, G.; Vinyals, O.; Dean, J. Distilling the knowledge in a neural network. *arXiv* **2015**, arXiv:1503.02531.
13. Han, K.; Wang, Y.; Tian, Q.; Guo, J.; Xu, C.; Xu, C. Ghostnet: More features from cheap operations. In Proceedings of the IEEE/CVF Conference on Computer Vision and Pattern Recognition, Seattle, WA, USA, 13–19 June 2020; pp. 1580–1589.
14. Iandola, F.N.; Han, S.; Moskewicz, M.W.; Ashraf, K.; Dally, W.J.; Keutzer, K. SqueezeNet: AlexNet-level precision with 50x fewer parameters and <0.5 MB model size. *arXiv* **2016**, arXiv:1602.07360.
15. Zhang, X.; Zhou, X.; Lin, M.; Sun, J. Shufflenet: An extremely efficient convolutional neural network for mobile devices. In Proceedings of the IEEE Conference on Computer Vision and Pattern Recognition, Salt Lake City, UT, USA, 18–23 June 2018; pp. 6848–6856.

16. Howard, A.G.; Zhu, M.; Chen, B.; Kalenichenko, D.; Wang, W.; Weyand, T.; Andreetto, M.; Adam, H. Mobilenets: Efficient convolutional neural networks for mobile vision applications. *arXiv* **2017**, arXiv:1704.04861.
17. Sandler, M.; Howard, A.; Zhu, M.; Zhmoginov, A.; Chen, L.C. Mobilenetv2: Inverted residuals and linear bottlenecks. In Proceedings of the IEEE Conference on Computer Vision and Pattern Recognition, Salt Lake City, UT, USA, 18–23 June 2018; pp. 4510–4520.
18. Hou, Q.; Zhou, D.; Feng, J. Coordinate attention for efficient mobile network design. In Proceedings of the IEEE/CVF Conference on Computer Vision and Pattern Recognition, Nashville, TN, USA, 20–25 June 2021; pp. 13713–13722.
19. He, K.; Zhang, X.; Ren, S.; Sun, J. Spatial pyramid pooling in deepwise convolutional networks for visual recognition. *IEEE Trans. Pattern Anal. Mach. Intell.* **2015**, *37*, 1904–1916. [CrossRef] [PubMed]
20. Liu, S.; Qi, L.; Qin, H.; Shi, J.; Jia, J. Path aggregation network for instance segmentation. In Proceedings of the IEEE Conference on Computer Vision and Pattern Recognition, Salt Lake City, UT, USA, 18 June–23 June 2018; IEEE Computer Society: Salt Lake City, UT, USA, 2018; pp. 8759–8768.
21. Bochkovskiy, A.; Wang, C.Y.; Liao HY, M. Yolov4: Optimal speed and precision of object detection. *arXiv* **2020**, arXiv:2004.10934.
22. Everingham, M.; Eslami, S.A.; Van Gool, L.; Williams, C.K.; Winn, J.; Zisserman, A. The pascal visual object classes challenge: A retrospective. *Int. J. Comput. Vis.* **2015**, *111*, 98–136. [CrossRef]
23. Hu, J.; Shen, L.; Sun, G. Squeeze-and-excitation networks. In Proceedings of the IEEE Conference on Computer Vision and Pattern Recognition, Salt Lake City, UT, USA, 18–23 June 2018; pp. 7132–7141.
24. Woo, S.; Park, J.; Lee, J.Y.; Kweon, I.S. Cbam: Convolutional block attention module. In Proceedings of the European Conference on Computer Vision (ECCV), Munich, Germany, 8–14 September 2018; pp. 3–19.
25. Geiger, A.; Lenz, P.; Urtasun, R. Are we ready for autonomous driving? The kitti vision benchmark suite. In Proceedings of the 2012 IEEE Conference on Computer Vision and Pattern Recognition; IEEE: Manhattan, NY, USA, 2012; pp. 3354–3361.
26. Deng, J.; Dong, W.; Socher, R.; Li, L.J.; Li, K.; Fei-Fei, L. Imagenet: A large-scale hierarchical image database. In Proceedings of the 2009 IEEE Conference on Computer Vision and Pattern Recognition, Miami, FL, USA, 20–25 June 2009; IEEE: Manhattan, NY, USA, 2009; pp. 248–255.
27. Kingma, D.P.; Ba, J. Adam: A method for stochastic optimization. *arXiv* **2014**, arXiv:1412.6980.
28. Zheng, Z.; Wang, P.; Ren, D.; Liu, W.; Ye, R.; Hu, Q.; Zuo, W. Enhancing geometric factors in model learning and inference for object detection and instance segmentation. *IEEE Trans. Cybern.* **2021**, *52*, 8574–8586. [CrossRef] [PubMed]
29. Glenn Jocher. Yolov5. 2021. 1, 2, 3, 5, 6. Available online: https://github.com/ultralytics/yolov5 (accessed on 5 February 2023).
30. Wang, C.Y.; Bochkovskiy, A.; Liao HY, M. Scaled-yolov4: Scaling cross stage partial network. In Proceedings of the IEEE/CVF Conference on Computer Vision and Pattern Recognition, Nashville, TN, USA, 20–25 June 2021; pp. 13029–13038.

Article

GDAL and PROJ Libraries Integrated with GRASS GIS for Terrain Modelling of the Georeferenced Raster Image

Polina Lemenkova * and Olivier Debeir

Laboratory of Image Synthesis and Analysis (LISA), École Polytechnique de Bruxelles, Campus du Solbosch, Université Libre de Bruxelles (ULB), Avenue Franklin Roosevelt 50, 1050 Brussels, Belgium; olivier.debeir@ulb.be
* Correspondence: polina.lemenkova@ulb.be; Tel.: +32-471-86-04-59

Abstract: Libraries with pre-written codes optimize the workflow in cartography and reduce labour intensive data processing by iteratively applying scripts to implementing mapping tasks. Most existing Geographic Information System (GIS) approaches are based on traditional software with a graphical user's interface which significantly limits their performance. Although plugins are proposed to improve the functionality of many GIS programs, they are usually ad hoc in finding specific mapping solutions, e.g., cartographic projections and data conversion. We address this limitation by applying the principled approach of Geospatial Data Abstraction Library (GDAL), library for conversions between cartographic projections (PROJ) and Geographic Resources Analysis Support System (GRASS) GIS for geospatial data processing and morphometric analysis. This research presents topographic analysis of the dataset using scripting methods which include several tools: (1) GDAL, a translator library for raster and vector geospatial data formats used for converting Earth Global Relief Model (ETOPO1) GeoTIFF in XY Cartesian coordinates into World Geodetic System 1984 (WGS84) by the 'gdalwarp' utility; (2) PROJ projection transformation library used for converting ETOPO1 WGS84 grid to cartographic projections (Cassini–Soldner equirectangular, Equal Area Cylindrical, Two-Point Equidistant Azimuthal, and Oblique Mercator); and (3) GRASS GIS by sequential use of the following modules: r.info, d.mon, d.rast, r.colors, d.rast.leg, d.legend, d.northarrow, d.grid, d.text, g.region, and r.contour. The depth frequency was analysed by the module 'd.histogram'. The proposed approach provided a systematic way for morphometric measuring of topographic data and combine the advantages of the GDAL, PROJ, and GRASS GIS tools that include the informativeness, effectiveness, and representativeness in spatial data processing. The morphometric analysis included the computed slope, aspect, profile, and tangential curvature of the study area. The data analysis revealed the distribution pattern in topographic data: 24% of data with elevations below 400 m, 13% of data with depths −5000 to −6000 m, 4% of depths have values −3000 to −4000 m, the least frequent data (−6000 to 7000 m) <1%, 2% of depths have values −2000 to 3000 m in the basin, while other values are distributed proportionally. Further, by incorporating the generic coordinate transformation software library PROJ, the raster grid was transformed into various cartographic projections to demonstrate distortions in shape and area. Scripting techniques of GRASS GIS are demonstrated for applications in topographic modelling and raster data processing. The GRASS GIS shows the effectiveness for mapping and visualization, compatibility with libraries (GDAL, PROJ), technical flexibility in combining Graphical User Interface (GUI), and command-line data processing. The research contributes to the technical cartographic development.

Keywords: GDAL; PROJ; GRASS GIS; terrain; DEM; elevation; ETOPO1; geomorphometry; cartography; mapping

PACS: 91.10.Da; 92.70.Pq; 93.30.Db; 91.10.Fc

MSC: 86A30; 86A04; 86A22; 53C22; 58E10; 97N60

JEL Classification: Y91; Q00; Q01; Q5; Q50; Q55; Q56; P18; P48; N50; N57

Citation: Lemenkova, P.; Debeir, O. GDAL and PROJ Libraries Integrated with GRASS GIS for Terrain Modelling of the Georeferenced Raster Image. *Technologies* **2023**, *11*, 46. https://doi.org/10.3390/technologies11020046

Academic Editors: Imran Ahmed and Gwanggil Jeon

Received: 30 January 2023
Revised: 19 February 2023
Accepted: 28 February 2023
Published: 22 March 2023

1. Introduction

1.1. Background

Topographic modelling is a central problem in cartography, land surface modelling, geodesy, and mapping [1]. A basis form of spatial analysis, topographic modelling consists of visualisation of the land surface and plotting contours that reveal variations in elevation and enables landform delineation [2,3]. Tasks such as geomorphic gradients, hydrological modelling, or stream profiling all require topographic analysis. While Geographic Information System (GIS) spatial analysis focuses on complex problems regarding modelling the Earth's shape [4,5], the technical aspects of spatial data handling consists of various tasks such as data capture [6–8], multi-source data formatting and processing [9–11], computational analysis and modelling [12–15], and visualization and mapping [16–18].

Maps became a successful form of geospatial data processing since they help exploit the complex phenomena of diverse processes and geographic objects on the Earth's surface [19]. For instance, features such as elevation points, contour isolines, and curvature gradients are extracted from the topographic maps and Digital Elevation Model (DEM) and matched against features of another map with thematic content (e.g., hydrological drainage network). Moreover, maps and georeferenced images enable one to learn and recognise regular patterns in visual spatio-temporal datasets for interpretation and analysis of the environmental dynamics and variability [20,21].

Therefore, a vital part of the spatial data interpretation is topographic visualization and mapping performed through a cartographic workflow. Topographic data modelling comprises two subtasks: (1) retrieval the information regarding the surface of the Earth from DEM, and (2) visualization and analysis of the key parameters of the elevation and objects depicted on the maps. Each of these subtasks can be solved using GIS. Many existing applications in GIS propose solutions to topographic modelling using their diverse functionality [22–24]. Due to the variety of the existing GIS (e.g., ArcGIS, QGIS, MapINFO, Environment for Visualizing Images (ENVI), Idrisi, System for Automated Geoscientific Analyses (SAGA) GIS, Erdas Imagine), the choice of tools may face questions regarding the technical functionality of the diversity of GIS. The difficulty arises when attempting to select the GIS software and find optimal solutions to mapping. Among the most essential factors of GIS are the open source availability and functionality providing a variety of tools and embedded computational algorithms [25–27]. Since the proprietary GIS, such as Erdas Imagine and ArcGIS, are a subject of access, open source tools present more effective solutions for mapping, geoinformation processing, and spatial data analysis.

In recent years, open source tools of computer vision and machine learning for data analysis have become increasingly successful in the analysis of spatial data and diverse cartographic tasks [28]. Free availability of the open source GIS has increased their popularity and demand with reported cases in geospatial analysis [29,30]. Much progress has been made, for instance, by applications of the programming scripts that aim to automate geoinformation processing using a given dataset. Examples include artificial intelligence and big data processing, pointed out in selected studies on geoinformatics [31,32]. The benefits of the open source GIS and scripting are widely discussed, mentioning easily data integration in a number of scenarios and geospatial research, handling 2D and 3D raster datasets [33], spatial data exchange, and processing [34–36].

Open source data are ubiquitous in the Earth sciences with numerous application across the field of topographical and geographical studies [37]. Common sources of open geospatial data include topographic and bathymetric grids [38,39] using General Bathymetric Chart of the Oceans (GEBCO), Shuttle Radar Topography Mission (SRTM) DEM, ETOPO1 or ETOPO5, geophysical datasets (Earth Gravitational Model of 1996 (EGM96), EGM2008), geological data, local data from the online repositories, embedded datasets, OpenStreetMap of Google Earth imagery, remote sensing data such as Landsat TM [40], or Sentinel-2A [41] satellite images. Further, many of the topographic datasets are available in rasters of vector digital formats, accessed via the online repositories, e.g., OpenStreetMaps or Google Earth. Increasingly, the geospatial data sources are developing online repositories for their data storage, such as the National Oceanic and Atmospheric Administration (NOAA), and GloVis, providing access to remote sensing datasets, and many more. Regular access to such resources enables their use in spatial analysis and raises the question of what effective methods are there for processing these data?

With ever larger geospatial datasets generating thousands of grids and thematic layers, two particular problems have become critical: (i) how to process these data with algorithms that use automation by scripts, and (ii) which GIS software to use for processing and visualising all the original data by an optimised workflow. Recently, there has been a rapid development of scripting approaches used in geographic research and Earth sciences aimed at addressing problem (i) [42]. The use of programming methods tend to become a progressively popular perspective in view of the rising amounts of spatial data and the need to process these data accurately and rapidly. For many types of spatial data, cartographic modelling approaches hereby have to address the problem of automation while handling these data. An ever-growing pool of spatial datasets encompass the multi-format data formats that vary in resolution, source, and scale which raise challenges in handling such big geospatial data [43,44]. These data can be used for the diverse purposes in Earth science such as hydrology and geomorphology [45], soils and landscape changes [46], environmental coastal studies [47], climate changes, land cover studies, geology [48], and many more. Various GIS software (e.g., ArcGIS, QGIS, and SAGA GIS) can be applied to process open source geospatial data available online with the aim of spatial analysis.

Programming applications address this problem by replicating scripts for the treatment of different images and selected scenes of the study area in the topographic map. Other approaches go one step further by automating the individual cartographic steps and explicitly model spatial data using a modular structure as in the Generic Mapping Tools (GMT). Others use extensions with the pre-written scripts in the American Standard Code for Information Interchange (ASCII) format as Unix shell scripts in the Geographic Resources Analysis Support System (GRASS) GIS. Such approaches allow for the mapping of the spatial datasets by scripts that are independent of their spatial position. Such performance is based on replicating the scripts for other extents of data using snippets of code and modified spatial parameters such as coordinates, data extent, and projections. However, the question of how to better process spatial data and which GIS program to select remains a challenging question in cartography due to the many advantages and drawbacks of various existing tools and proposed solutions.

1.2. Objectives and Motivation

The heterogeneity in the topographic structure of the Kuril–Kamchatka region reflects the cumulative effects from the complex geological setting and tectonic processes in the north-west Pacific Ocean, which requires advanced cartographic methods for geospatial data processing. To this end, this study presents the application of the GRASS GIS, an open source powerful software for topographical modelling, visualization, statistical analysis, and mapping with a case study of morphometric data analysis. Besides open source availability, GRASS GIS includes a variety of functional tools for processing both raster and vector data, using both a graphical user interface (GUI) and scripts for mapping. Leaving the GUI approach as a more trivial approach of geospatial data processing, this paper presents the implementation of the scripting methods using the GRASS GIS. We present a case of topographical modelling and data handling integrating GRASS GIS with such libraries as the Geospatial Data Abstraction Library (GDAL) and PROJ as key tools for transforming geospatial data formats and cartographic projections. The combination of these technical tools introduces major computational challenges for cartographic data processing and selection of the best projection to avoid distortions [49].

The objective of this study is to provide effective geoprocessing using architecture of the GRASS GIS, enabling data storage in one location of the 'mapset'. The 'mapset' in the GRASS GIS can be used for the same coordinate reference system and projection by interactive map design. The GRASS GIS modules were used to assess the morphometrics in the land-surface parameters over the study area based on the ETOPO1-derived basin morphometry: profile and tangential curvature, slope and aspect maps, and histograms of elevation distribution. As a result of the GRASS GIS application, the study revealed the effective coupling of scripting techniques and open source geospatial datasets in an optimised manner for modelling morphometric parameters and topographic mapping in a tectonically active region of the Pacific Ocean.

Current methods for geospatial data processing all focus on projecting the data according to the geospatial location. Thus, to enable data modelling, the QGIS and ArcGIS have a special modelling tool that allows one to create new tools with the use of sequential algorithms; while this can be a flexible solution for thematic maps, for topographical modelling this approach is unfeasible as placing different projections into separate folder locations is more optimal when dealing with several projections because it makes it possible to discriminate between the grids and process maps in distinct map sets. In contrast to traditional GIS software, the architecture of the GRASS GIS presents a collection of modules that can be stored as a sequence in a shell script and used for mapping using a console.

We instead focus on using the advantages of the GRASS GIS to demonstrate the effectiveness of scripts and data processing for mapping and visualization, as well as its compatibility with libraries GDAL, PROJ, and R [50–53]. The theoretical approaches in the morphometric studies [54] have been implemented in practical applications of GRASS GIS and developed in the GRASS GIS module r.slope.aspect [55], as demonstrated in this study. The topographical modelling and spatial analysis based on the continuous field of DEM assist in the analysis of slope stability in geologic risk assessment and its diverse applications in Earth sciences [56–58]. Selected features in the GRASS GIS function are comparable to the principles of other GIS. For instance, similar to QGIS, the technical flexibility of the GRASS GIS enables switching between the GUI and command-line mode of data processing. Similar to GMT, GRASS GIS uses a programming approach which enables the speeding up of repeated tasks through executable scripts.

Resembling the scripts in Unix shell, R and Python environments [59], GRASS GIS and GMT scripts are used to automate the workflow repeatedly performed for different maps using the same algorithm for various study areas. Other examples of automation of the cartographic workflow are diverse and include cases of rapid vectorisation of the isolines via machine learning [60], morphometric landscape analysis [61,62], geospatial analysis-based scripting libraries [63], topographic mapping using the GMT and QGIS [64], and fractal modelling quantifying the repetitive patterns [65].

Results from this study can be incorporated into a cartographic framework for spatial analysis of the morphometric structures in the west Pacific Ocean. The ability of GRASS GIS to smoothly handle large geospatial datasets, and improve and facilitate a cartographic workflow in various projects through automation is demonstrated in scripts. We present topographic mapping and morphometric surface modelling using several scripts of GRASS GIS. These results provide new perspectives for both cartographic workflow and topography-based spatial analysis in various fields such as applied geology. This paper also highlights the capabilities of mapping and modelling the bathymetric data by GRASS GIS for inaccessible deep marine and shelf areas. In contrast to terrain areas where fieldwork is possible, the submarine relief can only be evaluated and mapped by processing geospatial datasets using advanced cartographic methods.

2. Case Study

This study is focused on the Kuril–Kamchatka Trench and Kuril island arc that extend in the north-west Pacific Ocean, from the Kamchatka Peninsula to Hokkaido Island of the Japanese Archipelago, (Figure 1). This region is characterised by complex geological settings and high seismicity caused by the deep slab dynamics and morphology with high discontinuity of the subduction zone, as recently revealed in seismic cross-sections [66]. Thus, almost all of the islands from the Greater Kuril Chain are located along the submarine ridge of the Kuril island arc with depths of 100–300 m above them. Further, the morphological structure includes the submerged Vityaz ridge which extends from the ocean side of the Sea of Okhotsk. The Kuril–Kamchatka Trench generally has a flat seabed consisting of a chain of relatively wide and narrow segments representing short sections with depths of over −7000 m. However, it differs in the southern and northern parts with a maximum depth (−9717 m) recorded in its southwestern segment.

The basement surface of the Mesozoic structures near the northern coast of the Sea of Okhotsk forms a sub-latitudinal depression with depths reaching up to 3 km. Local minor trenches with depths of up to −2000 m extend from this depression in a sub-meridional direction, covering structural uplift in the centre of this sea. The southern part of the Sea of Okhotsk has a notable depression with depths exceeding −5000 m. The steep basement structures of the nearby shelf and the Kuril island arc descend towards the bottom of this depression.

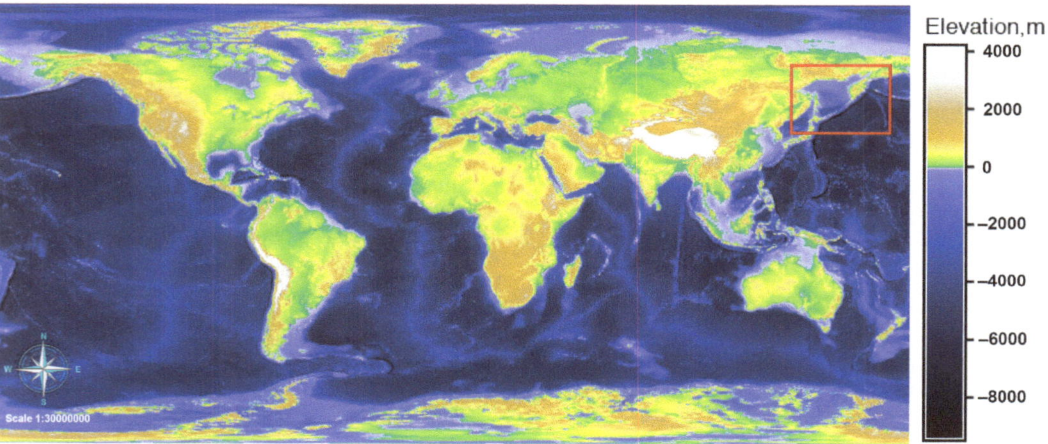

Figure 1. Study area on the Earth Global Relief Model (ETOPO1) GeoTIFF by National Oceanic and Atmospheric Administration (NOAA). Mapping: Geographic Resources Analysis Support System Geographic Information System (GRASS GIS).

The surface morphology of the seafloor basement in the Kuril–Kamchatka Trench correlates well with its modern relief. However, the depths of the seabed trenches are about −1000 to −2000 m deeper. According to the age of the constituent rocks, the Kuril–Kamchatka island arc was formed during the Oligocene period with some of the most ancient seamounts, mostly located on the slopes of the trenches, formed on the Mesozoic crust. Following the subduction of the tectonic plates under the island arcs, these seamounts were displaced from the basement surface and imprinted in the structures of the island arcs and coastal structures of the active continental margins. Such volcanic mountains are found, for instance, in the modern structures of the submarine slopes of the Kamchatka Peninsula.

The geophysical settings in the study area have a zone of linear positive and negative magnetic anomalies with an isometric magnetic field in the Kuril basin, which extends along the Kuril island arc and continues further within the Kamchatka Peninsula. The upper edges of the anomaly-forming bodies, associated with the volcanic rocks of the Pliocene-Quaternary age, are deposited closer to the surface. The geological complexity of the region of the Kuril–Kamchatka Trench and Kuril island arc, as briefly described above, illustrates the existing connectivity and correlation between the submarine and terrestrial morphology of the study area.

3. Materials and Methods

The research was implemented in the GRASS GIS [67], an open source software and image and graphics production system [68,69]. The major advantage of the GRASS GIS, similar to the QGIS, is that it combines the GUI with command-line interface (CLI) run from the console and enables operation with cartographic projections by GDAL. This makes it convenient to switch between the two modes of the data processing since it allows both cartographic and analytical experimentation with or without a traditional GUI menu. All the maps made in the GRASS GIS were stored using the Universal Transverse Mercator (UTM) map coordinate system, Zone 57, within the geographic extent of the study area. The full scripts used in this study for topographic mapping by the GRASS GIS are available in the GitHub repository of the authors with a provided open access using the following link: https://github.com/paulinelemenkova/Mapping_KKT_GRASS_GIS_Scripts (accessed on 18 February 2022).

The GRASS GIS uses a modular system organised by the name according to their functionality class (raster or vector) with the first letter indicating a function class, separated by dots and describing the specific task performed by this module. This feature is specific for GRASS GIS, yet differs from the GMT, a scripting cartographic toolset [70]. The console-based approach has distinct scripting toolsets from the GIS which mostly use the GUI as a main workflow environment. In contrast, the GRASS GIS enables the use of both GUI with a standard menu and a scripting approach, one of its advantages demonstrating its flexibility between the fully console-based GMT and the conventional GIS. This enables cartographic work to be performed in both ways, depending on the needs, preferences and profits from the combination of both approaches.

3.1. Data Capture and Organisation

The data used in this project include the open source datasets available online. The georeferenced Tag Image File Format (TIFF) file of the ETOPO1 grid by NOAA (https://www.ngdc.noaa.gov/mgg/global/ (accessed on 18 February 2022)) was used as a base map from the ETOPO1 Bedrock: Grid of Earth's surface depicting the bedrock underneath the ice sheets, (Figure 1). The 'gdalinfo' module of the GDAL library was used to display raster metadata and the 'gdalwarp' was applied to project the initial raster file in the XY Cartesian coordinate into the WGS84. Likewise, the GDAL library 'gdal_translate' utility was used to subset target regions from the global grid; the GRASS GIS command-line modules r.info, d.mon, and d.rast were used for visualising the output raster. Metadata checking for raster maps was performed by the 'g.list rast' GRASS GIS module, (Figure 2). The information on the data grid was checked by the command 'r.info' module which returns the min/max coordinates of the study area, resolution, number of classes, carto-graphic projection, scale, geometry with the number of centroids, points, lines, boundaries and islands, and the type of data (3D or 2D).

The command 'd.mon' displays a raster map in a selected monitor. The monitors are used in the GRASS GIS system for visualization of various grids by the commands 'd.mon wx0' and 'd.mon wx1'. The project management in GRASS GIS has a certain sim-ilarity to the ArcGIS in terms of data storage. Thus, each raster or vector map consists of several auxiliary files which include various data types, categories, header, and other specific information. The information on the projection of the current mapset was checked up by the commands: 'g.proj-p' (general description), 'g.proj-w' (here: the area of inter-est (AOI) extension and mathematical approximation are as follows: standard parallels, latitude of origin, central meridian, and false easting and northing). A summary of the algorithm used for plotting is presented in Listing 1:

Listing 1. GRASS GIS script for mapping ETOPO1 raster file.

```
1 #!/bin/sh
2 # raster file imported to the Location folder 'Kamchatka'
3 r.in.gdal -o input = /Users/pauline/ETOPO1_KKT_WGS84.tif output =
ETOPO1_KKT_WGS84.tif
4 # checked up the list of the raster files in the working directory
5 g.list rast
6 #starting graphical display (or 'monitor') in which the maps were
displayed
7 d.mon wx0
8 # visualizing raster by d.rast utility which displays grid cell maps in
the current graphics region
9 d.rast ETOPO1_Bed_g_geotiff
10 # changing color palette, from default viridis to srtm_plus
11 r.colors ETOPO1_KKT_WGS84 col = srtm_plus
12 # check of the range of the elevation values (depths/heights) of the map
13 r.info ETOPO1_KKT_WGS84
14 # display a legend on a map
15 d.rast.leg map = ETOPO1_KKT_WGS84
16 # providing numerical output of category numbers or range of values in a
raster
17 r.describe ETOPO1_KKT_WGS84
18 # print a list of category numbers and associated labels
19 r.cats
20 # check up and print the current region extent (in this case:   NS:
20.000000; EW: 45.000000).
21 g.region -e
22 # Removing auxiliary files
23 g.remove type = raster name = ETOPO1_KKT_WGS84.tiff -f
```

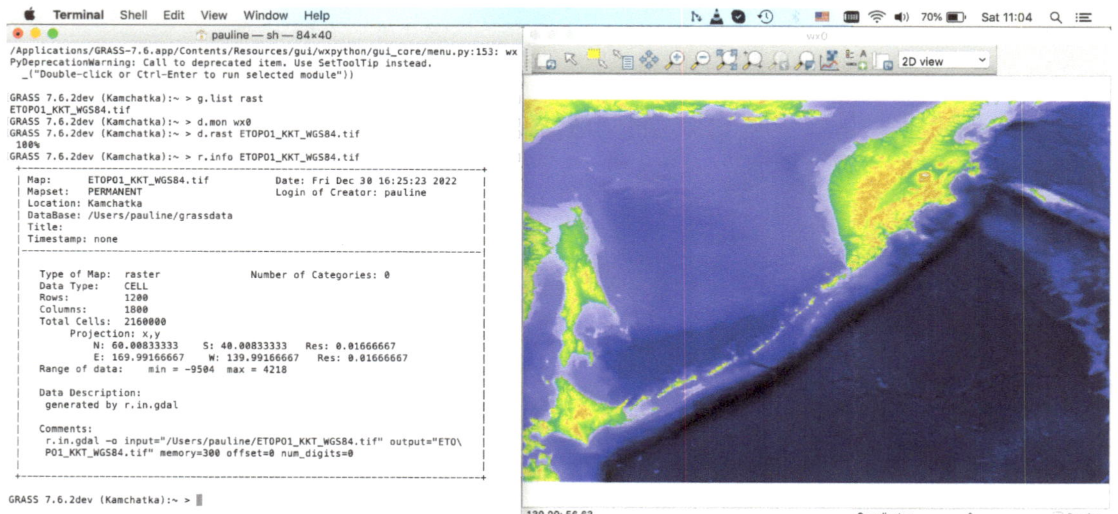

Figure 2. Subset of the study area (Kuril–Kamchatka region), in the Cartesian XY coordinates (metadata: **left**) visualized on a display (GRASS GIS GUI: **right**). Source: authors.

The trial directories were removed from the Unix utility 'rm' from a console by commands in two steps: (1) *cd /Users/<...>/grassdata* (changing to the current folder); (2) *rm -r testKKT* (here: removing testing folder for the Kuril–Kamchatka Trench with all files). The 'g.region-p' command was used to retrieve the information on the data extension and selected region (West, East, South, or North (WESN); number of rows; columns and cells; as well as additional cartographic data: ellipsoid, datum, and zone).

3.2. Data Processing by GDAL

The initial operations with raster files were performed by the Geospatial Data Abstraction Library (GDAL), a translator library for raster and vector geospatial data formats with available GDAL documentation [71]. The input to the reference list [71] was run from the GMT console. In general, the GDAL was used to re-project raster files to the supported cartographic projection, as well as to perform image mosaicking and transformations. In this study, we used three GDAL utilities using the commands of the following utilities: *'gdalwarp'*, *'gdal_translate'*, and *'gdalinfo'*. The 'gdalwarp' utility was used for the re-projection and warping of the initial original file (ETOPO1_Bed_g_geotiff.tif) from the XY Cartesian coordinates to the World Geodetic System 1984 (WGS84) [72].

The PROJ library, used for transforming the projections, was installed via the Homebrew package manager using the following commands: 'brew install proj' and 'brew upgrade proj' for installing and updating the package, respectively. The transformation uses the European Petroleum Survey Group (EPSG) number, which in this case is 4326 (corresponds to the WGS84). The warping was performed by the command for spatial reference system (SRS) as follows:

```
gdalwarp -t_srs EPSG:4326 ETOPO1_Bed_g_geotiff.tif ETOPO1_WGS84.tif
```

The next step entails running the utility 'gdal_translate' with the argument '-*projwin_srs* $< srs\_def >$', which specifies the projwin SRS GDAL, in which it interprets the coordinates (here, given coordinates in the EPSG system).

Using the gdal_translate, the file was subset, or clipped, to the target square area from the global extent with clockwise coordinates in degrees as follows: W = 140 N = 60 E = 170 S = 40. The re-projection was performed by the 'gdalwarp' utility and is illustrated in Figure 3. The subsetting of the target area was performed by the following command:

```
gdal_translate -projwin 140.0000 60.0000 170.0000 40.0000 ETOPO1_WGS84.tif
ETOPO1_KKT_WGS84.tif
```

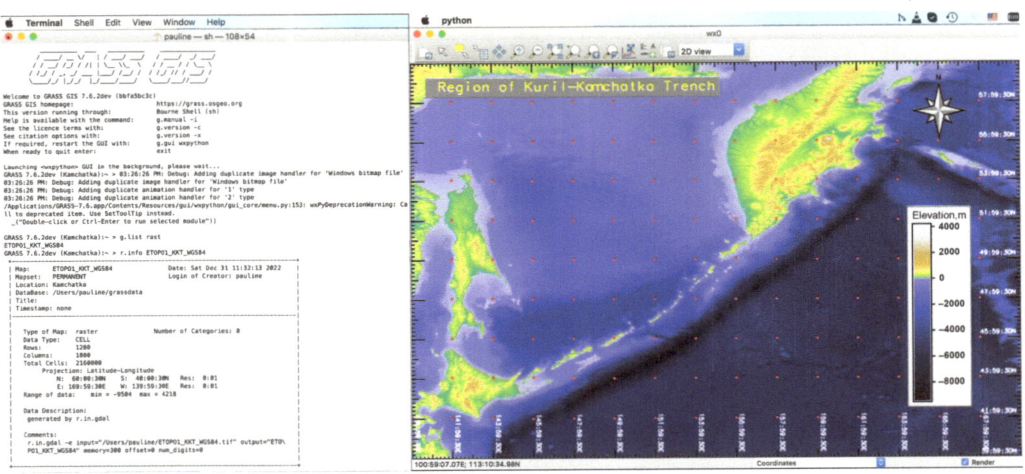

Figure 3. Raster projected from XY to WGS84 Lat/Lon coordinates with cartographic elements. Source: authors.

Now that all the components of the code have been explained, we provide a summary in Listing 2.

Listing 2. GRASS GIS script for adding cartographic elements.

```
1 #!/bin/sh
2 g.region rast = ETOPO1_KKT_WGS84.tif -p
3 d.mon wx0 4 d.rast ETOPO1_KKT_WGS84
5 r.colors ETOPO1_KKT_WGS84 col = srtm_plus
6 d.legend raster = ETOPO1_KKT_WGS84 title = Elevation,
m title_fontsize = 12
7 font = Helvetica fontsize = 8 -t -b bgcolor = white label_step = 2000
border_color = gray -f thin = 10
8 d.northarrow style = fancy_compass rotation = 0 label = N -w
color = black fill_color = gray fontsize = 10
9 d.grid size = 02:00:00 -a -d color = red width = 0.1 fontsize = 8
text_color = white
10 d.text text = "Region of Kuril-Kamchatka Trench" color = yellow bgcolor
= gray size = 8
```

The method presentation is shown by the projection transformation by GDAL utility 'gralwarp' using the PROJ library. It demonstrates the transformation of the TIFF file in the WGS84 geoid datum warped to a two-point equidistant azimuthal projection using the PROJ arguments:

```
gdalwarp -t_srs '+proj=tpeqd +lat_1=45 +lat_2=55 +lon_1=100 +lon_2=120'
ETOPO1_KKT_WGS84.tif ETOPO1_KKT_2p.tif -overwrite
```

Now the file was reprojected to the WGS84 and subset (cut-off) from the initial raster grid with global extent. A gdalinfo utility was used to check the correctness of the performed operations and the output data as follows: *gdalinfoETOPO1_KKT_WGS84.tif*. The standard parallels for the study area are 45° N and 55° N. Next, we present the code in Listing 3:

Listing 3. GRASS GIS script for plotting two-point equidistant azimuthal projection.

```
1 # stored in GeoTIFF in WGS84 warped to a Two-Point Equidistant Azimuthal
projection:
2 gdalwarp -t_srs '+proj = tpeqd +lat_1 = 45 +lat_2 = 55 +lon_1 = 100
+lon_2 = 120' ETOPO1_KKT_WGS84.tif ETOPO1_KKT_2p.tif -overwrite
3 #!/bin/sh
4 g.list rast
5 r.info ETOPO1_KKT_2p
6 d.mon wx0
7 g.region raster = ETOPO1_KKT_2p -p
8 r.colors ETOPO1_KKT_2p col = grey.eq
9 d.rast ETOPO1_KKT_2p
10 d.redraw
11 r.contour ETOPO1_KKT_2p out = Bathymetry1000 step = 1000
12 d.vect Bathymetry1000 color = 'blue' width = 0
13 d.grid -g size = 2.5 color = '255:0:0'
14 d.text text = "2 Point Equidist. Azimuth." color = '0:0:51' bgcolor =
'224:224:224' size = 3
```

The next step was continued in the GRASS GIS version 7.6 and included coordinate transformation, performed by 'proj' with several tested projections and their visualization using GRASS GIS scripts presented in Listings 4 for the Cassini–Soldner projection, Listing 5 for the oblique Mercator projection, and Listing 6 for the equal-area cylindrical projection. The 'proj' is a cartographic projection filter of the PROJ software which allows for converting between the geographic and projection coordinates within one datum WGS84. The maps projections were tested in PROJ via the 'gdalwarp' utility of GDAL before importing them into the GRASS GIS. The 'gdalwarp' is specifically designed to reproject maps to the target projections, coordinate systems, ellipsoid, or geodetic datum. In this case, we selected the datum as a constant WGS84 and kept it constant for this project.

The ellipsoidal Cassini–Soldner projection is well suited for the topographic mapping of areas with restricted extent along the meridian since the distortion of the areas along the central meridian is minimised (in this case, the meridian is 155° E and latitude 50° N). Thus, due to the more accurate computations in the ellipsoidal formulae, it provides a compromise between the conformal and equal-area projections keeping the distortions minimal in the longitudinal direction. The definition of this projection is based on the meridian and parallel of the central point of the study area. The plotting has been performed using the script provided in Listing 4.

Listing 4. GRASS GIS script for plotting map in Cassini–Soldner projection.

```
1 # stored in GeoTIFF in WGS84 warped to an Cassini (Cassini-Soldner)
equirectangular projection:
2 gdalwarp -t_srs '+proj = cass lat_0 = 50 lon_0 = 155'
ETOPO1_KKT_WGS84.tif ETOPO1_KKT_C.tif
3 #!/bin/sh
4 g.list rast
5 r.info ETOPO1_KKT_C
6 d.mon wx0 7 g.region raster = ETOPO1_KKT_C -p
8 r.colors ETOPO1_KKT_C col = grey.eq
9 d.rast ETOPO1_KKT_C
10 d.redraw
11 r.contour ETOPO1_KKT_C out = Bathymetry1000C step = 1000
12 d.vect Bathymetry1000C color = 'blue' width = 0
13 d.grid -g size = 2.5 color = '255:0:0'
14 d.text text = "Cassini-Soldner prj" color = '0:0:51' bgcolor =
'224:224:224' size = 3
```

The geodetic transformations used to plot the map in the conformal oblique Mercator projection are expressed as a series of operations which represent the parallels and meridians as complex curves. These include the definition of the key parameters such as longitude and latitude of the projection centre, azimuth of the oblique equator with a scale along it representing the oblique aspect of the Mercator projection. The advantages of the oblique Mercator projection is that it provides a high accuracy in zones close to the longitude and latitude of the projection centre and a constant scale on the line of tangency, that is, the central longitude and preserves local shapes in regions of large lateral extent such as the Kuril–Kamchatka and Kuril islands. The map in the oblique Mercator projection was plotted using Listing 5.

Listing 5. GRASS GIS script for plotting map in oblique Mercator projection.

```
1 # stored in GeoTIFF in WGS84 warped to a Oblique Mercator projection:
2 gdalwarp -t_srs '+proj = omerc +lat_1 = 45 +lat_2 = 55 lon_1 = 177 lon_2
= 210 +ellps = GRS80' ETOPO1_KKT_WGS84.tif ETOPO1_KKT_OM.tif -overwrite
3 #!/bin/sh
4 g.list rast
5 r.info ETOPO1_KKT_OM
6 d.mon wx0
7 g.region raster = ETOPO1_KKT_OM -p
8 r.colors ETOPO1_KKT_OM col = grey.eq
9 d.rast ETOPO1_KKT_OM
10 r.contour ETOPO1_KKT_OM out = Bathymetry10000M step = 1000
11 d.vect Bathymetry10000M color = 'blue' width = 0
12 d.grid -g size = 2.5 color = '255:0:0'
13 d.text text = "Oblique Mercator prj" color = '0:0:51' bgcolor =
'224:224:224' size = 3
```

The key parameters for defining the equal-area cylindrical projection include the latitude (the meridian) of the true scale or standard parallels (in this case, 50° N and 155° E). The distortions along the central meridian and parallels are minimised since they provide a centre point of the extent of the study area. The projection has been defined by the 'gdalwarp' and then plotted using the scripting in Listing 6:

Listing 6. GRASS GIS script for plotting map in Equal-Area Cylindrical projection.

```
1 # stored in GeoTIFF in WGS84 warped to an Equal Area Cylindrical
projection:
2 gdalwarp -t_srs '+proj = cea lat_ts = 50 lon_0 = 155'
ETOPO1_KKT_WGS84.tif ETOPO1_KKT_EAC.tif
3 #!/bin/sh
4 g.list rast
5 r.info ETOPO1_KKT_EAC
6 d.mon wx0
7 g.region raster = ETOPO1_KKT_EAC -p
8 r.colors ETOPO1_KKT_EAC col = grey.eq
9 d.rast ETOPO1_KKT_EAC
10 d.redraw
11 r.contour ETOPO1_KKT_EAC out = Bathymetry1000 step = 1000
12 d.vect Bathymetry1000 color = 'blue' width = 0
13 d.grid -g size = 2.5 color = '255:0:0'
14 d.text text = "Equal Area Cylindrical prj" color = '0:0:51' bgcolor =
'224:224:224' size = 3
```

3.3. Displaying Raster Grids in GRASS GIS

Upon re-projection of the file, it was imported to the GRASS GIS by the 'r.in.gdal' utility. Displaying the dataset was performed using the sequence of the GRASS modules shown in Listing 1, based on the GRASS GIS modular techniques [73] and visualized in Figure 2. The visualized cartographic elements include grid ticks, legend and the title, which were placed on the map in Figure 3 using the code in Listing 2. Generating the continuous fields of the bathymetric contours (isolines) were derived from the raster ETOPO1_KKT_WGS84 file using the GRASS GIS module 'r.contour' by script. In this example the step parameter was set to 750 m and is visualized in Figure 4. The script of the GRASS GIS used for plotting the topographic isolines is shown in Listing 7:

Listing 7. GRASS GIS script for visualising raster grid and adding contour isolines.

```sh
1 #!/bin/sh
2 d.mon wx0
3 r.info ETOPO1_KKT_WGS84
4 g.region raster = ETOPO1_KKT_WGS84 -p
5 d.rast ETOPO1_KKT_WGS84
6 d.grid size = 03:00:00 -a -d color = red width = 0.1 fontsize = 8
text_color = white
7 d.text text = "Kuril-Kamchatka Area" color = '255:255:255' size = 5
8 r.contour ETOPO1_KKT_WGS84 out = Bathymetry750 step = 750 -overwrite
9 d.vect Bathymetry750 color = '102:37:4' width = 0
```

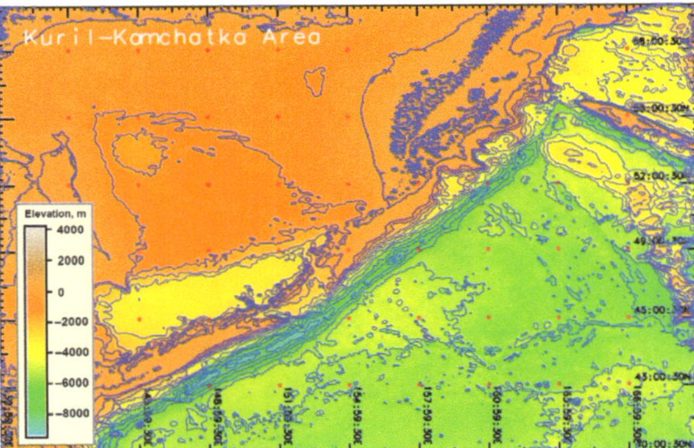

Figure 4. Plotting isolines by the 'r.contour' module and visualising the raster in the 'elevation' colour table. Source: authors.

3.4. Calculating Morphometric Parameters

The correlation between the profile and tangential curvature reflects the methodological approaches of the algorithm which concerns the mathematical approximation of the morphological profile of the slopes and inflation of the terrain heights. The differences in calculus are well-illustrated in Equation (1) for the profile curvature K_P [74] and Equation (2) for the tangential curvature K_T [74].

$$K_P = \frac{\left(\frac{\delta^2 z}{\delta x^2}\right)\left(\frac{\delta z}{\delta x}\right)^2 + 2\left(\frac{\delta^2 z}{\delta x \delta y}\frac{\delta z}{\delta x}\frac{\delta z}{\delta y}\right) + \frac{\delta^2 z}{\delta y^2}\left(\frac{\delta z}{\delta x}\right)^2}{pq^{\frac{3}{2}}} \tag{1}$$

$$K_T = \frac{\left(\frac{\delta^2 z}{\delta x^2}\right)\left(\frac{\delta z}{\delta y}\right)^2 - 2\left(\frac{\delta^2 z}{\delta x \delta y}\frac{\delta z}{\delta x}\frac{\delta z}{\delta y}\right) + \frac{\delta^2 z}{\delta y^2}\left(\frac{\delta z}{\delta x}\right)^2}{pq^{\frac{1}{2}}} \tag{2}$$

where $p = \left(\frac{\delta z}{\delta x}\right)^2 + \left(\frac{\delta z}{\delta y}\right)^2$ and $q = 1 + p$ for both cases.

The plotting of the morphometric parameters was performed using the code:
`'r.slope.aspect elevation=ETOPO1_KKT_WGS84 slope=slope aspect=aspect pcurvature=pcurv tcurvature=tcurv'`

where the input elevation raster map is the ETOPO1_KKT_WGS84, and the output raster map is 'slope'. The curvature shows the rate of variations of the slope, which makes it the second function of the elevation variable related to the surface heights [75]. This process technically includes both cases: whether the purpose is profile or tangential curvature.

However, the difference is in the approaches of the mathematical algorithms that are embedded in the GRASS GIS. Hence, the difference between the profile and tangential curvature consists of the method of calculation. For the tangential curvature, it is computed as perpendicular to the slope gradient, while for profile curvature, it measures a curvature of the surface in the direction of the gradient [76]. The slope shows a steepness or the degree in the topographic surface covering the study area. The aspect, or exposure, is a compass direction that a slope faces (East, West, South, or North). Geographically, the aspect of a slope has a significant impact on the local microclimate which affects physical and often botanical features of a mountain slope, i.e., a slope effect on the terrestrial areas.

Such effects are owing to the strong correlation between the mountain slope and temperature, which is in turn affected by the angle of sun. Hence, both parameters are derivative from the slope of the elevation with certain variations in the mathematical approach and angle direction of the gradient degree (direct and perpendicular). The theoretical foundations of the slope and angle modelling in geospatial research can be found in existing works reporting, for instance, on the effects from the cell size [77], gradient angles [78], numerical terrain analysis [79], or quantifying land surface and slope curvatures [80–83]. The increase in values of a specific depth was analysed by plotting the histograms as bars and pie charts using the GRASS GIS 'd.histogram' module for calculating depths distribution [84]. The visualized images are displayed using various colour map palettes (r.colors) specifically intended for the cartographic visualization of the terrain raster grids.

4. Results and Discussion

The GRASS GIS algorithms were implementedfor cartographic data visualization and map compilation in various projections. The topographic data visualization and a morphometric analysis were based on the GRASS GIS modules. The input cartographic data based on the ETOPO1 raster grid comprised the derived variables: slope, aspect, curvature, and hillshade. The data analysis included a plotted histogram and pie chart showing data distribution. Several projections were tested and visualized using the PROJ library. The data transformation was performed using GDAL integrated with the GRASS GIS.

4.1. Cartographic Projections by PROJ

The PROJ, a generic coordinate transformation software library [85], was used to transform the raster grid files in Lat/Lon geospatial coordinates from one coordinate reference system (CRS) to another. This method includes the PROJ transformation which performs a variety of operations with cartographic data, starting from the simple and well-known projections such as cylindrical Mercator projection to the very complex transformations across many reference coordinate frames. The PROJ was used for testing various projections (conic, cylindrical, and azimuthal) selected from a variety of available projections in the library: Cassini–Soldner equirectangular; equal area cylindrical; two-point equidistant azimuthal; and oblique Mercator [86].

Originally made as a technical tool for cartographic projections, PROJ gradually transformed into being a functional generic coordinate transformation engine. The supported projections were checked by the command 'cs2cs -lp' from the GRASS GIS console [87]. The PROJ was applied for the scale cartographic projections and coordinate transformation (Figure 5), thus presenting the geodetic transformations of varying complexity. In this study, the visualization of various cartographic projections in GRASS GIS (Figure 5) was performed by a combination and sequential use of the following modules of GRASS GIS: 'd.rast', 'r.contour', 'd.vect', 'd.grid', and 'd.text'. The script is based on the developed methodology of data processing by GRASS GIS [88]. It was used for plotting the contour maps with an example of Cassini–Soldner projection. Mapping the study area in various cartographic projections was recorded in separate plots in the GRASS GIS and is presented in Figure 5 for comparison: Cassini–Soldner equirectangular; equal area cylindrical; two-point equidistant azimuthal; and oblique Mercator.

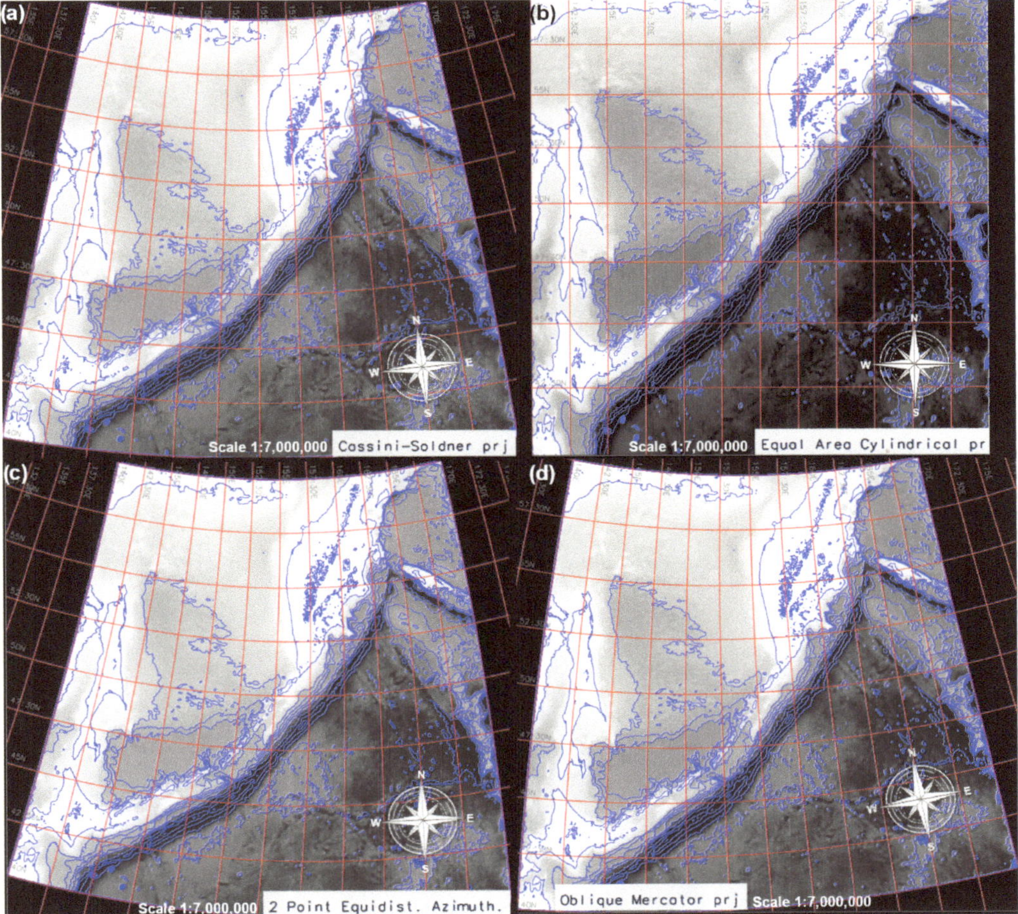

Figure 5. Comparison of the four maps visualized in GRASS GIS in four different cartographic projections transformed by the 'gdalwarp' GDAL utility and PROJ parameters. The projections from left to right, top to bottom are as follows: (**a**) Cassini–Soldner equirectangular; (**b**) equal area cylindrical; (**c**) two-point equidistant azimuthal; (**d**) oblique Mercator. Source: authors.

4.2. Morphometric Parameters

The input elevation data were derived from the ETOPO1 raster grid containing the elevation values in a digital format covering the study area. The 'r.slope.aspect' module of GRASS GIS was used to plot raster maps using the developed methodology [55,89,90]. The resulting maps of slope, aspect (computed counterclockwise from the east), profile curvature, and tangential curvature from the ETOPO1 raster grid are shown in Figure 6. The results include the morphometric elements of slope, aspect, profile curvature, and tangential curvature (Figure 6).

The results of the morphometric modelling show that the terrain slope in the Kuril–Kamchatka region ranges from 0° to 39°, as shown on the slope map calculated from the ETOPO1_KKT_WGS84 grid. The slope in the study area is shown in gradually increasing degrees starting from 0° (flat horizontal) to 39° (maximal vertical). The profile curvature shows modelled terrain surfaces in the gradient direction which reflects the variations in the slope steepness angle within the range of -87×10^{-5} to 68×10^{-5}. The largest area coloured light green on the lower left map (Figure 6) has values of -10×10^{-5} to 0. Practically, it indicates that the velocity of mass falls downwards along the slope curve. The tangential curvature reflects the change in the aspect angle and influences the divergence, or convergence, of the water flow, respectively.

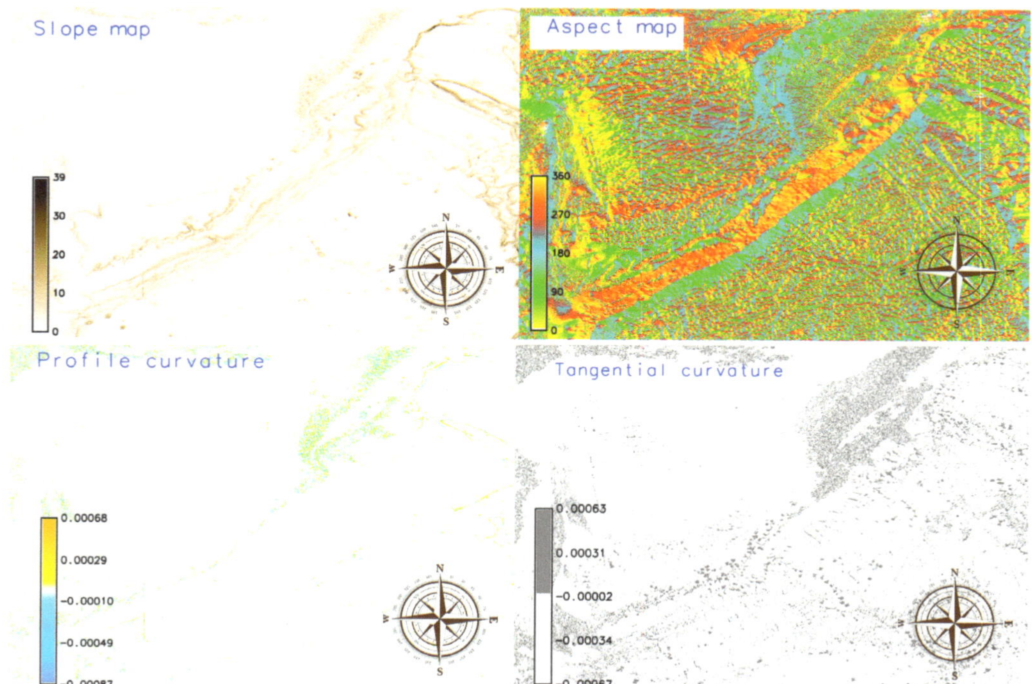

Figure 6. Slope, aspect, profile, and tangential curvature maps, Kuril–Kamchatka region, based on raster grid ETOPO1. Source: authors.

The tangential curvature shows values of -67×10^{-5} to 63×10^{-5}. The terrestrial areas of the Kamchatka Peninsula and Greater Kuril Chain show higher positive values, depicted by the dark grey colours, in the lower right map (Figure 6) compared to the water areas of the Sea of Okhotsk (white and light grey colours) which shows variations in the land surface steepness over the terrain compared to water areas. Numerically, the aspect map shows the orientation in degrees varying from $0°$ to $360°$ (upper right map in Figure 6). It shows a clear correlation between the aspect and topographic area of the Kuril–Kamchatka Trench ($130°$–$150°$, emerald green colour on the map) and the Greater Kuril Chain which extends in parallel to the trench both in the southeastern orientation (beige colour from the selected 'aspectcolr' colour palette, $300°$–$330°$,). A clear delineation of the Sakhalin Island can be noted as the north-western orientation (yellow-green colours, $0°$–$30°$,) and the Kamchatka Peninsula (NW, greens, $60°$–$120°$).

The produced maps show morphometric variables over the target area performed by the GRASS GIS techniques which include data conversion, warping, transforming projections, and adding cartographic elements (ticks, legend, and isolines) on the final layout. Each map sources a range of new visualizations of the data relevant to the morphology of the study region. Similar to the GMT techniques, the scripting methods of the GRASS GIS have a command-line approach of data processing. This is convenient for controlling the cartographic design and enables the creation of print-quality maps.

Further, we demonstrated the effectiveness of the GRASS GIS methods in visualising raster grids and performing raster data analysis in a geologically complex region, such as the northwestern sector of the Pacific Ocean. The computed geomorphic indices included slope, aspect, profile, and tangential curvature maps, derived from the topographic grid using the embedded algorithms in GRASS GIS.

4.3. Analysis of the Elevation Frequency

The next research step included the analysis of the distribution of the elevation over the study area. Using computational techniques of GRASS GIS, we calculated the frequency of the data in various range diapason and the repeatability of corresponding depths. The most frequent depths or elevations over the study area were recorded and summarised for the analysis of spatially distributed data [91–93]. The increase in values of a specific depth reflects the frequency of the elevation recorded on a bathymetric and topographic grid in a subsequent region of the study area. The analysis of the spatial data distribution aimed to demonstrate a range of variations in data, including skewness, the variation of values within intervals of the topography, organisation of categorical data, selecting relevant peaks in the interpretation and description of the dataset, or other resources showing the correlation among the morphometric parameters, such as the aspect and slope with relief. The distribution of topographic data has been evaluated using statistical tools embedded in GRASS GIS and presented in a histogram and a pie chart (Figure 7).

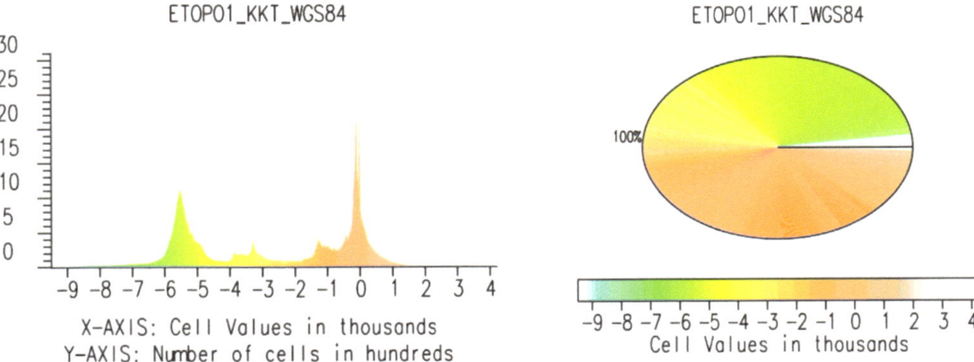

Figure 7. Bar graph histogram for elevation map (**left**); pie graph of height distribution based on the raster grid ETOPO1 (**right**). Source: authors.

The critical synthesis and analysis of a geomorphological setting of the study area is largely based on the experiential techniques of cartographic data visualization using GRASS GIS. A range of its modules utilised in scripts for cartographic transformations were used to visualise various projection and colour map palettes. The topographic data analysis demonstrated variations in the percentage of data distribution and morphometry in the study area, as follows. The depths in the interval below 400 to −2000 m covered 24% of the total points; the depths in the interval from −2000 to −3000 m summed to 2%; the depth from −3000 to −4000 m took up 4% of the total points; the depths in the range between −5000 and −6000 m were 13%; and the depth in the diapason from −6000 to 7000 m were less than 1%. The rest of the elevation points are covered by the data with the diverse heights and topographic values.

The extent of the data range is summarised and visualized in the histogram and pie chart (Figure 7). The bar plot and the pie chart show the frequencies of the elevation levels in a raster representing morphometric unevenness of the terrain. For each topographic range, the number of pixels in the raster image is counted and drawn on the diagram in the two views of representation. The number of pixels corresponding to the each value of depth/height contributes to the height of this particular bar on the chart (or a radius of a segment in the pie). The x-axis of the bar chart diagram represents the topographic/bathymetric values (heights and depths), while the vertical y-axis shows the number of pixels counted at a given step.

The morphometry is an essential approach of quantitative GIS-based terrain analysis. It is increasingly used in diverse geological applications for relief visualization. Based on elevation derived from DEM, the landforms can be visualised in the target area for morphological modelling and mapping at regional scale. The morphometric terrain analysis may further include extended approaches such as analysis and interpretation of the land forms or seafloor surface as a continuous field, and upscaling the morphological shapes for a closer analysis of the terrain structure. The slope gradient indicates the terrain roughness and represents a base for the land surface discreteness. A slope gradient can be used as an indication of terrain roughness, while the aspect can be applied for mapping slope distribution, depending on the compass orientation of the terrain.

The findings revealed that the topography of the Kuril–Kamchatka Trench varied significantly reflecting its regional geologic evolution and tectonic setting which largely affects the structure of the relief. Our study presented a case for displaying topographic data to analyse height differences using cartographic and statistical data processing. We have shown a specific example of bathymetric data processing for the north Pacific Ocean area performed using a scripting approach by GRASS GIS syntax with its specific language operated from the command line. The scripts of the GRASS GIS were used to rapidly plot the topographic data as a series of morphometric variables. The results can be extended to similar datasets using scripts presented for the repeatability of data processing in GRASS GIS: visualization of the slope, aspect, curvature, and statistical data distribution of elevation data.

5. Conclusions

Advancing insights into the complexities of the heterogeneous bathymetry and topography in coastal areas can be enhanced through the application of advanced cartographic tools. We applied the ETOPO1 World Global Relief Model data on the western segment of the Pacific Ocean to quantify morphometric terrain parameters and evaluate the performance of various cartographic projections by spatial data transformation in the GRASS GIS environment. ETOPO1 data provided detailed information on the variations in the topographic structure over the target region. We modelled slope, aspect, profile curvature and tangential curvature using a combination of the GRASS GIS modules presented in scripts coupled with the GDAL library for data transformation and formatting. The results of the topographic modelling were statistically processed for analysis of the depth and heights distributions over the study area to evaluate the heterogeneity of landform elements.

The integration of the GRASS GIS scripts with geospatial data allowed us to characterise the distribution of elevation values throughout the region. Using such an approach, this paper revealed the value of spatial data processing to model morphometric parameters by GRASS GIS scripts in semi-automatic topographic mapping. As a result, we found and visualised the high complexity of land surface structures and adjacent bathymetry in the region, linked to the geological and tectonic processes that shaped the current landforms. The variations in the slope and aspect maps were evaluated along with the profile and tangential curvature of the topographic dataset to model the curvature of the relief in the Kuril–Kamchatka region. Topographic mapping and morphometric modelling rely on spatial data processing and representation. However, real-world spatial data are generated from DEM involving may factors. These include the geologic drivers and tectonic processes in the past that affect modern relief of the Earth.

Such complexity of the relief is reflected in the diverse datasets with various resolution and details of the land surface. Processing these data and visualising the topography of the Earth with variability of the morphometric patterns requires the advanced tools of cartographic representation and automated approaches of data handling. As a result, recent efforts have been made to integrate the programming tools into the GRASS GIS for geospatial studies. To this end, we propose the use of the GRASS GIS modules for morphometric modelling and cartographic reprojecting the raster grids on selected area of the north-west Pacific Ocean. We demonstrate that the GRASS GIS is effective in terms of cartographic visualization and accuracy of spatial data processing. In addition, we provide a series of scripts used for plotting and data visualization as a reference in similar studies.

The application of the GRASS GIS algorithms aimed at topographic mapping and morphometric modelling that uses a script-based approach for cartographic data visualization, transformation, and analysis. The selected region of the tested area includes the north-western segment in the Far East based on the ETOPO1 grid. The extensive experimental results on mapping of the raster images by scripts show that GRASS GIS outperforms the traditional software in terms of automating data processing which increases the speed and accuracy of cartographic plotting and, specifically, modelling the morphometric parameters. The latter include slope, aspect, and curvature in both the down-slope and across-slope directions, which is presented in this case study based on the GRASS GIS modules.

The results show the effective application of the GRASS GIS techniques for cartographic data processing. Using the console in scripting mode, we identified topographic variability in the target area through spatial analysis of the processed data and visualised raster grids in various cartographic projections. We have also shown that the geodetic transformation of the dataset in various cartographic projections performs well using the GDAL utility and PROJ library, while the choice of the cartographic projections is wholly due to the spatial location of the study area controlled by the possible distortions in sizes, angles, distances, and directions of the visualized objects. We demonstrated the method to be flexible in changing projections using the integration of GRASS GIS and GDAL.

With regard to addressing the cartographic plotting, GRASS GIS shows robustness and effectiveness of data handling both using the GUI and from the console in a series of listings. We believe that techniques of GRASS GIS for handling geospatial data can be further extended to deal with spatial datasets using scripts presented in this study. Thus, the advantage of this study is that it provides the scripts in the GitHub open repository that can be reused in similar studies for repeatability. These scripts are sequential used for algorithms from the GRASS GIS and composed with the use of several modules for topographic modelling and mapping.

Author Contributions: Supervision, conceptualization, methodology, software, resources, funding acquisition, and project administration, O.D.; writing—original draft preparation, methodology, software, data curation, visualization, formal analysis, validation, writing—review and editing, and investigation, P.L. All authors have read and agreed to the published version of the manuscript.

Funding: The publication was funded by the Editorial Office of *Technologies*, Multidisciplinary Digital Publishing Institute (MDPI), by providing 100% discount for the APC of this manuscript. This project was supported by the Federal Public Planning Service Science Policy or Belgian Science Policy Office, Federal Science Policy—BELSPO (B2/202/P2/SEISMOSTORM).

Institutional Review Board Statement: Not applicable.

Informed Consent Statement: Not applicable.

Data Availability Statement: Not applicable.

Acknowledgments: The authors thank the anonymous reviewers of *Land* for reading, their suggestions, and their comments that improved an earlier version of this manuscript.

Conflicts of Interest: The authors declare no conflicts of interest.

Abbreviations

The following abbreviations are used in this manuscript:

ASCII	American Standard Code for Information Interchange
AOI	Area of Interest
CLI	Command Line Interface
DEM	Digital Elevation Model
EGM96	Earth Gravitational Model of 1996
EGM2008	Earth Gravitational Model of 2008
ENVI	Environment for Visualizing Images
EPSG	European Petroleum Survey Group
ETOPO1	Earth's surface topography
GDAL	Geospatial Data Abstraction Library
GEBCO	General Bathymetric Chart of the Oceans
GIS	Geographic Information System
GMT	Generic Mapping Tools
GRASS	Geographic Resources Analysis Support System
GUI	Graphical User Interface
NOAA	National Oceanic and Atmospheric Administration
PROJ	a library for performing conversions between cartographic projections
QGIS	Quantum GIS
SAGA	System for Automated Geoscientific Analyses
SRS	Spatial Reference System
SRTM	Shuttle Radar Topography Mission
TIFF	Tag Image File Format
UTM	Universal Transverse Mercator
WGS84	World Geodetic System 1984
WESN	West East South North

References

1. Wang, J.; Wu, F. *Advances in Cartography and Geographic Information Engineering*; Springer: Singapore, 2022; Volume 638. [CrossRef]
2. Minár, J.; Evans, I.S.; Jenčo, M. A comprehensive system of definitions of land surface (topographic) curvatures, with implications for their application in geoscience modelling and prediction. *Earth-Sci. Rev.* **2020**, *211*, 103414. [CrossRef]
3. Maxwell, A.E.; Shobe, C.M. Land-surface parameters for spatial predictive mapping and modeling. *Earth-Sci. Rev.* **2022**, *226*, 103944. [CrossRef]
4. Zhou, W. GIS for Earth Sciences. In *Encyclopedia of Geology*, 2nd ed.; Alderton, D., Elias, S.A., Eds.; Academic Press: Oxford, UK, 2021; pp. 281–293. [CrossRef]
5. Ruzickova, K.; Ruzicka, J.; Bitta, J. A new GIS-compatible methodology for visibility analysis in digital surface models of earth sites. *Geosci. Front.* **2021**, *12*, 101109. [CrossRef]
6. Deseilligny, M.; Le Men, H.; Stamon, G. Map understanding for GIS data capture: Algorithms for road network graph reconstruction. In Proceedings of the 2nd International Conference on Document Analysis and Recognition (ICDAR '93), Tsukuba, Japan, 20–22 October 1993; pp. 676–679. [CrossRef]
7. Sasso, D.; Biles, W.E. An object-oriented programming approach for a GIS data-driven simulation model of traffic on an inland waterway. In Proceedings of the 2008 Winter Simulation Conference, Miami, FL, USA, 7–10 December 2008; pp. 2590–2594. [CrossRef]
8. Rathod, N.; Subramanian, R.; Sundaresan, R. Data-Driven and GIS-Based Coverage Estimation in a Heterogeneous Propagation Environment. In Proceedings of the 2018 IEEE Global Communications Conference (GLOBECOM), Abu Dhabi, United Arab Emirates, 9–13 December 2018; pp. 1–6. [CrossRef]
9. Shrestha, B.; Devarakonda, R.; Palanisamy, G. An open source framework to add spatial extent and geospatial visibility to Big Data. In Proceedings of the 2014 IEEE International Conference on Big Data (Big Data), Washington, DC, USA, 27–30 October 2014; pp. 64–66. [CrossRef]
10. Scott, G.J.; Angelov, G.A.; Reinig, M.L.; Gaudiello, E.C.; England, M.R. cvTile: Multilevel parallel geospatial data processing with OpenCV and CUDA. In Proceedings of the 2015 IEEE International Geoscience and Remote Sensing Symposium (IGARSS), Milan, Italy, 26–31 July 2015; pp. 139–142. [CrossRef]
11. Scott, G.J.; Backus, K.; Anderson, D.T. A multilevel parallel and scalable single-host GPU cluster framework for large-scale geospatial data processing. In Proceedings of the 2014 IEEE Geoscience and Remote Sensing Symposium, Quebec City, QC, Canada, 13–18 July 2014; pp. 2475–2478. [CrossRef]

12. Chen, T.; Yuan, H.y.; Yang, R.; Chen, J. Integration of GIS and Computational Models for Emergency Management. In Proceedings of the 2008 International Conference on Intelligent Computation Technology and Automation (ICICTA), Changsha, China, 20–22 October 2008; Volume 2, pp. 255–258. [CrossRef]
13. Li, G.; Zhang, J.; Wang, N. Construction and Implementation of Spatial Analysis Model Based on Geographic Information System (GIS)—A Case Study of Simulation for Urban Thermal Field. In Proceedings of the 2008 International Conference on Computational Intelligence for Modelling Control & Automation, Vienna, Austria, 10–12 December 2008; pp. 1095–1098. [CrossRef]
14. Zhang, J.; Luo, W.; Yuan, L.; Mei, W. Shortest path algorithm in GIS network analysis based on Clifford algebra. In Proceedings of the 2010 2nd International Conference on Future Computer and Communication, Wuhan, China, 21–24 May 2010; Volume 1, pp. 432–436. [CrossRef]
15. Huang, Z.; Fang, Y. A novel approach for geospatial computational task processing in Grid environment. In Proceedings of the 2010 IEEE International Geoscience and Remote Sensing Symposium, Honolulu, HI, USA, 25–30 July 2010; pp. 3980–3982. [CrossRef]
16. Franges, S.; Zupan, R. New encouragement on cartographic visualisation. In Proceedings of the ISPA 2001—Proceedings of the 2nd International Symposium on Image and Signal Processing and Analysis—In conjunction with 23rd International Conference on Information Technology Interfaces (IEEE Cat.), Pula, Croatia, 19–21 June 2001; pp. 368–372. . [CrossRef]
17. Zhang, B.; Ran, H.; Yu, J. Visualizaiton of water system based on the cartographic presentation. In Proceedings of the 2012 International Symposium on Geomatics for Integrated Water Resource Management, Lanzhou, China, 19–21 October 2012; pp. 1–4. . [CrossRef]
18. Yamaguchi, N. Visualizing states in autoregressive hidden Markov models using generative topographic mapping. In Proceedings of the 2012 8th International Conference on Natural Computation, Chongqing, China, 29–31 May 2012; pp. 138–142. [CrossRef]
19. Pazouki, E. A smart surface irrigation design based on the topographical and geometrical shape characteristics of the land. *Agric. Water Manag.* **2023**, *275*, 108046. [CrossRef]
20. Moreira, E.P.; Valeriano, M.M. Application and evaluation of topographic correction methods to improve land cover mapping using object-based classification. *Int. J. Appl. Earth Obs. Geoinf.* **2014**, *32*, 208–217. [CrossRef]
21. Chowdhury, M.S. Modelling hydrological factors from DEM using GIS. *MethodsX* **2023**, *10*, 102062. [CrossRef] [PubMed]
22. Boulton, S.J.; Stokes, M. Which DEM is best for analyzing fluvial landscape development in mountainous terrains? *Geomorphology* **2018**, *310*, 168–187. [CrossRef]
23. Buterez, C.; Olariu, B.; Mihai, B.; Rujoiu-Mare, M.; Cruceru, I. General topography of Prahova County, Romania. *J. Maps* **2016**, *12*, 541–545. [CrossRef]
24. Duncan, D.T.; Regan, S.D. Mapping multi-day GPS data: A cartographic study in NYC. *J. Maps* **2016**, *12*, 668–670. [CrossRef]
25. Duarte, L.; Teodoro, A.C.; Sousa, J.J.; Pádua, L. QVigourMap: A GIS Open Source Application for the Creation of Canopy Vigour Maps. *Agronomy* **2021**, *11*, 952. [CrossRef]
26. Armstrong, M.P. Geography and computational science. *Ann. Am. Assoc. Geogr.* **2000**, *90*, 146–156. [CrossRef]
27. Duarte, L.; Teodoro, A.C.; Maia, D.; Barbosa, D. Radio Astronomy Demonstrator: Assessment of the Appropriate Sites through a GIS Open Source Application. *ISPRS Int. J. Geo-Inf.* **2016**, *5*, 209. [CrossRef]
28. Brovelli, M.A.; Mitasova, H.; Neteler, M.; Raghavan, V. Free and open source desktop and Web GIS solutions. *Appl. Geomat.* **2012**, *4*, 65–66. [CrossRef]
29. Christl, A. Free Software and Open Source Business Models. In *Open Source Approaches in Spatial Data Handling*; Advances in Geographic Information Science; Hall G.B., Leahy M.G., Eds.; Springer: Berlin, Germany, 2008; Chapter 2, pp. 21–48. [CrossRef]
30. Turton, I. Geo Tools. In *Open Source Approaches in Spatial Data Handling*; Springer: Berlin/Heidelberg, Germany, 2008; pp. 153–169. [CrossRef]
31. Barnes, N. Publish your computer code: It is good enough. *Nature* **2010**, *467*, 753–753. [CrossRef] [PubMed]
32. Issaka, Y.; Kumi-Boateng, B. Artificial intelligence techniques for predicting tidal effects based on geographic locations in Ghana. *Geod. Cartogr.* **2020**, *46*, 1–7. [CrossRef]
33. Lemenkova, P.; Debeir, O. Quantitative Morphometric 3D Terrain Analysis of Japan Using Scripts of GMT and R. *Land* **2023**, *12*, 261. [CrossRef]
34. Ajvazi, B.; Czimber, K. A comparative analysis of different DEM interpolation methods in GIS: Case study of Rahovec, Kosovo. *Geod. Cartogr.* **2019**, *45*, 43–48. [CrossRef]
35. Guan, Q.; Hu, S.; Liu, Y.; Yun, S., High-Performance GeoComputation with the Parallel Raster Processing Library. In *GeoComputational Analysis and Modeling of Regional Systems*; Springer International Publishing: Cham, Switzerland, 2018; pp. 55–74. [CrossRef]
36. Kralidis, A.T., Geospatial Open Source and Open Standards Convergences. In *Open Source Approaches in Spatial Data Handling*; Springer: Berlin/Heidelberg, Germany, 2008; pp. 1–20. [CrossRef]
37. Wagemann, J.; Siemen, S.; Seeger, B.; Bendix, J. Users of open Big Earth data—An analysis of the current state. *Comput. Geosci.* **2021**, *157*, 104916. [CrossRef]
38. Basith, A.; Prastyani, R. Evaluating ACOMP, FLAASH and QUAC on Worldview-3 for satellite derived bathymetry (SDB) in shallow water. *Geod. Cartogr.* **2020**, *46*, 151–158. [CrossRef]

39. Habib, M.; Alfugara, A.; Pradhan, B. A low-cost spatial tool for transforming feature positions of CAD-based topographic mapping. *Geod. Cartogr.* **2019**, *45*, 161–168. [CrossRef]

40. Yin, Z.; Li, J.; Liu, Y.; Xie, Y.; Zhang, F.; Wang, S.; Sun, X.; Zhang, B. Water clarity changes in Lake Taihu over 36 years based on Landsat TM and OLI observations. *Int. J. Appl. Earth Obs. Geoinf.* **2021**, *102*, 102457. [CrossRef]

41. Chen, L.; Ma, Y.; Lian, Y.; Zhang, H.; Yu, Y.; Lin, Y. Radiometric Normalization Using a Pseudo-Invariant Polygon Features-Based Algorithm with Contemporaneous Sentinel-2A and Landsat-8 OLI Imagery. *Appl. Sci.* **2023**, *13*, 2525. . [CrossRef]

42. Rey, S.J., Code as Text: Open Source Lessons for Geospatial Research and Education. In *GeoComputational Analysis and Modeling of Regional Systems*; Springer International Publishing: Cham, Switzerland, 2018; pp. 7–21. . [CrossRef]

43. Zhang, Y.; Li, Q.; Tu, W.; Mai, K.; Yao, Y.; Chen, Y. Functional urban land use recognition integrating multi-source geospatial data and cross-correlations. *Comput. Environ. Urban Syst.* **2019**, *78*, 101374. [CrossRef]

44. Kim, S.; Hoang, Y.; Yu, T.T.; Kanwar, Y.S. GeoYCSB: A Benchmark Framework for the Performance and Scalability Evaluation of Geospatial NoSQL Databases. *Big Data Res.* **2023**, *31*, 100368. [CrossRef]

45. Teixeira, J.; Chaminé, H.I.; Carvalho, J.M.; Pérez-Alberti, A.; Rocha, F. Hydrogeomorphological mapping as a tool in groundwater exploration. *J. Maps* **2013**, *9*, 263–273. [CrossRef]

46. Sâvulescu, I.; Mihai, B. Mapping forest landscape change in Iezer Mountains, Romanian Carpathians. A GIS approach based on cartographic heritage, forestry data and remote sensing imagery. *J. Maps* **2011**, *7*, 429–446. [CrossRef]

47. Hernández-Santana, J.R.; Méndez-Linares, A.P.; López-Portillo, J.A.; Preciado-López, J.C. Coastal geomorphological cartography of Veracruz State, Mexico. *J. Maps* **2016**, *12*, 316–323. [CrossRef]

48. Ribeiro, J.; Viveiros, D.; Ferreira, J.; Lopez-Gil, A.; Dominguez-Lopez, A.; Martins, H.F.; Perez-Herrera, R.; Lopez-Aldaba, A.; Duarte, L.; Pinto, A.; et al. ECOAL Project—Delivering Solutions for Integrated Monitoring of Coal-Related Fires Supported on Optical Fiber Sensing Technology. *Appl. Sci.* **2017**, *7*, 956. [CrossRef]

49. Kiani, M.; Chegini, N.; Safari, A.; Nazari, B. Spheroidal spline interpolation and its application in geodesy. *Geod. Cartogr.* **2020**, *46*, 123–135. [CrossRef]

50. Bivand, R. Using the R statistical data analysis language on GRASS 5.0 GIS database files. *Comput. Geosci.* **2000**, *26*, 1043–1052. [CrossRef]

51. Bivand, R.S. *Integrating GRASS 5.0 and R: GIS and Modern Statistics for Data Analysis*; Technical Report 228; University of Bergen, Department of Geography: Bergen, Norway, 1999.

52. Lemenkova, P.; Debeir, O. R Libraries for Remote Sensing Data Classification by K-Means Clustering and NDVI Computation in Congo River Basin, DRC. *Appl. Sci.* **2022**, *12*, 12554. [CrossRef]

53. Grohmann, C.H. Morphometric analysis in Geographic Information Systems: Applications of free software GRASS and R. *Comput. Geosci.* **2004**, *30*, 1055–1067. [CrossRef]

54. Chapman, C.A. A new quantitative method of topographic analysis. *Am. J. Sci.* **1952**, *250*, 428–452. . [CrossRef]

55. Hofierka, J.; Mitasova, H.; Neteler, M., Geomorphometry in GRASS GIS. In *Geomorphometry: Concepts, Software, Applications—Developments in Soil Science*; Elsevier: Amsterdam, The Netherlands, 2009; Volume 33, pp. 387–410. [CrossRef]

56. Hofierka, J.; Súri, M. The solar radiation model for Open Source GIS: Implementation and applications. In Proceedings of Open Source Free Software GIS—GRASS Users Conference, Trento, Italy, 11–13 September 2002; pp. 11–13.

57. Mitas, L.; Mitasova, H., Spatial interpolation. In *Geographical Information Systems: Principles, Techniques, Management and Applications*; Longley, P., Goodchild, M., Maguire, D., Rhind, D., Eds.; Wiley: New York, NY, USA, 1999; pp. 481–492.

58. Mitasova, H.; Hofierka, J.; Zlocha, M.; Iverson, L. Modeling topographic potential for erosion and deposition using GIS. *Int. J. Geogr. Inf. Sci.* **1996**, *10*, 629–641. [CrossRef]

59. Lemenkova, P.; Debeir, O. Satellite Image Processing by Python and R Using Landsat 9 OLI/TIRS and SRTM DEM Data on Côte d'Ivoire, West Africa. *J. Imaging* **2022**, *8*, 317. [CrossRef]

60. Lemenkova, P. Geodynamic setting of Scotia Sea and its effects on geomorphology of South Sandwich Trench, Southern Ocean. *Pol. Polar Res.* **2020**, *42*, 1–23. [CrossRef]

61. Clarke, K.C. Computation of the fractal dimension of topographic surfaces using the triangular prism surface area method. *Comput. Geosci.* **1986**, *12*, 713–722. [CrossRef]

62. Sofia, G. Combining geomorphometry, feature extraction techniques and Earth-surface processes research: The way forward. *Geomorphology* **2020**, *355*, 107055. [CrossRef]

63. Lemenkova, P.; Debeir, O. Seismotectonics of Shallow-Focus Earthquakes in Venezuela with Links to Gravity Anomalies and Geologic Heterogeneity Mapped by a GMT Scripting Language. *Sustainability* **2022**, *14*, 15966. . [CrossRef]

64. Lemenkova, P.; Debeir, O. Satellite Altimetry and Gravimetry Data for Mapping Marine Geodetic and Geophysical Setting of the Seychelles and the Somali Sea, Indian Ocean. *J. Appl. Eng. Sci.* **2022**, *12*, 191–202. [CrossRef]

65. Malinverno, A., Fractals and Ocean Floor Topography: A Review and a Model. In *Fractals in the Earth Sciences*; Springer US: Boston, MA, USA, 1995; pp. 107–130. [CrossRef]

66. Cui, Q.; Zhou, Y.; Liu, L.; Gao, Y.; Li, G.; Zhang, S. The topography of the 660-km discontinuity beneath the Kuril-Kamchatka: Implication for morphology and dynamics of the northwestern Pacific slab. *Earth Planet. Sci. Lett.* **2023**, *602*, 117967. [CrossRef]

67. Neteler, M.; Mitasova, H. *Open Source GIS—A GRASS GIS Approach*, 3rd ed.; Springer: New York, NY, USA, 2008.

68. Neteler, M.; Beaudette, D.; Cavallini, P.; Lami, L.; Cepicky, J., GRASS GIS. In *Open Source Approaches in Spatial Data Handling*; Springer: Berlin/Heidelberg, Germany, 2008; pp. 171–199. [CrossRef]

69. Neteler, M. Geosynthesis 11: Der Praktische Leitfaden zum Geographischen Informationssystem GRASS. In *GRASS-Handbuch*; University of Hannover: Hannover, Germany, 2000.
70. Wessel, P.; Luis, J.F.; Uieda, L.; Scharroo, R.; Wobbe, F.; Smith, W.H.F.; Tian, D. The Generic Mapping Tools version 6. *Geochem. Geophys. Geosystems* **2019**, *20*, 5556–5564. [CrossRef]
71. GDAL/OGR Contributors. *GDAL/OGR Geospatial Data Abstraction Software Library*; Open Source Geospatial Foundation: Chicago, IL, USA, 2020.
72. Defense Mapping Agency. *Department of Defense World Geodetic System 1984: Its Definition and Relationships with Local Geodetic Systems: Technical Report*; Technical Report 8350; Defense Mapping Agency: Fairfax, VA, USA, 1991.
73. Brown, W.; Astley, M.; Baker, T.; Mitasova, H. GRASS as an integrated GIS and visualization environment for spatio-temporal modeling. In Proceedings of the Auto-Carto XII, ACSM/ASPRS, Charlotte, NC, USA, 27 February–2 March 1995; Volume 1, pp. 89–99.
74. Golden Software, Inc. *Full User's Guide: Surfer 12—Powerful Contouring, Gridding, and Surface Mapping*; Golden Software, Inc.: Golden, CO, USA, 2014.
75. Schmidt, J.; Evans, I.S.; Brinkmann, J. Comparison of polynomial models for land surface curvature calculation. *Int. J. Geogr. Inf. Sci.* **2003**, *17*, 797–814. [CrossRef]
76. Gallant, J.C.; Moore, I.D.; Hutchinson, M.F.; Gessler, P. Estimating fractal dimension of profiles: A comparison of methods. *Math. Geol.* **1994**, *26*, 455–481. [CrossRef]
77. Hodgson, M.E. What cell size does the computed slope/aspect angle represent? *Photogramm. Eng. Remote Sens.* **1995**, *61*, 513–517.
78. Hodgson, M.E. Comparison of Angles from Surface Slope/Aspect Algorithms. *Cartogr. Geogr. Inf. Syst.* **1998**, *25*, 173–185. [CrossRef]
79. Zevenbegen, L.W.; Thorne, C.R. Quantitative analysis of land surface topography. *Earth Surf. Process. Landforms* **1987**, *12*, 47–56. [CrossRef]
80. Dikau, R., The application of a digital relief model to landform analysis in geomorphology. In *Three Dimensional Applications in Geographic Information Systems*; Taylor & Francis: London, UK, 1989; pp. 51–77. [CrossRef]
81. Dunn, M.; Hickey, R. The Effect of Slope Algorithms on Slope estimates within a GIS. *Cartography* **1998**, *27*, 9–15. [CrossRef]
82. Guth, P.L. Slope and aspect calculations on gridded digital elevation models: Examples from a geomorphometric toolbox for personal computers. *Z. Geomorphol.* **1995**, *101*, 31–52.
83. Heerdegan, R.G..; Beran, M.A. Quantifying source areas through land surface curvature and shape. *J. Hydrol.* **1982**, *57*, 359–373. [CrossRef]
84. Mitasova, H.; Mitas, L.; Brown, W.; Gerdes, D.; Kosinovsky, I.; Baker, T. Modeling spatially and temporally distributed phenomena: New methods and tools for GRASS GIS. *Int. J. Geogr. Inf. Sci.* **1995**, *9*, 433–446. [CrossRef]
85. Evers, K.; Knudsen, T. Transformation pipelines for PROJ.4. In *Surveying the World of Tomorrow—From Digitalisation to Augmented Reality, Prceedings of the FIG Working Week 2017, Helsinki, Finland, 29 May–2 June 2017*; Gim International: Latina, Italy, 2017; pp. 1–13.
86. Snyder, J.P. *Map Projections—A Working Manual*; 1. U.S. Geological Professional Paper; U.S. Government Printing Office: Washington, DC, USA, 1987; 385p.
87. Evenden, G.I. *Cartographic Projection Procedures for the UNIX Environment—A User's Manual*; Technical Report 90-284, USGS Open-File Report; USGS: Reston, VA, USA, 1990.
88. Dassau, O.; Holl, S.; Neteler, M.; Redslob, M. *An Introduction to the Practical Use of the Free Geographical Information System GRASS 6.0. version 1.2*; GDF Hannover: Hannover, Germany, 2005.
89. Mitasova, H. Cartographic Aspects of Computer Surface Modeling. Ph.D. Thesis, Slovak Technical University, Bratislava, Slovakia, 1985.
90. Horn, B.K.P. Hill Shading and the Reflectance Map. *Proc. IEEE* **1981**, *69*, 14–47. [CrossRef]
91. Antrop, M.; De Maeyer, P.; Neutens, T.; Van de Weghe, N. *Geografische Informatiesystemen*; Academia Press: Gent, Belgium, 2013.
92. Bailey, T.; Gatrell, A. *Interactive Spatial Data Analysis*; Longman Scientific: Harlow, UK; John Wiley & Sons: New York, NY, USA, 1995.
93. Burrough, P.A.; McDonnell, R.A. *Principles of Geographical Information Systems*; Oxford University Press: Oxford, UK, 1998.

 technologies

Article

Identifying Historic Buildings over Time through Image Matching

Kyriaki A. Tychola [1,*], **Stamatis Chatzistamatis** [1,2], **Eleni Vrochidou** [1], **George E. Tsekouras** [2] and **George A. Papakostas** [1,*]

[1] MLV Research Group, Department of Computer Science, International Hellenic University, 65404 Kavala, Greece
[2] Department of Cultural Technology and Communications, University of the Aegean, 81100 Mytilene, Greece
* Correspondence: kytzcho@cs.ihu.gr (K.A.T.); gpapak@cs.ihu.gr (G.A.P.); Tel.: +30-2510462361 (G.A.P.)

Abstract: The buildings in a city are of great importance. Certain historic buildings are landmarks and indicate the city's architecture and culture. The buildings over time undergo changes because of various factors, such as structural changes, natural disaster damages, and aesthetic interventions. The form of buildings in each period is perceived and understood by people of each generation, through photography. Nevertheless, each photograph has its own characteristics depending on the camera (analog or digital) used for capturing it. Any photo, even depicting the same object, is impossible to capture in the same way in terms of illumination, viewing angle, and scale. Hence, to study two or more photographs depicting the same object, first they should be identified and then properly matched. Nowadays, computer vision contributes to this process by providing useful tools. In particular, for this purpose, several feature detection and description algorithms of homologous points have been developed. In this study, the identification of historic buildings over time through feature correspondence techniques and methods is investigated. Especially, photographs from landmarks of Drama city, in Greece, on different dates and conditions (weather, light, rotation, scale, etc.), were gathered and experiments on 2D pairs of images, implementing traditional feature detectors and descriptors algorithms, such as SIFT, ORB, and BRISK, were carried out. This study aims to evaluate the feature matching procedure focusing on both the algorithms' performance (accuracy, efficiency, and robustness) and the identification of the buildings. SIFT and BRISK are the most accurate algorithms while ORB and BRISK are the most efficient.

Keywords: computer vision; feature extraction; building identification; culture heritage; architectural heritage; image descriptor; image matching; image analysis

1. Introduction

A city is continuously evolving, and this implies large and unforeseen changes. As part of the urban landscape, historic buildings are the epitome and reflection of civilization. Over time, the urban net grows rapidly, and this entails large and unexpected spatial–temporal changes in the buildings. Some buildings have historical importance and are considered landmarks of a city [1]. Architectural heritage is a wealth of cultural heritage expression and an invaluable testimony to the past. Hence, it should be protected [2]. This is achieved with new technologies. Nowadays, the contribution of computer vision with various tools available plays a catalytic role in building protection. For instance, a comparative study of the building evolution over different periods has positive effects on the cultural heritage of the city and for anyone interested. In addition, comparing images of historic buildings helps those in charge of better decision-making in the management of heritage [3]. For this purpose, various suitable matching techniques are applied aiming at the correct correspondence between images and their further processing [4].

An image is a result of a camera recording and is a way to capture reality. The images, which capture an entity over time, such as a building, were not captured under the same

Citation: Tychola, K.A.; Chatzistamatis, S.; Vrochidou, E.; Tsekouras, G.E.; Papakostas, G.A. Identifying Historic Buildings over Time through Image Matching. *Technologies* **2023**, *11*, 32. https://doi.org/10.3390/technologies11010032

Academic Editor: Gwanggil Jeon

Received: 14 January 2023
Revised: 4 February 2023
Accepted: 11 February 2023
Published: 17 February 2023

conditions. Therefore, for their comparison, feature extraction and matching techniques should be applied. Specifically, for feature extraction and matching from two or more images of the same scene but from different viewing angles and captured from the same or different camera, homologous point detection techniques are used. This is an important task in image analysis [5]. Especially, it is the primary preprocessing image step for further processing, and it has been widely applied in various computer vision applications, such as pattern recognition, robot navigation, image stitching and mosaicking [6], visual odometry [7], pose estimation, object classification, and 3D reconstruction in the case of 3D images [8,9]. In the case of buildings, this procedure is of major importance, as in a further analysis of the images they can be compared and identify possible lesions and deformations. To achieve strong feature matching, the invariant properties of the images should be utilized in order that the extracted features not vary with respect to lighting changes, scale, position, rotation, and viewing angle [10].

Feature extraction from images is achieved by feature detectors, which detect feature invariants when the image undergoes different transformations. The term 'invariant feature' refers to those features that remain invariant when rotation, scaling, illumination, and affine transformation are applied. Detection is achieved by scaling the image to extract distinct features across various scales of objects to identify. After this process, various descriptors are applied to describe the features extracted with repeatability, compatibility, accuracy, and efficient representations, which are also invariant to scale, rotation, affine transformation, occlusion, and illumination [2,11–13].

After feature detection and description follows the feature matching. The matching procedure finds and matches identical points between image pairs by calculating the displacement from the changes in the pixels [14]. There are two main image-matching methods: the area-based methods, where detectors are applied to find the similarity of the pixels between source and target images followed by optimization algorithms [15–18], and the feature-based methods, where the features are extracted directly from images without calculating the intensity values. The latter category is suitable for images with complex geometric distortions and lighting changes [17,18]. Historical images of buildings are an important source of information for researchers. Finding images of specific objects, visually comparing different constructions, or estimating proportions, is important. These tasks are largely related to metadata (relative position, orientation of images, and download time), and since the quality of metadata varies, it poses a problem in detection and identification of entities in an image [19]. To date, there are reliable image-matching techniques for matching vector- or binary-based images to modern images. Nevertheless, they present inaccuracies in historical images [20–22].

Although in the literature, there are many studies on building management and conservation, such as building disaster prediction [23], there are very few studies about feature-matching methods and techniques for historical building 2D images. For instance, Agarwal et al. [24] have presented edge-detection techniques for matching images, based on graphs. Surapog et al. [25] proposed the fusion of historical and modern photographs with a database-based indexing technique; however, their original purpose was not the matching. Heider and Whitehead [20] applied vector-based, binary-based, and hybrid techniques, such as ORB detector/SURF descriptor, on historical and modern buildings of the same landmarks to examine the correspondence problem. Hybrid techniques have proved the most efficient, whereas in the same period they used feature descriptors based on the gradient of two historical images that converted them into a matrix. The results were incorrect [26]. Wolf et al. [27] proposed an innovative approach for feature matching using image regions. The expectation was that using regions instead of corners and edges would lead to greater precision in the matching process. The initial results were mixed, and no concrete solution was found. Wu et al. [28] have proposed a matching methodology for historical buildings based on contours. They based it on the Canny edge-detection algorithm and replaced the Sobel filter with a modified Scharr filter, and they automatically adjusted the local limit, instead of using the Canny algorithm settings. Their algorithm was sensitive to multidirectional gradient change but was very efficient at detailed building corners. Kabir S. R. et al. [2] proposed four computational methods to feature detection (Canny edge detection, Hough line transform, find contours, and Harris corner detector) on historic and modern buildings. Then, they evaluated their algorithms' performances, concluding that these detectors were best suited for this purpose. Samaun Hasan et al. [29] applied the Canny edge detector to a dataset of modern and old Indian buildings and then designed a neural CNN model able to distinguish two different time periods (Sultans and Mongols). Maiwald F. et al. [30] used exclusively geometrical features and semantic constraints (windows, material, and overlays) to match two or more images. As long as strong features were taken into account, the linear structures of objects were easily detected as quadrilaterals. Yue L. and Zheng X. [31] proposed a method for the distorted images of buildings but not of historical ones. They applied the TILT algorithm to correct the image, and an automated detection method to low-textured photographs, while for image matching the ORB algorithm was used to remove the outliers. Si, L. et al. [32] implemented a genetic image-matching algorithm for associating two building images but not historical ones to solve the problem of optimization and proposed techniques for quickly associating homologous points. Edward J. and Yang G.Z. [33] applied the RANSAC algorithm to pairs of a building's images in changing environments for learning invariant features, during the day and different seasons. The authors emphasized the geometry of similar buildings and the calculation of distances and corners between all the matched features. Finally, Avrithis, Y. and Tolias, G. [34] introduced a methodology of histogram pyramids based on Hough voting, where the votes were derived from the single feature matches, for accelerated large-scale image retrieval. It is noted that this investigation was not only limited to buildings. In addition, the authors applied the SURF algorithm to extract and describe the image features.

This study attempts to investigate the feature extraction and matching from various aspects such as comparing historical buildings over time between them, as well as with buildings of other architectures. In addition, it studies the performance of three algorithms and the possible changes on buildings over different time periods. More specifically, this work is focused on the employment of well-established image-matching algorithms to the problem of historical buildings' identification over time, given their indisputable performance and considering the inherent limited data available for the problem under study. To accomplish that task, three algorithms are compared, namely, the SIFT [35], the ORB [36], and the BRISK [37] algorithm. Alternative methods such as deep learning could be investigated; however, they are in need of large datasets, which in the case of historical buildings over time are challenging and time-intensive to collect.

The current research focuses on studying the correspondence problem through experiments, related to the identification of historical buildings in Drama city, Greece. Feature extraction and matching techniques on 2D images were applied, which were captured in different conditions of lighting, viewing angle, scale, season, weather conditions, and 24 h period. Then, the results were evaluated by focusing on both the performance and reliability of used algorithms, and any changes on buildings were reflected through the results.

The rest of the paper is organized as follows: Section 2 presents the motivation and contribution of this work. Section 3 presents and provides our research strategy, including a theoretical background of features extraction and image matching. In addition, the correspondence problem is highlighted and its importance in the computer vision is discussed. Section 4 describes the dataset and presents the simulation experiments, explaining the parameters that were implemented. Section 5 illustrates the experimental results. Finally, Section 6 report the discussion and conclusions of this study.

2. Motivation and Contribution

Our motivation for conducting a comprehensive survey on identifying historic buildings over time through image matching is twofold. First, the significance of the subject. Simply stated, the ability of algorithms to extract and match features between two images is the essence of most computer vision tasks. Second, the previous reviews on the subject (see next section) carry out experiments with traditional algorithms, either focusing on their performance on historic and/or modern buildings or aiming to identify and retrieve the images through databases (image retrieval).

Undoubtedly, the previous works offer a significant contribution. Most of them analyze the algorithm performance for feature matching at the level of feature extracted from two or more images. Nevertheless, none of them provides a comprehensive study gathering, presenting, and analyzing in detail both the algorithm performance and identifying historic buildings over time among them and with other buildings, through experiments. This encourages us to carry out a thorough investigation through experiments studying all factors and parameters that are crucial in identifying buildings. Hence, the contribution and innovation of our work lie in that we present a comprehensive analysis of classical detectors and description algorithms such as SIFT, ORB, and BRISK, we conduct experiments on image pairs of the same building and on different buildings with the best parameters of these algorithms, and we analyze their capabilities and limitations. In addition, we provide solutions and identify open issues and research trends, thereby providing directions for future work. Hence, our unique contribution is a focused and detailed analysis of identifying historic buildings over time through image matching.

Moreover, we try to answer various research questions about identifying historic buildings over time through image matching. The research questions are summarized below:

- Does time affect the identity of the building?
- To what extent is the validity of the identification affected?

3. Literature Analysis

The research methodology that has been followed in this paper included an initial search on Scopus with the following query: TITLE-ABS-KEY (feature AND detectors AND descriptors). This search, conducted in November 2022, yielded 1832 papers, but only journal articles, conference papers, and book chapters were selected. To pursue a more exhaustive exploration of the literature we went beyond Scopus and coupled it with a search in Google Scholar, using the same terms. Then, our probing branched out in two parts. First, we conducted a series of secondary, more detailed Google Scholar searches, using terms specific to the phrases "feature matching methods" and "feature extraction algorithms". Then, we sought and gathered any relevant works referenced in the papers from our primary Scopus search, examined and properly evaluated them. This wide-reaching reconnaissance brought 87 papers under our scrutiny, which we screened for relevancy. Figure 1 shows the percentage of each type of publication in the papers selected, while Figure 2 shows the progress of the published papers per year. Early in our exploration of the bibliography, we noticed that this field is quite old. The first publications date back to 1983, with the volume of research steadily increasing over the following years.

According to the above chart, the studies start "timidly" only in 1983, and for the first 9 years, there are few posts on this topic. From 2004 to 2016, researchers' interest seems to be increased. From 2016 to date, a slight drop in publications is noted. This phenomenon can be explained by the fact that deep learning methods have been gaining more and more ground in recent years. The correspondence problem is still being investigated and seems to be attracting interest from the scientific community.

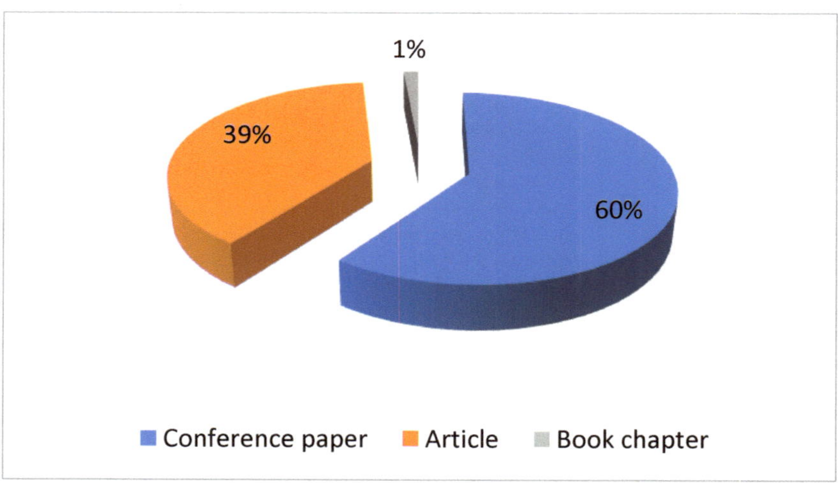

Figure 1. The pie chart illustrates the proportion of papers from different types of publications.

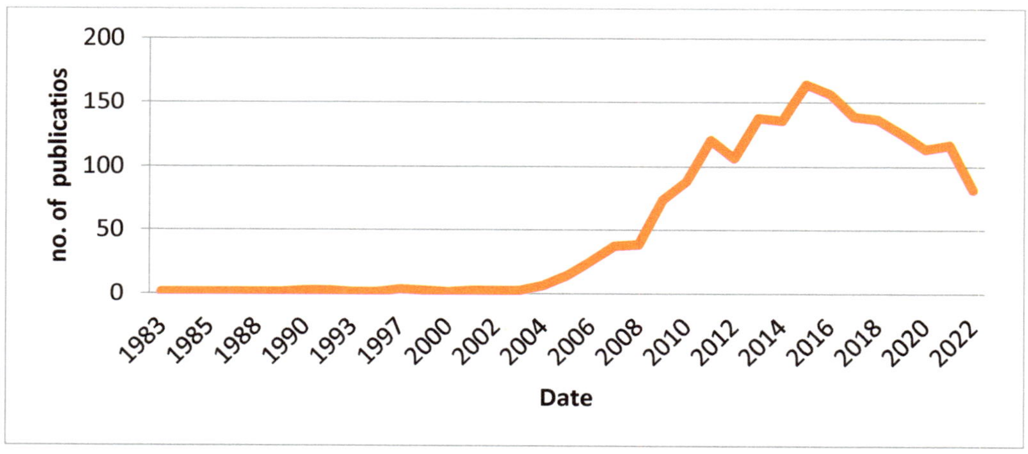

Figure 2. Number of relevant publications per year, about feature detectors and descriptors (statistics from November 2022).

4. Materials and Methods

In image processing, feature extraction is the process of transforming raw data into numeric attributes that can be further processed, preserving the information in the original dataset [38–42]. This procedure aims to reduce the number of attributes of a dataset and to create new from existing ones [43]. The characteristics to be selected should be representative so that they contain as much information as possible about the original set of features. The usefulness of this process is to increase the accuracy of the results [44,45]. Nowadays, we have at our disposal a wide variety of images depicting the same landmarks over time, which are either captured by analog cameras with film or digital cameras. It is understood that a collection of photographs that contains the same object, e.g., a building, cannot be identical (different viewing angle, different material, various weather conditions, lighting, scale, or date). Therefore, feature matching between historic and modern images is a complex process and has constraints such as the correspondence problem.

4.1. Correspondence Problem

After feature extraction follows feature matching, which is undoubtedly the main element in various applications such as optical flow stereo vision and structure from motion (SfM) [46,47]. Feature matching refers to finding matching attributes from two similar images, based on a search-distance algorithm. One of the images is considered the source (reference image) and the other the target, and the feature-matching technique is applied either for detection or for extraction and transport of a feature from reference to target image. In particular, feature matching analyzes the topology of images, detects feature patterns, and matches with features within localized patterns. The accuracy of matching features depends on the similarity, complexity, and quality of the image. In addition, feature selection plays an important role. For this purpose, various algorithms have been developed; however, none of them can be universally accepted as to high efficiency and accuracy [48]. The idea of feature matching arose in the early 1980s, when it was realized that the human brain associates (i.e., matches) entities altogether based on their characteristics and not just the tones of hue. Matching becomes less sensitive to radiometric and geometric distortions of images and takes into account space structure information, ensuring more powerful and reliable solutions. There are two stages for matching: the first stage includes feature extraction such as points, lines, edges, corners, and regions on each image, which is linked to descriptive properties (attributes) in the form of descriptive vectors (descriptors), and the second relates to feature matching after calculating a measure of similarity between their properties. In this case, the search for homologous points is restricted to the index and to descriptive features and not to the entire image, reducing managed information and thereby increasing the computational cost. When Takeo Kanade was asked, "What are the three most important problems in computer vision?" he replied, "Correspondence, correspondence, correspondence" [49]. Successful features matching allows us to create matches between pairs of points and interpret the visual world (Figure 3).

Figure 3. Feature detection and matching homologous points.

The correspondence problem remains one open issue. Given two or more images of the same scene captured from different perspectives, the correspondence problem refers to the problem of determining the parts of an image to which parts of another image correspond, i.e., the points of features in an image are the same with another image and create homologous points. In addition, the problem intensifies further because of other factors that are mentioned above [50,51]. Figure 4 shows an example of the correspondence problem.

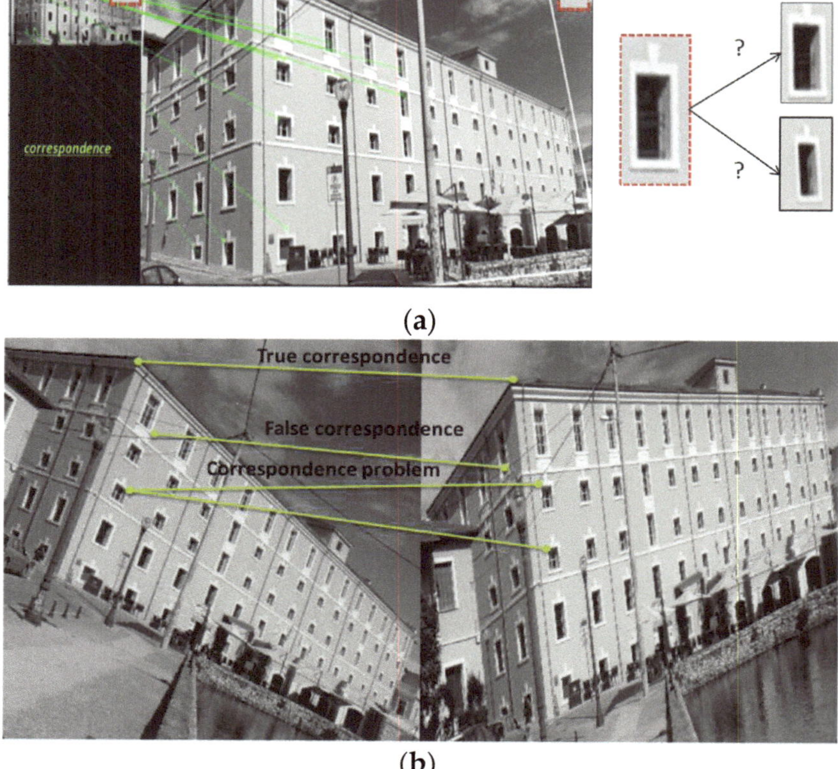

Figure 4. Correspondence problem: (**a**) feature matching between two images with the same building; (**b**) clear indication of the correspondence problem. There are false and true correspondences but in the third case (i.e., + correspondence problem), the algorithm is having trouble locating the position of the point between similar characteristics.

4.2. Feature Detectors and Descriptors for Feature Matching

Feature detectors and descriptors have been powerful tools over the last few decades in many applications in the computer vision field and pattern recognition. Feature is a piece of information extracted from an image representing a more detailed region and can be divided into *global*, which provides information about the entire image, and *local*, which focuses on a specific part of an image [52]. A feature detector to be qualified should fulfill the following criteria [5]:

- *Stability*: the locations of the features detected should be independent of different geometric transformations, scaling, rotation, translation, photometric distortions, compression errors, and noise;
- *Repeatability*: detectors should be able to detect the same features of the same scene or object repeatedly under various viewing conditions;
- *Generality*: detectors should be able to detect features that can be used in different applications;
- *Accuracy*: the feature detection should be the same accuracy localized both in image location;
- *Efficiency*: fast detection to support applications in real time;

- *Quantity*: the number of detected features should be sufficiently large, such that a reasonable number of features are detected even on small objects.

After detection follows the description of interest points. The terminology "interesting feature" includes several interchangeable terms, such as keypoint, landmark, interest point, or anchor point, all of which refer to features such as corners, edges, or patterns that can be found repeatedly with high likelihood [53]. A feature descriptor can be computed at each key point to provide more information about the pixel region surrounding the keypoint and it is a representation of transforming the located features into a new space called the feature description space, where the feature matching is more easily distinguished [54–56]. Feature description is foundational to feature matching, leading to image understanding scene analysis. The central problems in feature matching include how to determine if a feature is differentiated from other similar features, and if the feature is part of a larger object. The method of determining a feature match is critical, for many reasons such as computational cost, memory size, repeatability, accuracy, and robustness, while a perfect match is ideal. In practice, a relative match is determined by a distance function, where the incoming set of feature descriptors is compared with known feature descriptors [57,58]. Figure 5 illustrates the steps of feature matching.

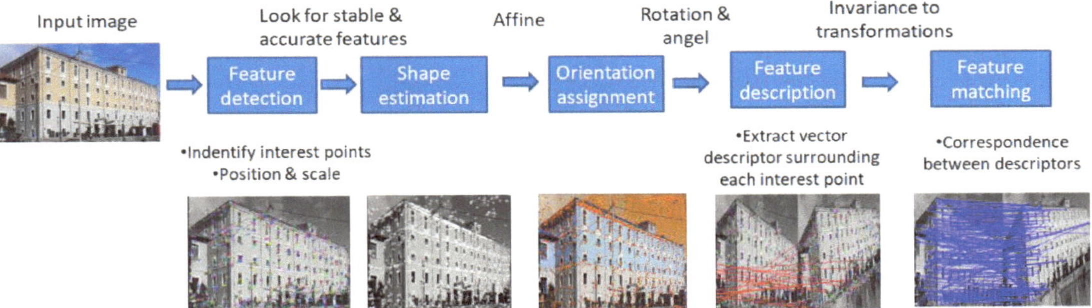

Figure 5. Feature detection and matching process.

According to Figure 5, scaling differences exist between the same images, especially in regions where the distances are close. First, the features detected, such as corners, detect interest points in a space scale (identify position and scale) around each image while subsequently local descriptions are extracted from the adjacent areas of these points. Nevertheless, feature extraction may not extract an adequate number of points of interest because of the bad texture of repeated patterns, different viewing angles, lighting, and blur [59,60]. Second, affine transformation is applied to reduce asymmetry and different scales on the axes of the two images, and their orientation is assessed, considering the different rotations of the two images around the detection point. From the above procedure is determined the characteristics' position, scale, affine shape, and rotation. Therefore, a high-dimensional vector represents features detected, descriptors are derived independently for each image, and the matching of the two images is achieved by searching the neighborhood [55,60]. Image-matching methods can be divided into two major categories: area-based methods emphasizing the step of matching and work directly with the image intensity value, and feature-based methods that are based on feature extraction of important structures such as regions (forest and lakes), lines (area boundaries, coastlines, roads, and rivers) or points (area corners, line intersections, and high curves) [61–65]. In recent years, hybrid approaches have been applied at the same time, i.e., methods based on areas and features [66].

To identify historic buildings over time through image matching, we carry out experiments applying various classical feature detection and description algorithms such as SIFT [35], ORB [36], and BRISK [37] using the trial-and-error method for feature matching.

This section can be divided by subheadings. First, we describe our dataset and discuss the buildings' history. Second, we discuss the background of the algorithms, focusing on their parameters, and we analyze, in detail, the parameters selected for our experiments. Finally, we provide a concise comparison among algorithms.

4.3. Study Area

Drama city is part of the East Macedonia and Thrace region (Figure 6) and is located in northeastern Greece. Drama has experienced a great economic boom mainly during the Ottoman era with tobacco production and trade. This area was chosen, as it is a tourist attraction because of the various historical events over time and is considered a cultural center of great interest with milestones of various architectural orders.

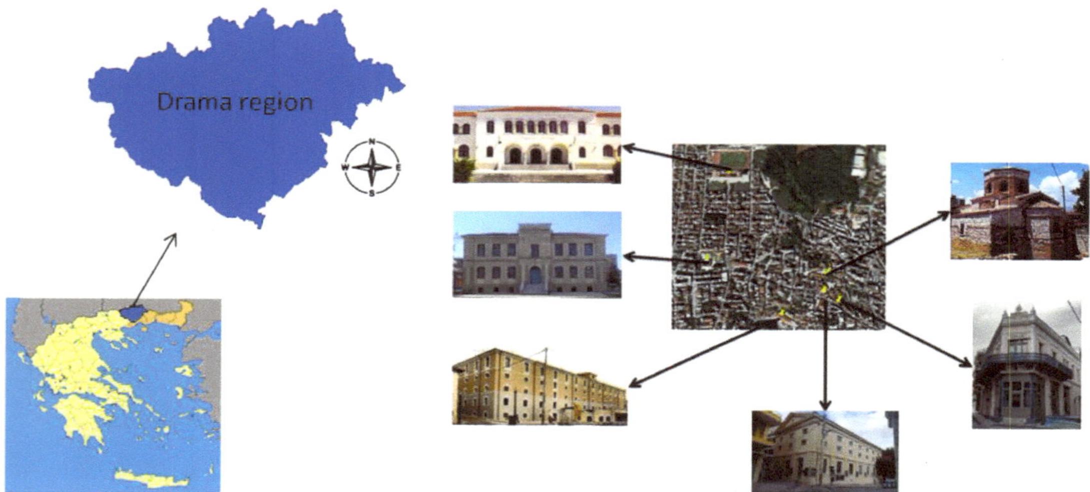

Figure 6. The study area is a part of Drama city. In the smaller map of Greece is the location of Drama, while the yellow pins represent the four landmarks that were used in this study.

4.4. Data Acquisition and Description

For the needs of the experiments, various images, with landmarks at various years, under different condition shots, and with different lenses for analog cameras and digital cameras, of Drama city in Greece were collected from the site https://gr.pinterest.com/ (accessed on 13 January 2023) and books. In addition, we used images from personal files received on mobile phones such as Samsung A21 and Huawei P20 Lite. Furthermore, historical photo collectors have offered us some images from their collections. Note that the collection of images of historical buildings over time requires very extensive and systematic research in a large volume of files and different sources that may not be easily accessible. Finally, our dataset includes four different landmarks: tobacco industry (1990–2022), elementary school (1908–2020), Hagia Sofia church (1994–2022), and café "Eleutheria" (1940–2022), where each landmark contains 28, 7, 39, and 17 images, respectively. Figure 7 shows a sample of our dataset.

The tobacco industry was built in 1874 and belongs to I. Anastasiades and was a sample of economic development of Drama city. The school was built in 1908 on the initiative of the Metropolitan of Drama in the plans of the architect Chrysostomos Hatzimichalis. The family of Pavlos Melas offered a part of the cost, while artisans of the region aided in the construction. Nowadays, it operates as an elementary school (12th elementary school). The Church of Hagia Sophia is the oldest preserved building in Drama. It was built in the highest part of the city during the 10th century, along with the city's old walls. The conditions for the erection of the monument are unknown, but its construction coincides with a Byzantine period of great prosperity. In addition, the church was probably dedicated to Theotokos and it was built on the ruins of a larger three-aisled church. An early Christian basilica, where debris was revealed during the landscaping work, and in terms of architecture, the temple keeps similarities with iconic monuments of Byzantine church architecture, such as the Hagia Sophia of Thessalonica and the Assumption of the Virgin Mary in Nicaea, Bithynia, Asia Minor. The historical café "Eleftheria" was built in the early 20th century (1906–1907) by the Greek community of Drama at the intersection of the present Venizelou and Kountouriotou streets. It got its name "Eleutheria" ("Freedom" in Greek) after the liberation of the city. The last few years it operated as a coffee shop, whereas, sometimes, it offers hospitality to painting exhibitions.

Figure 7. Landmarks samples from the dataset of Drama city.

4.5. Experiments

In this work, we carry out experiments using handcrafted matching methods to investigate feature extraction and matching for identifying historic buildings over time that the tobacco industry is the basic building of this study. Specifically, we implement a trial-and-error method to find the best parameters of algorithms that make them efficient. It should be noted here that this is not a generic approach, and thus, the extracted parameters may need to be recalculated for large deviations in the buildings' architecture. The focus of this work is on the employment of well-established algorithms in the field of image matching to be systematically applied to the problem of historical buildings' identification over time, given their indisputable performance and considering the limited data at our disposal. To this end, we apply classical algorithms to pairs of images, such as SIFT, ORB, and BRISK. These algorithms were selected because they present some advantages. For instance, SIFT (scale-invariant feature transform) is perfectly suitable because it detects features that are invariant to image scale and rotation. Moreover, features are robust to lighting changes, noise, and minor changes in viewpoint. ORB features are also invariant to scale and rotation, while the BRISK algorithm is robust to noise and invariant to scale and rotation. All the experiments were designed and implemented in Open CV. Figure 8 shows the workflow chart of our methodology.

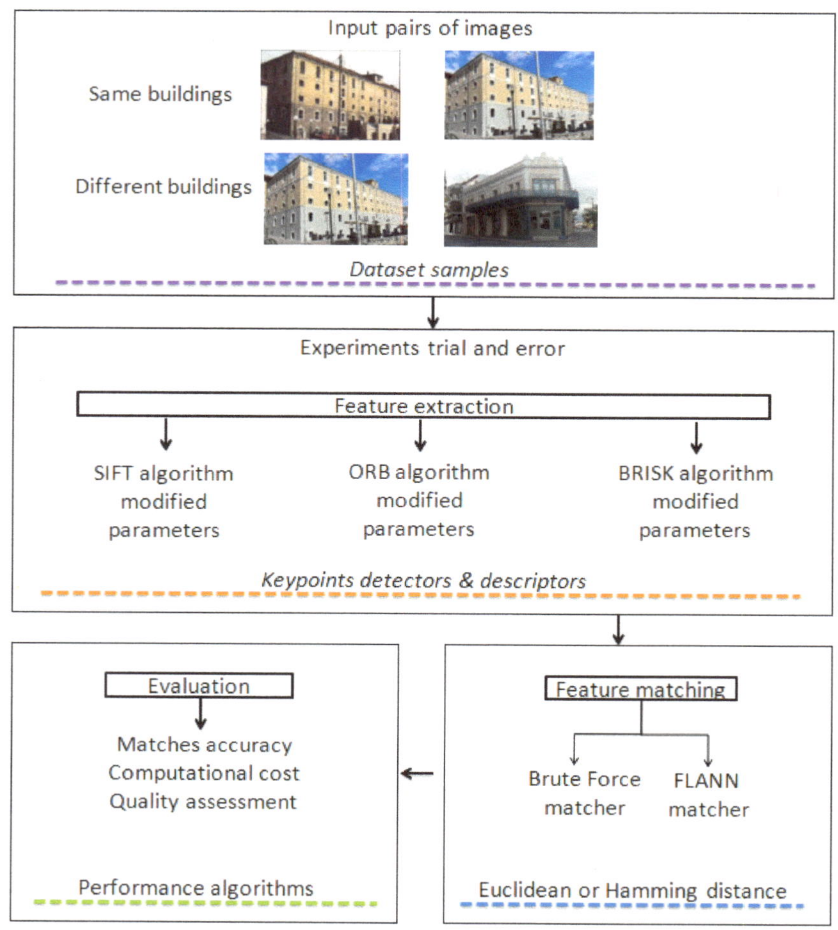

Figure 8. The flowchart of experiments.

We divided the whole procedure into four stages. Specifically, first, we input pairs of images of the tobacco industry from 1990 until 2022 (pair 9–pair 28). It should be noted that images of 2022 captured the building from different aspects, scales, and rotations. In addition, we consider that buildings from 2021 (pair 6–pair 28) and beyond are modern buildings, whereas the rest are historic (pair 1–pair 5). Besides, we compared each date of the building to all the rest, while then we tested the tobacco industry landmark with three other buildings based on the architectural orders, structure materials, and date. In the second step, we applied feature detectors and descriptions, such as SIFT, ORB, and BRISK algorithms to extract features from pairs of images. In this step, after tests, we modified the algorithm parameters in such a way as to produce the best possible results without interfering with any extra optimization method. In the third stage, we adopted the strategy of nearest neighbor distance ratio (NNDR), which was proposed by K. Mikolajczyk in [67] and D.G. Lowe in [68]. In this case, a ratio of the nearest neighbor to the second nearest neighbor is calculated for each feature descriptor, and a certain threshold ratio is set to filter out the preferred matches. We implemented the brute force matcher [68,69], i.e., a matching descriptor that compares two sets of descriptors' keypoints and produces a list of matches, and then the FLANN (FAST library for approximate nearest neighbors) matcher [70,71], which is used in two different forms, depending on the algorithm. Finally, in the next step, the matching pairs are evaluated by various quantitative and qualitative techniques. It is understood that because of the different applications of algorithms, it is difficult to judge their performance with a unified and commonly accepted system. Hence, various indicators should be found for evaluation. Commonly, precision and best score matching for each pair of images based on the distance (Euclidean or Hamming) are evaluated as algorithm performance benchmarks [72]. The closer the homologous points between two images, the better and more valid the match is. However, since only these evaluations were not sufficient, we applied additional measures, including the computational cost.

In what follows, our evaluation measures are listed:

- The total number of keypoints from each pair of images.
- The total number and the rate of the best matches (good matches). The percentage of best matches was calculated by dividing the best matches by the smallest number of keypoints extracted from the first or second image (good matches/min no, keypoints*100).
- Precision is a performance measure of the best matches (best matches/total matches), while then, we evaluated optically the best matches, aiming to find the false positive matches.
- Effectiveness (%) measure (total matches/total keypoints) to evaluate the actual number of keypoints used for matching.

4.6. SIFT, ORB, BRISK Trial and Error Methods

4.6.1. SIFT Algorithm

SIFT (scale-invariant feature transform) algorithm has been proposed by D. G. Lowe [35] and extracts invariant features. Today, it is globally accepted and is considered the best and most accurate algorithm. The algorithm detects local features (keypoints) and localizes them on the image. Its greatest advantage is that SIFT is robustly invariant to image rotations, scale, and limited affine variations, while its main drawback is the high computational cost. Figure 9 illustrates a SIFT workflow chart.

SIFT parameters are number of features, nOctaveLayers, contrast, edge Threshold, and Gaussian sigma during the parameter test. Since the desired effect was to extract the keypoints number as large as possible, the number of features was not set with value. As the number of octavelayers increases, the number of keypoints increases; however, it only increased by one point because a further increase did not favor the increase in matches partially, i.e., during the test, with a decrease of 1, the difference is also matches and keypoints decreased. ContrastThreshold is a parameter that specifies under which threshold the values of keypoints should be discarded. To be accepted, there must be points with contrast above the threshold. Increasing the value of filtered, more and more points are discarded. In short, as the value of the parameter increases, the process tightens, but at the same time, the success rate of the mappings increases. The ideal is many keypoints and a high success rate. The edgeThreshold parameter is a limit for the ratio of two identical vectors given by filtering keypoints on edges to discard them. The higher the limit, the more difficult it is to discard possible keypoints. As the threshold value increases, the number of keypoints increases. In the context of the test with the price fluctuation, the difference is not obvious and does not offer any improvement. The success rate decreases as the value increases. Because "pass" keypoints are not completely valid, the reduction is therefore negligible. The value of Gaussian sigma depends more on the quality of the images. Since most pictures have a good resolution, it was not considered necessary to dramatically change the value of the original. Increasing the value observed a significant decrease in keypoints and a large percentage of matching that was false, while a slight value increase decreases dramatically both the matching and the number of keypoints but not proportionally. Note that if the images were clear and largely blurred, a small value would be required. Therefore, the value selected was 1 point less than the default value. Finally, we also increased the ratio.

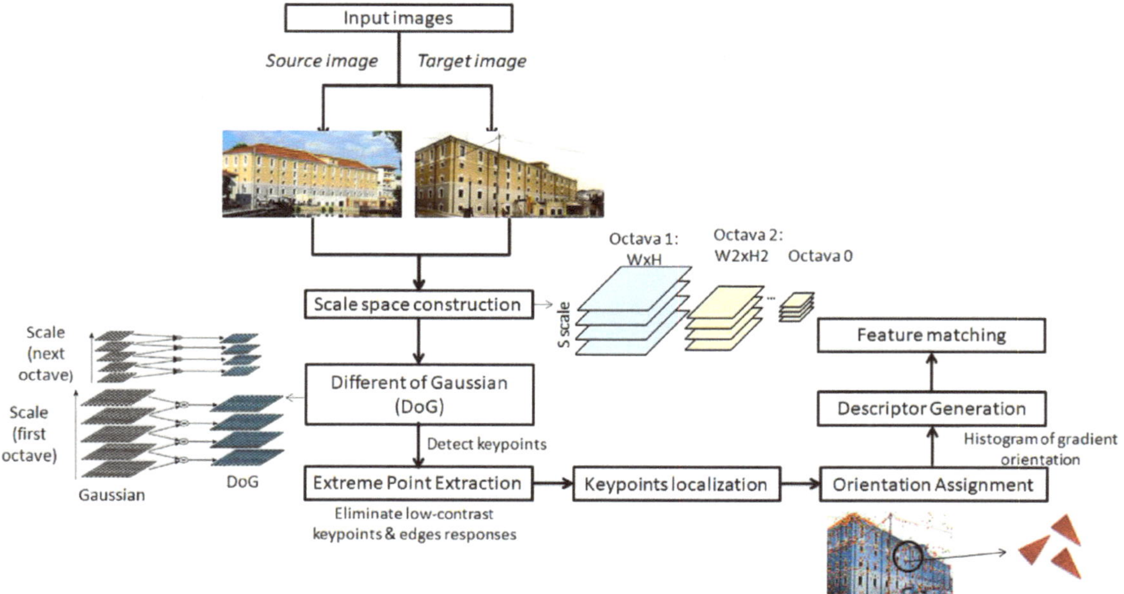

Figure 9. Workflow chart of SIFT algorithm. SIFT detector is based on the difference-of-Gaussians (DoG) operator, which is an approximation of Laplacian-of-Gaussian (LoG). Feature points are detected by searching local maxima using DoG at various scales of the subject images. The description method extracts a 16×16 neighborhood around each detected feature and further segments the region into sub-blocks, rendering a total of 128 bin values.

4.6.2. ORB Algorithm

The ORB (oriented FAST and rotated BRIEF) algorithm [36] is a combination of FAST (features from accelerated segment test) [73] and BRIEF (binary robust independent elementary features) algorithms [74,75], with some changes to improve accuracy. Figure 10 illustrates the workflow chart of ORB.

ORB parameters are the number of features, scaleFactor, number of octaves level, edgeThreshold, firstLevel, WTA_K and patchSize, and FastThreshold. The max number of keypoints has been set to 5000. The scale factor increased from 1.2 to 1.5. As a result, keypoints and good matches decreased significantly, while total matches decreased slightly. Moreover, it was observed that a scale factor above 1.5 gives reduced results but a slightly higher percentage of better matches than the default value of 1.2. The next test involved the modification of edgeThreshold and sizepatches. From 31, which is the default value, it has been reduced to 28, and at the same time it has been slightly increased and the scalefactor price has increased from 1.2 to 1.5. As a result, keypoints, total matches, and good matches were reduced by half as compared to the previous modifications, while if the scale factor remains at 1.2 and edgeThreshold and size patches will be reduced then, the same number of keypoints is produced in pairs of images; however, we have low total matches and good matches. Then we increased the scale factor to 1.5 and decreased edgeThreshold to 29 and FASTThreshold from 20 to 8. The results were dramatically reduced in both the number of keypoints and matches. Furthermore, we discovered that it makes no sense to increase or decrease edgeThreshold and sizepatch because good matches are increasing; nevertheless, important points on the images are ignored. Finally, FASTThreshold does not offer any significant information or improvement of results. Finally, the ratio was modified to 0.85.

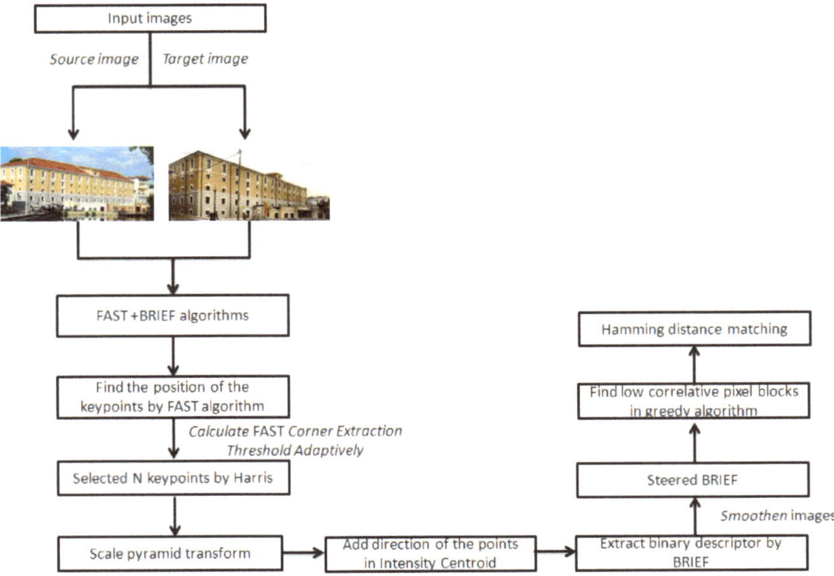

Figure 10. Workflow chart of ORB. The ORB detector takes the intensity threshold between the center pixel and those in a circular ring about the center, and FAST uses a simple measure of corner orientation, the intensity centroid that assumes that a corner's intensity is offset from its center, and this vector is used to refer to an orientation.

4.6.3. BRISK Algorithm

The BRISK (binary robust invariant scalable keypoints) algorithm [37,76] was developed in 2011 as a free alternative to SIFT. It is a robust salient point detection, description, and matching algorithm. The BRISK algorithm utilizes a 16-point FAST detector to identify potential salient points in octaves and intra-octaves of scale-space pyramid. The FAST detector also calculates FAST score S, which is the threshold that still considers an image point as a corner. Then, a non-maxima suppression routine is applied to detect salient points. Figure 11 shows the workflow of BRISK. To identify a test point in the scale-space pyramid as a salient point, the point should satisfy the following conditions:

1. At least nine consecutive pixels within the 16-pixel circle centered on the test point must be sufficiently brighter or darker than the test point;
2. The test point needs to fulfill the maximum condition with respect to its eight neighboring FAST scores S in the same octave;
3. The score of the test point must be larger than the scores of corresponding pixels in the above and below layers [77].

It detects and describes features with invariant scaling and rotational variability. BRISK constructs the description of local image attributes through the grayscale relationship of random pairs of points in the neighborhood of the local image and obtains the description of binary attributes. Two types of pairs are used for sampling: short where the distance is below a set threshold distmax and long which has a distance above distmin pairs. Moreover, its descriptor has a predefined sampling pattern as compared to ORB. Long pairs are used for orientation and short pairs are used for calculating the descriptor by comparing intensities and pixels are sampled over concentric rings.

BRISK parameters are threshold, number of octaves, and patternScale. Threshold is used between the center pixel density and the pixels within the circle around this pixel while the patternScale applies this scale to the pattern used for sampling the neighborhood of a keypoint. Initially, the tests were conducted only on the octaves. If the default value is increased then only the number of keypoints decreased is negligible, whereas if the value decreases by 1 then the number of keypoints and total matches are decreased. However, the good matches and the best matching rate are increased. Then we tried to modify the value of patternScale. A value greater than 1 (double, i.e., 2) led to a decrease in all results, while a half-doubling of the value (i.e., 0.5) increased all results, but the results are almost similar to those when default parameters are used. In case all parameters decrease, an increase in total matches is observed; however, it is virtual, as all other parameters remain at the same levels as the default parameters. On the other hand, if only threshold is increased and the other two parameters are reduced, then the number of keypoints and total matches is reduced and the success rate of best matching is increased twice. Finally, reducing only the threshold increases all results except the percentage of success of the matching remaining at the same levels, while increasing the threshold reduces all results dramatically. Finally, the ratio was also modified to 0.85. In Table 1, we present in brief the SIFT, ORB, and BRISK properties and characteristics.

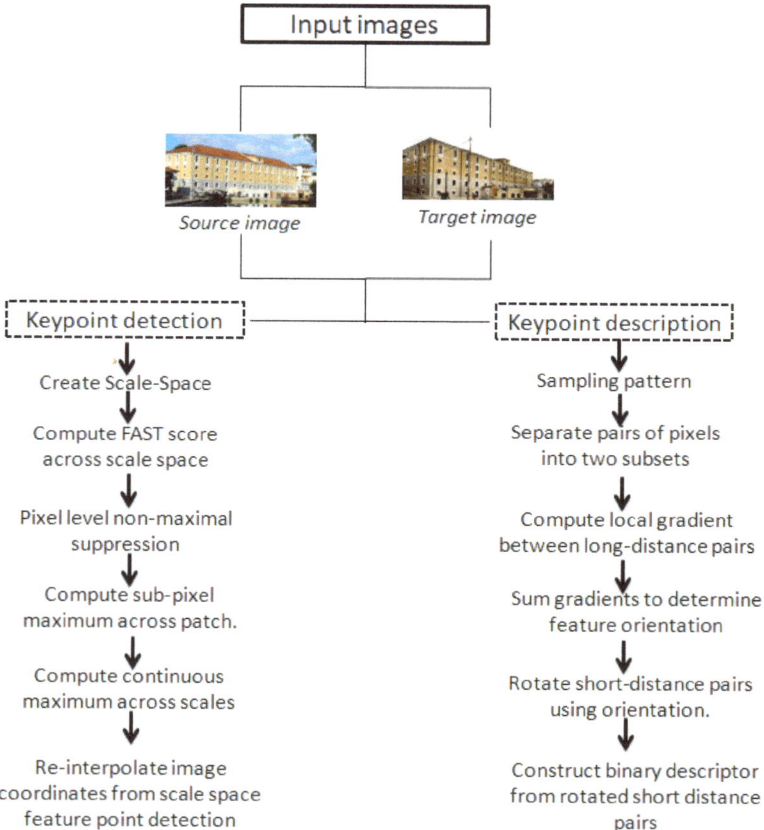

Figure 11. Workflow of BRISK algorithm scale-space keypoint detection: points of interest are identified across both the image and scale dimensions using a saliency criterion. Keypoints are detected in octave layers of the image pyramid. Keypoint description: a sampling pattern consisting of points lying on appropriately scaled concentric circles is applied at the neighborhood of each keypoint to retrieve gray values (processing local intensity gradients and the feature characteristic direction is determined).

Table 1. Illustration of a brief comparison of SIFT, ORB, and BRISK algorithms.

Properties	SIFT (2004)	ORB (2011)	BRISK (2011)
Operators	(a) Detecting keypoints from the multi-scale image space, presented by difference-of-Gaussians (DoG) operator (i.e., approximation of Laplacian-of-Gaussian (LoG)), (b) Keypoint point localization by removing low-contrast and those on edge, (c) assigning orientations to each keypoint based on an orientation histogram, weighted by gradient magnitude and Gaussian-weighted circular window, and (d) providing a unique and robust keypoint descriptor by considering the neighborhood around the keypoint and its orientation histogram [33]	It is a combination of modified FAST (Features from Accelerated Segment Test) which detects corner objects as candidate points, and the Harris Corner score is then utilized to refine them from low-quality points. [73], detection and BRIEF (binary robust independent elementary features) descriptor [74]	It first extracts corners as feature point candidates using the AGAST algorithm [74] and then refines them with the FAST corner score in each scale-space pyramid layer. The illumination robust and rotation invariant descriptor has been generated based on each feature's characteristic direction and simple brightness tests [37]
Keypoints	DoG [57,78]	FAST [79]	AGAST [80]
Detectors type	Blob [81]	Corner [82]	Corner [82]
Descriptors			
Descriptors type and length	Integer vector 128 bytes [35]	Binary string 256 bits [36]	Binary string 64 bytes [37]
Encode information	Gradient-based descriptor [83,84]	Intensity-based descriptor [16]	Intensity-based descriptor [16]
Scale invariant	Yes	Yes (achievable via an image pyramid)	Yes
Rotation invariant	Yes	Yes (achievable via intensity centroid)	Yes
Distance matching	Euclidean [85]	Hamming [86]	Hamming [86]
Constraints	Limited affine changes, high computational cost	Limited affine changes	Limited affine changes, error rate does not exist
Strong points	Robust to illumination fluctuations, noise, Partial occlusion, and minor viewpoint changes in the images [87]	Very Fast, Reduce sensitivity to noise	Robust to noise and affine performance

5. Experimental Results

In this section, we provide a concise and precise description of the experimental results. We itemize the algorithms' parameters and we discuss the keypoints detected and matched, and present our evaluation methods through charts.

5.1. Algorithms Parameters Trial and Error

Table 2 illustrates the default and the modified parameters of algorithms after our trial-and-error tests.

Table 2. Default and modified parameters of algorithms.

Algorithms	Default Parameters	Modified Parameters Trial and Error
SIFT	nfeatures = 0, nOctaveLayers = 3, contrastThreshold = 0.04, edgeThreshold = 10, sigma = 1.6, ratio = 0.7	nfeatures = 0, nOctaveLayers = 4, contrastThreshold = 0.05, edgeThreshold = 8, sigma = 1.5, ratio = 0.85
ORB	nfeatures = 500, scaleFactor = 1.2, nlevels = 8, scoreType = cv.ORB_HARRIS_SCORE, edgeThreshold = 31, firstlevel = 0, scoreType:Harris_score, patchsize = 31, WTA_K = 2, FastThreshold = 20, ratio = 0.7	nfeatures = 5000, scaleFactor = 1.5, nlevels = 8, scoreType = cv.ORB_HARRIS_SCORE, edgeThreshold = 31, firstLevel = 0, WTA_K = 2, patchSize = 31, FastThreshold = 20, ratio = 0.85
BRISK	threshold = 30, octaves = 3, patternScale = 1.0, ratio = 0.7	thresh = 40, octaves = 2, patternScale = 1.0, ratio = 0.85

5.2. Features Detection and Matching

In this subsection, the visualized results from representative images of features extracted and matched by algorithms, both of the same building over time and between different buildings based on certain criteria, are presented. Figure 12 provides some examples of the tobacco industry from keypoints detected between historic buildings, between historic and modern buildings, and between modern buildings. Keypoints were obtained by SIFT (green), ORB (blue), and BRISK (red) detectors. Figure 13 compares the tobacco industry building with different buildings from various aspects such as date, material, and architectural order.

Figure 12. Tobacco industry building: detection of keypoints. The first pair of buildings is historic (1990–1992), the second pair has 20 years of distance (2000–2021), while the third pair of buildings is modern and of the same year (2022). Green, blue, and red keypoints indicate SIFT, ORB, and BRISK algorithms, respectively.

Figure 13. Keypoints detected by SIFT (green), ORB (blue), and BRISK (red) algorithms between different buildings. The first column includes two modern buildings. The second column includes two stone buildings, and the third column includes two buildings of the same order and at the same time; one is historical and the other modern.

Between historic buildings, the SIFT algorithm detected more keypoints around the point of interest, i.e., windows; however, it did not manage to detect additional features such as walls and vegetation. The ORB algorithm detected more keypoints than SIFT, but they are in scattered form and dense, while the BRISK algorithm detected the least number of keypoints as compared to SIFT and ORB but are more concentrated on corners. Between historic and modern building keypoints were detected by SIFT, mainly on material that divides the texture of the building line under each windows row but not around the interest points such as windows edges and corners. Obviously, the algorithm recognizes the small texture or pattern changes. ORB extracted more and dense keypoints to the entire building, while BRISK detected fewer and sparse keypoints; however, it has more keypoints on all sides of the building. Between modern buildings with different rotation and scales, the SIFT algorithm confirms its advantages as keypoints are detected on all strong interest points; however, it is partially affected by other elements such as water. On the contrary, ORB detected more keypoints, but it seems to present a weakness in the region where the texture or material changes. For instance, there are keypoints denser and scattered near the water. BRISK has the same behavior; however, it has better and more uniform results than ORB.

In Figure 13, the first pair of images includes modern and historic buildings (tobacco industry 2022 and café "Eleutheria" 1940). The second pair contains two buildings with the same structure material (tobacco industry 2016 and Hagia Sofia 2019), while the third pair includes two buildings with almost of the same architecture—however, from different periods (tobacco industry 2022 and elementary school 1990). In the first case, all algorithms provide sparse detections concentrated on corners and at the points of pattern changes. ORB excels related to the density of keypoints, which is higher, while SIFT detects fewer and sparse feature points in the second pair of images. Thus, we notice that SIFT has a weakness to find more features when the texture changes, and we suppose that the spatial range of the feature points is constrained to the buildings with a non-rectangular geometric shape. BRISK detected sparse and scattered keypoints; however, it managed to detect more keypoints on corners than SIFT. Between different buildings by the same material (stone), all algorithms detected keypoints on the interest points; however, ORB and SIFT are affected more by other elements such as vegetation compared to BRISK. In addition, ORB and BRISK detected more dense keypoints in Hagia Sofia compared to SIFT. SIFT recognized better edges and corners on buildings with rectangle regions. In the third case, all the algorithms seem to have the same detection. From this case are seen clearly the advantages of algorithms, despite the diversity of buildings.

Then, we examined the feature matching results from the above buildings and the feature matching of the tobacco industry pairs of images over time (1990–2022). Figure 14 shows the feature matching from tobacco industry pairs. Between historic and modern buildings, SIFT and BRISK algorithms have fewer matches than ORB and confirmed intensively the correspondence problem. In addition, SIFT is also affected by vegetation, while ORB by additional elements such as small walls. Between historic and modern buildings, SIFT and BRISK produced almost the same results; however, the BRISK algorithm presents a weakness, although keypoints extracted were better uniformly distributed, while ORB has more dense matches mainly on the strong interest of points, and it seems to recognize all the textures of buildings. Between modern buildings, the feature matching is more organized and correct, visually. Specifically, SIFT and BRISK managed better performance than ORB and recognized also all the viewing angles of the building.

Figure 15 presents the feature matching results by the tobacco industry building compared to different buildings. The ORB algorithm gave the best results from the similar building shapes aspect while BRISK and SIFT failed to recognize strong points, although of the keypoint extraction phase, they detected them.

Figure 16 presents the feature matching results of the tobacco industry building over time by three algorithms. The building is between 1990–2022. Green, blue, and red colors are matches of the SIFT, ORB, and BRISK algorithms, respectively. SIFT and BRISK algorithms extract fewer matches both between historic and modern buildings, with an exception. Between modern images where there is no great rotation and scale, the feature matches are dense and concentrated on all buildings. On the contrary, ORB maintains density and uniformity of all pairs of images (i.e., historic and modern) and recognizes new additions and minor changes, such as on walls of historic buildings. Thus, the ORB algorithm is superior to the others, showing a robust visualized performance.

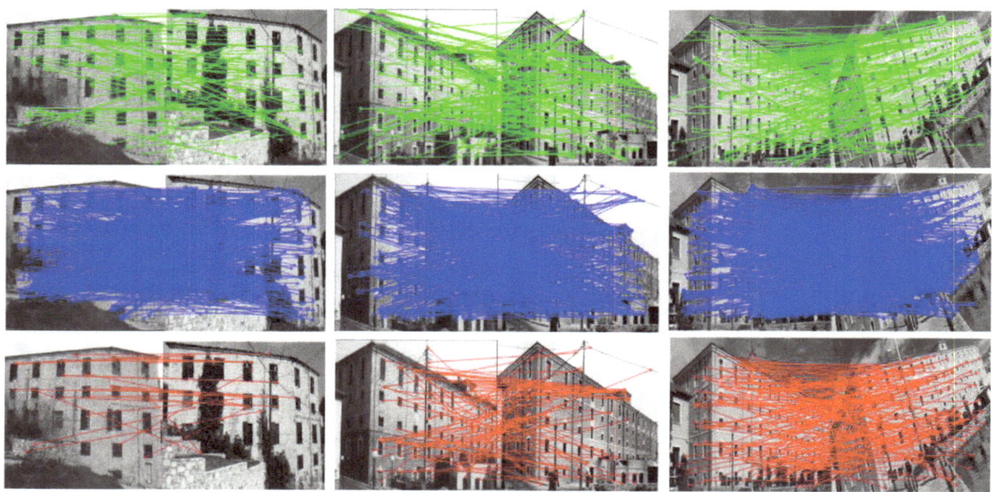

Figure 14. Feature matching. The first column is a pair of historic buildings, the second includes modern and historic buildings, while the third column includes two modern buildings by SIFT (green), ORB (blue), and BRISK (red) by trial-and-error-modified parameters.

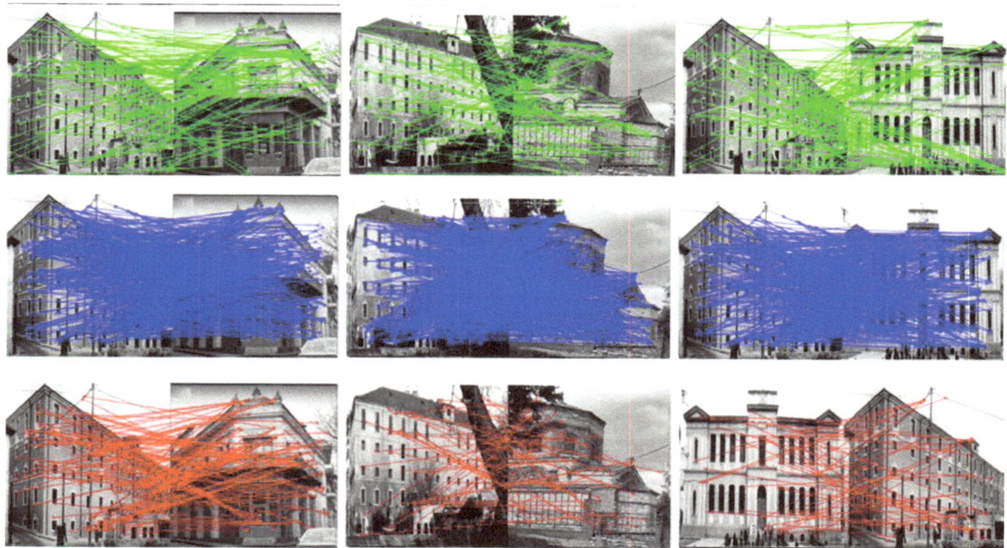

Figure 15. Feature matching of different buildings by SIFT, ORB, and BRISK algorithms with modified parameters.

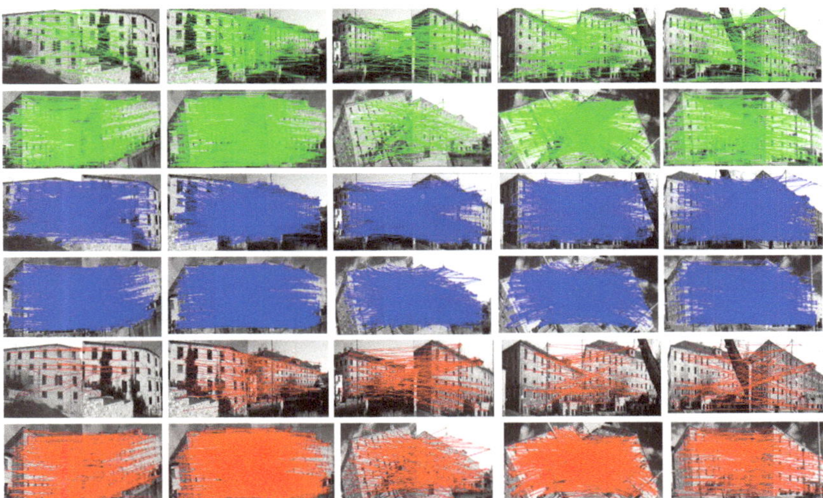

Figure 16. Tobacco industry building. Feature matching historic buildings over time. Green, blue, and red matches are by SIFT, ORB, and BRISK algorithms, respectively.

5.3. Feature Performance Evaluation

In this subsection, the quantified features detection and matching results are presented. Specifically, the number of keypoints detected by the SIFT, ORB, and BRISK algorithms, including the computational cost, are evaluated and compared. Figure 17 shows the total of keypoints from the tobacco industry pairs by algorithms with trial-and-error parameters. The ORB algorithm detects the highest number of features in all pairs of images compared to BRISK, which detects the greater number of features in modern buildings and fewer in historic buildings, while the SIFT algorithm is the opposite of BRISK, i.e., it detected fewer features in modern buildings and more in historic buildings. However, SIFT, ORB, and BRISK detected more keypoints on historic buildings (pair 2), while SIFT and BRISK detected them on modern buildings (pairs 18 and 21) with rotation of the greater number of keypoints, while ORB detected many keypoints on modern buildings without scale and intense rotation as observed in Figure 17.

The next evaluation concerns the evaluation of computational cost, which is an important factor during the feature-matching procedure. The computation efforts of each technique for dealing with modern to historic images can be estimated by recording and analyzing the runtime taken by each technique on each image-pair matching of all test operations. Table 3 shows the keypoints and total matches of the tobacco industry, all pairs (1990–2022), and the runtime results. The BRISK algorithm is the most efficient detector providing the fastest image matching, while SIFT has the greatest computational cost. However, it is normal because it presents also the greatest number of total matches. On the other hand, ORB has less cost compared to SIFT, although the number of total matches is not very different, while the number of total keypoints is almost double.

Table 3. The sum results obtained from pairs of images after applying the three algorithms to datasets, which contain sequences of images for the same building.

Algorithms	Total Keypoints	Total Matches	Runtime (s)
SIFT	76,192	25,246	44.18
ORB	169,233	21,017	31.64
BRISK	101,074	14,881	10.93

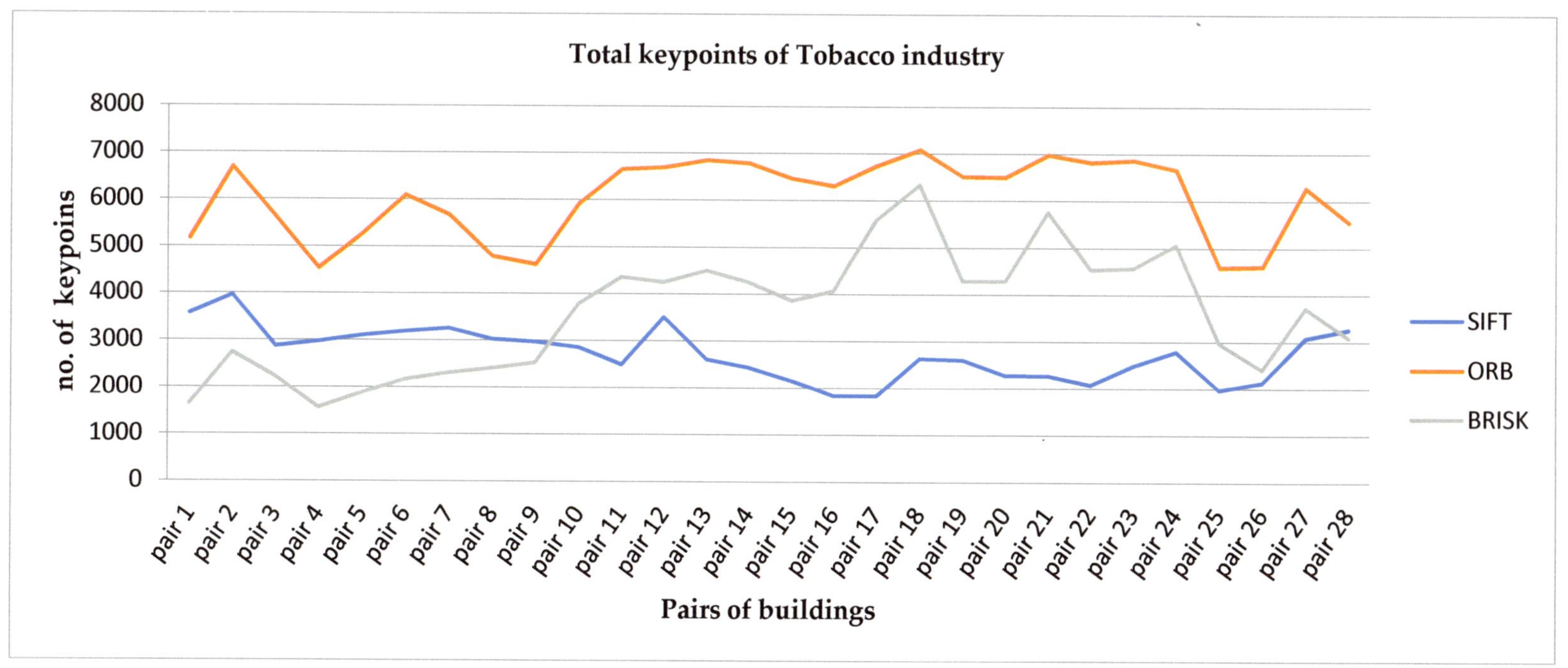

Figure 17. Total keypoints from each pair of images over time using modified parameters of algorithms.

Table 4 shows the number of keypoints detected from each pair of different buildings of the total matches and the computational cost. In this case of different buildings, it is observed that the number of features detected on each pair has significant differences. This is an indication that there are no common points between the images. Nevertheless, BRISK presents less computational cost, while SIFT higher, and ORB comes in second place. It is worth mentioning that the results are also justified based on the levels of human reflection. One may visually conclude whether buildings are similar or not. However, human reflection is also verified by the results. The latter is attributed to the fact that the extracted keypoints in all cases belong to the main structure of the building that does not alter over time. From a detailed observation of the keypoints located in the images, it can be clearly seen that they belong to basic structural elements of the building (windows, definition of floors, etc.) that remain unchanged and not to keypoints on plaster or cracks that are temporary and can be changed over time.

Table 4. Quantitative comparison and computational cost by algorithms for pairs of tobacco industry with other buildings.

Buildings	Algorithms	Keypoints		Total Matches	Runtime (s)
		Image 1	Image 2		
Pair 1	SIFT	1428	900	297	2.00
	ORB	2011	3672	606	2.70
	BRISK	1242	2824	375	0.72
Pair 2	SIFT	1586	1345	320	3.16
	ORB	2514	1978	623	2.10
	BRISK	1282	586	265	0.43
Pair 3	SIFT	2532	1497	177	2.34
	ORB	3424	2462	754	1.93
	BRISK	1329	894	282	0.91
Pair 4	SIFT	1579	2000	372	1.78
	ORB	3096	3527	880	1.54
	BRISK	1126	3382	502	0.75
Pair 5	SIFT	1586	1158	268	1.40
	ORB	2514	3184	674	1.76
	BRISK	1282	2958	492	0.65
Pair 6	SIFT	1042	1010	183	1.12
	ORB	1743	1824	434	1.93
	BRISK	309	549	109	0.42
Pair 7	SIFT	1586	1010	240	1.22
	ORB	2514	1824	553	1.75
	BRISK	1282	549	236	0.85

5.4. Efficient Keypoints Matching

In this subsection, we evaluate the feature matching of the tobacco industry building over time (Figure 18) and the tobacco industry building with different buildings (Figure 19), while then we investigate the effectiveness of keypoints to examine the number of keypoints that were actually used for matching.

The greatest rate of the best matches arose from the pairs with modern buildings that did not have rotation or differences at scale. All the algorithms have similar rates; however, BRISK and ORB present high rates between buildings with intensive scale differences (pairs 21 and 25). On the other hand, BRISK yields the least rates of the best matches mainly to historic buildings, while SIFT follows. Among the pairs, there are also buildings at night (pairs 8 and 9). In this case, the rate of best matches is reduced as compared to the during the day. The low illumination probably is a limiting factor for the satisfactory number of feature detection and matching.

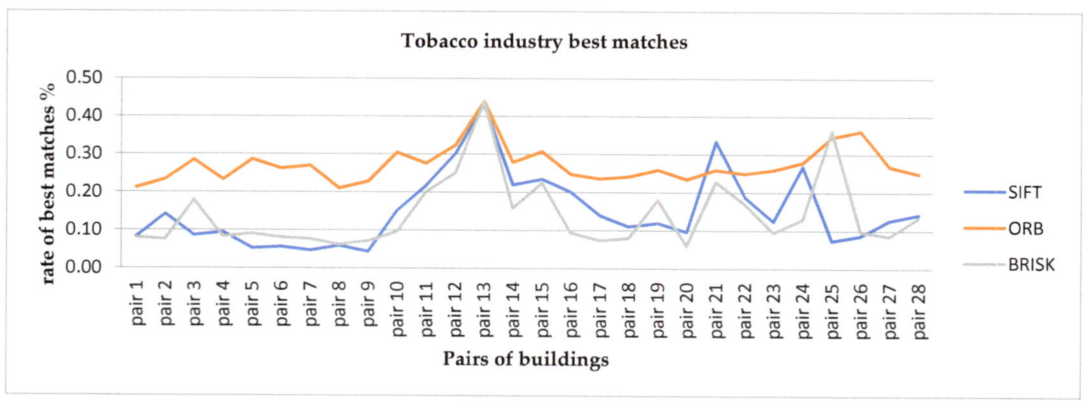

Figure 18. Tobacco industry: rate of the best matches from pairs of buildings over time.

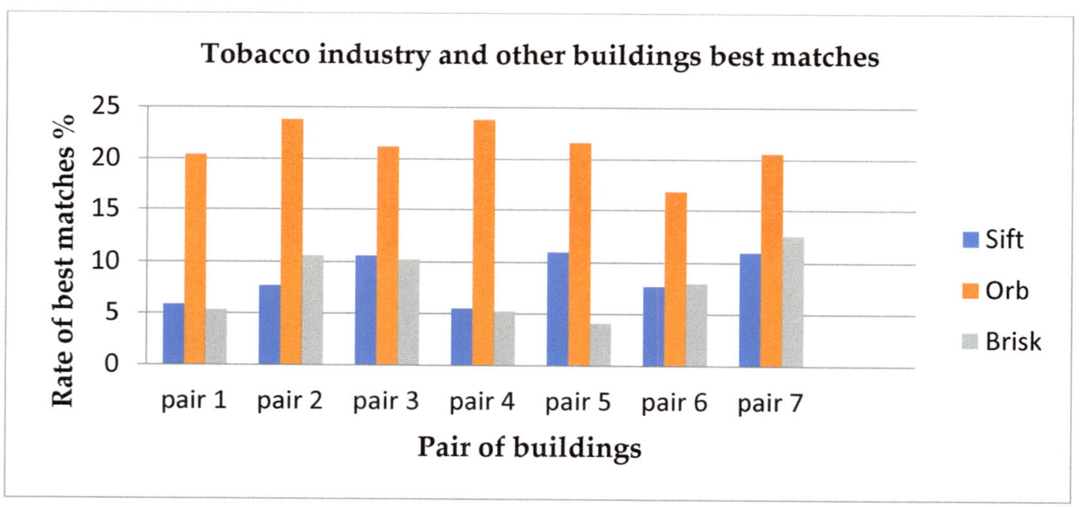

Figure 19. Rate of the best matches of different buildings by algorithms.

According to Figure 19, the ORB algorithm presents the highest rate of best matches while BRISK and SIFT present almost the same results. However, it is noticed that BRISK has the least rate of best matches (pair 5), i.e., between buildings with different architectural styles. (the tobacco industry has a rectangular shape and Hagia Sofia church an irregular shape).

Figure 20 shows the effectiveness of the tobacco industry building over time, while Figure 21 shows the effectiveness between the tobacco industry and other buildings. For the tobacco industry building over time, the BRISK algorithm has the least rate of keypoints matching both historic (1999–2000) and modern (pairs 21 and 25) buildings. On the other hand, the peak rate is presented by all algorithms to modern buildings without rotation (pair 13) and scale or with intensive rotation (pair 21), while BRISK presents the least rate on historic buildings (pair 3) and SIFT on different view-angle modern building (pair 7). Finally, all buildings have a tendency to decrease the rate of the keypoints actually used.

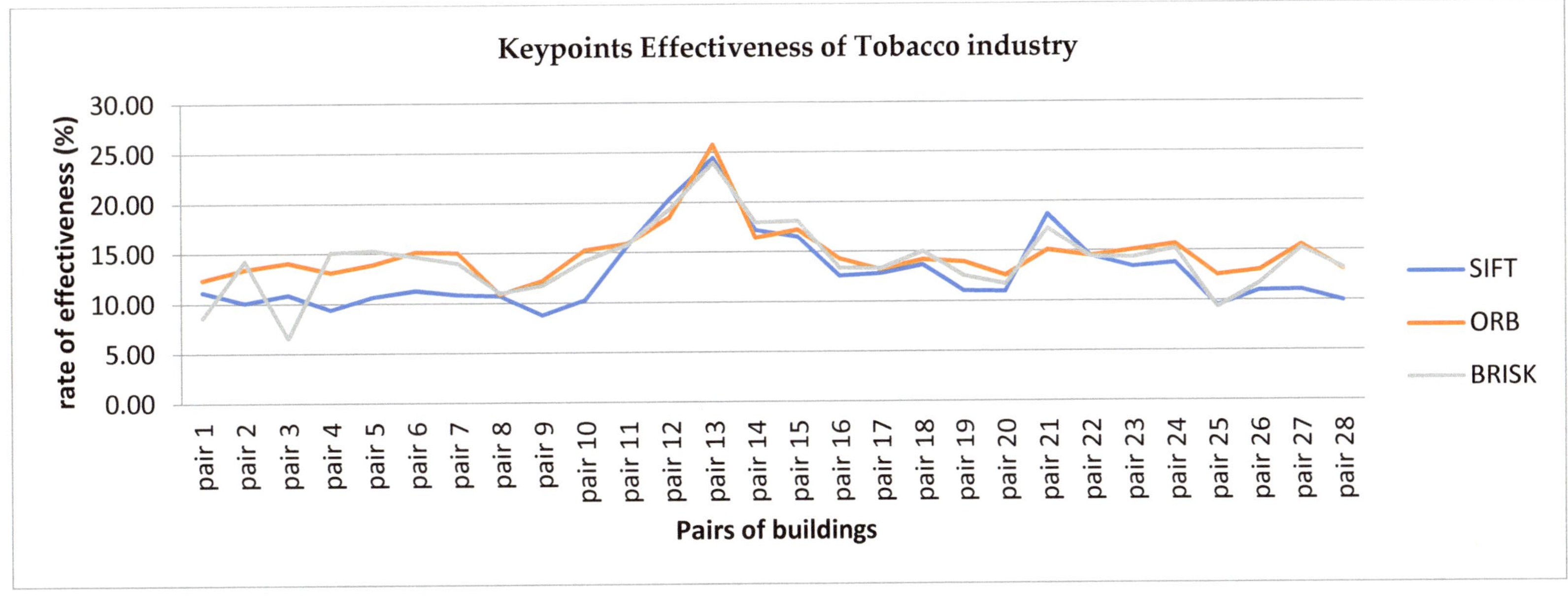

Figure 20. Effectiveness of keypoint matching in pairs of the same building over time.

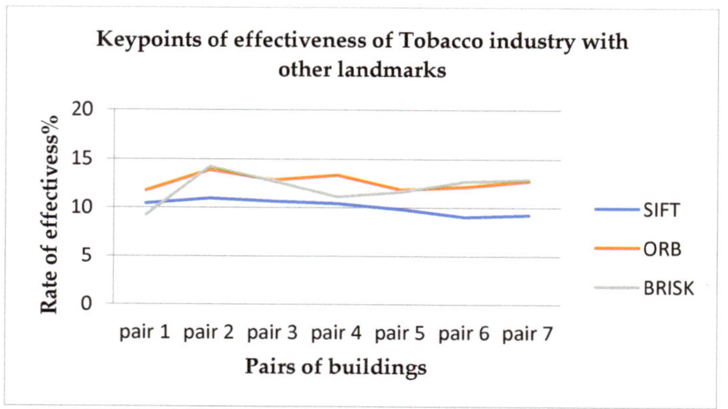

Figure 21. Effectiveness of keypoint matching in pairs of different building over time.

In Figure 21, BRISK and ORB present similar results to all pairs of images, while the rate by SIFT is similar to all pairs of buildings. However, in pair 4, while ORB presents an increase, SIFT presents a decrease. This is justified by the fact that between two buildings with different materials, SIFT fails to find similarities. Nevertheless, all results are virtual because there are not in reality common elements among the buildings.

Image Matching Accuracy

In this subsection, we investigate the final matching accuracy. First, we use an average precision index to evaluate coarsely the ratio of best matches/total matches, while since none of the algorithms completely guarantees correct feature matching results, we evaluate false positive matches from the best matches. The assessment is done visually without the threshold value because in such a case the results will probably not be unbiased. Figure 22 presents the average precision of good matches from the tobacco industry pairs over time. ORB provides the highest precision, contrary to BRISK, while SIFT comes in second place. In general, algorithms provide results from moderate to high precision. This is reasonable because the precision is affected by matching criteria. The stricter the matching criteria are, the greater the number of correct matches and the higher the accuracy.

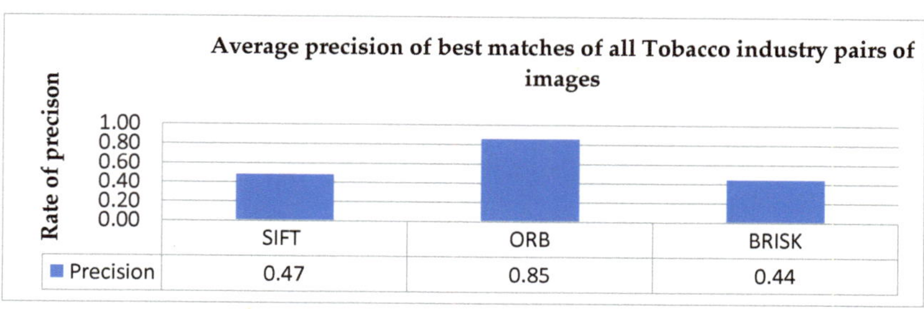

Figure 22. Average precision matching from all pairs of images.

Figure 23 illustrates the average precision of the tobacco industry building from each date with all the others. All algorithms provide similarities to their results. The ORB algorithm is proved to have higher precision of best matches on modern buildings that are under similar conditions and come up to 100% (pair 15). The SIFT algorithm is following; however, it presents rates over 50% for both historic (pair 2) and modern buildings (pairs 10, 16, 24, and 28), i.e., buildings without differences. The BRISK algorithm has the fewer rates; however, on modern buildings precision is similar under all conditions.

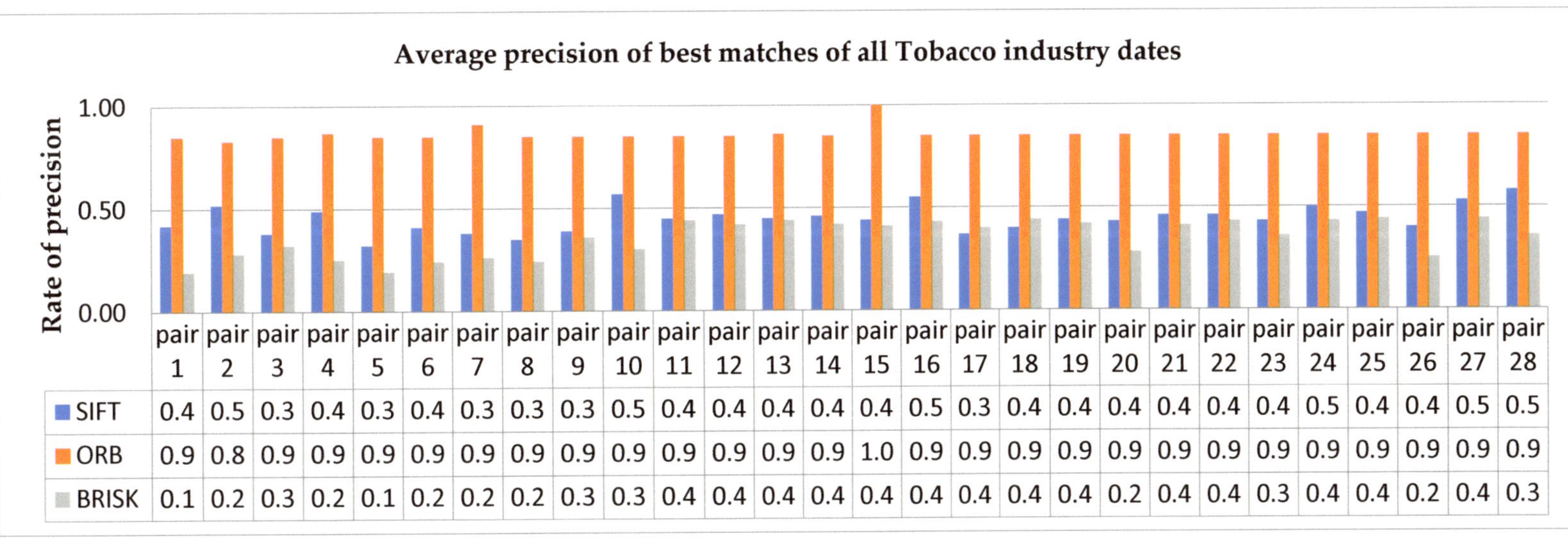

	pair 1	pair 2	pair 3	pair 4	pair 5	pair 6	pair 7	pair 8	pair 9	pair 10	pair 11	pair 12	pair 13	pair 14	pair 15	pair 16	pair 17	pair 18	pair 19	pair 20	pair 21	pair 22	pair 23	pair 24	pair 25	pair 26	pair 27	pair 28
SIFT	0.4	0.5	0.3	0.4	0.3	0.4	0.3	0.3	0.3	0.5	0.4	0.4	0.4	0.4	0.4	0.5	0.3	0.4	0.4	0.4	0.4	0.4	0.4	0.5	0.4	0.4	0.5	0.5
ORB	0.9	0.8	0.9	0.9	0.9	0.9	0.9	0.9	0.9	0.9	0.9	0.9	0.9	0.9	1.0	0.9	0.9	0.9	0.9	0.9	0.9	0.9	0.9	0.9	0.9	0.9	0.9	0.9
BRISK	0.1	0.2	0.3	0.2	0.1	0.2	0.2	0.2	0.3	0.3	0.4	0.4	0.4	0.4	0.4	0.4	0.4	0.4	0.4	0.2	0.4	0.4	0.3	0.4	0.4	0.2	0.4	0.3

Figure 23. Tobacco industry. Average precision each year compared to all the others.

Figure 24 shows the tobacco industry precision with other landmarks' precision of the best matches. The SIFT algorithm has almost half the precision of BRISK. In particular, SIFT provides lower rates mainly on buildings with similar dates but different architectural styles. The ORB algorithm has similar results, while BRISK presents less precision to modern and different architectural-style buildings.

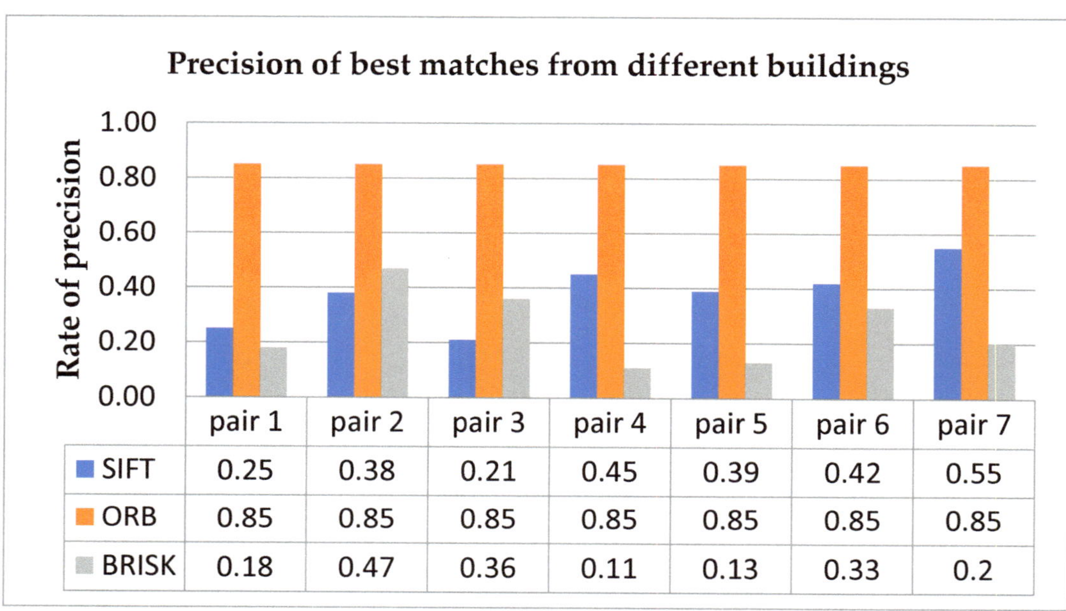

Precision of best matches from different buildings

Rate of precision	pair 1	pair 2	pair 3	pair 4	pair 5	pair 6	pair 7
SIFT	0.25	0.38	0.21	0.45	0.39	0.42	0.55
ORB	0.85	0.85	0.85	0.85	0.85	0.85	0.85
BRISK	0.18	0.47	0.36	0.11	0.13	0.33	0.2

Figure 24. Rate of precision among pairs of different buildings.

Finally, the last evaluation is the false positive of incorrectly detected feature matching. Table 5 shows pairs of the tobacco industry building over time results. According to the results on the buildings over a 15-year time (pair 1–pair 5) the correct matches fail at 100% by all algorithms except ORB, which rises near half. The reason for the failure is partly justified because the viewpoint of images is completely different, i.e., images (pair 1) illustrate only the facade of the building while the texture remains the same. In images where the building is captured only from the facade, all algorithms except SIFT succeed at half rates compared to the previous. In general, between modern buildings, the rate of false positives ranges in low percentages by all algorithms with some exceptions. For instance, the SIFT and BRISK algorithms have high rates of mismatching (pair 10). This pair of buildings has different illumination. In addition, SIFT, ORB, and BRISK appear with equally high rates of false matching in buildings with steep rotation and different scales (pairs 19, 20, 25, and 27). Finally, in fact, the evaluation of false positive matches is not undertaken among different landmarks because there are no common features.

Table 5. Rates (%) of false positives of feature matching.

Buildings	SIFT	ORB	BRISK
Pair 1	97.67	51.47	68.00
Pair 2	100.00	58.58	100.00
Pair 3	93.55	74.74	93.00
Pair 4	100.00	96.63	100.00

Table 5. *Cont.*

Buildings	SIFT	ORB	BRISK
Pair 5	100.00	97.75	97.06
Pair 6	98.88	99.23	92.90
Pair 7	94.74	82.05	26.14
Pair 8	100.00	91.63	100.00
Pair 9	100.00	99.17	100.00
Pair 9	84.90	55.41	86.99
Pair 11	21.80	42.16	33.86
Pair 12	7.57	20.95	44.14
Pair 13	10.18	18.21	37.59
Pair 14	15.48	10.77	35.01
Pair 15	24.90	10.33	51.67
Pair 16	47.44	52.35	77.30
Pair 17	71.56	40.37	79.19
Pair 18	53.39	11.48	39.09
Pair 19	62.90	44.23	38.36
Pair 20	68.32	67.82	94.52
Pair 21	25.51	12.96	34.48
Pair 22	27.69	8.57	24.28
Pair 23	32.06	39.25	70.62
Pair 24	23.62	13.85	64.47
Pair 25	100.00	85.89	98.71
Pair 26	94.12	71.76	92.06
Pair 27	41.45	60.92	96.08
Pair 28	90.86	95.95	90.85

6. Discussion—Conclusions

In this paper, we applied feature extraction and matching techniques to pairs of images, using SIFT, ORB, and BRISK detectors and descriptors to identify historic buildings over time. Comparing historic and modern buildings, the SIFT algorithm detected fewer numbers of features on modern buildings and higher numbers on historic buildings with rotations contrary to BRISK, which detected more features on modern buildings, also without rotation. In addition, although SIFT detected keypoints to window corners, small texture and pattern changes are not recognized, while the ORB algorithm detected more features on modern buildings without scale and rotation changes and are not concentrated on strong points such as corners, compared to the BRISK algorithm. This means that ORB also equally takes into account other elements in the scene.

On the feature-matching level, SIFT presented fewer rates of best matches to historical buildings after BRISK, as compared to ORB, which performed the highest rate of best matching. It is notable that it recognizes all the viewpoints of buildings. Moreover, ORB and BRISK had also a high rate of matching between buildings with various scales.

On the basis of the computational cost, BRISK was more efficient, providing fast matching to all cases, while SIFT had the greatest computational cost but also the greatest number of total matches. ORB had a high computational cost—however, fewer than SIFT. BRISK might be the correct choice in simpler cases that require speed.

The actual number of keypoints was used for feature matching. SIFT provided fewer rates, in particular to modern buildings with different view angles, while ORB had similar results to all pairs of images and BRISK the least. Image-matching precision results showed that from greater to less, ORB presented the best performance and BRISK and SIFT followed, while comparing each date to all others, ORB retained first place, whereas SIFT and BRISK followed, mainly on historical buildings. The same results were also found among pairs of different buildings. Finally, on the basis of false positives, SIFT and BRISK presented higher rates than BRISK, which had half.

According to the above results, SIFT performs the best on modern buildings under rotations, while ORB under scale changes. In addition, SIFT and BRISK had higher accuracy for image rotation changes. On the other hand, ORB moves stable and high detection rates on match tests for feature extraction and matching. Besides, it presents high rates of isolated, aged buildings. In general, ORB is more robust and reliable than SIFT, although it is less scale-invariant, while BRISK seems to be competitively related to SIFT as it performs under rotation and scale changes.

Before algorithm characteristics, we took into account some constraints that were unavoidable. For instance, some old images have been converted into digital format. This had an impact on the loss of basic and valuable information. In addition, since some data are derived from books, the quality of paper degradation is also an important factor.

The buildings in the urban scene have various complex patterns, structural materials, and different shapes. In addition, in urban environments exist other elements, such as buildings, vegetation, or water, and these lead to the appearance of shadows or hidden parts of study buildings. On the other hand, the different conditions of captured buildings make the identification of buildings more difficult. However, our algorithms in many cases found homologous points on the same material. The experimental results confirmed that our methods, up to a point, are capable of identifying the minor changes in buildings over time as the feature matches are more sparse or dense and the feature detection caused by the various factors are not always on the strong points such as (corners and edges). Finally, it is notable that although SIFT is considered the best for many applications, our experiments showed that the ORB and BRISK algorithms are also efficient. However, these standard methods are not sufficient but are satisfactory for identifying landmarks over time.

Future work will consider the computational optimization of the algorithms' parameters to examine the trade-off between enhanced performance and processing time. An extended dataset is also undergoing, with the aim to further investigate the performance of data-intensive techniques, such as deep learning, to the problem under study. Although this study replied to the initial research questions, new challenges are emerging. Deep learning methods to learn dense feature representations gain more and more ground. In the future, it is expected that modern methods such as photogrammetry and remote sensing will contribute to a better understanding of scenery, including urban historic buildings, using dense point clouds and orthophotomaps that provide further metric information, as only from geometry or texture it is difficult to extract conclusions.

Author Contributions: Conceptualization, K.A.T. and G.A.P.; methodology, K.A.T. and G.A.P.; software, K.A.T.; validation, G.A.P., S.C. and G.E.T.; formal analysis, K.A.T.; investigation, K.A.T.; resources, K.A.T.; data curation, K.A.T.; writing—original draft preparation, K.A.T.; writing—review and editing, E.V., S.C. and G.E.T.; visualization, K.A.T. and S.C.; supervision, G.A.P.; project administration, G.A.P., S.C. and G.E.T.; funding acquisition, G.A.P. All authors have read and agreed to the published version of the manuscript.

Funding: This research received no external funding.

Institutional Review Board Statement: Not applicable.

Informed Consent Statement: Not applicable.

Data Availability Statement: Not applicable.

Acknowledgments: This work was supported by the MPhil program "Advanced Technologies in Informatics and Computers", hosted by the Department of Computer Science, International Hellenic University, Kavala, Greece.

Conflicts of Interest: The authors declare no conflict of interest.

References

1. Whitehead, A.; Opp, J. Timescapes: Putting History in Your Hip Pocket. In Proceedings of the Computers and Their Applications Conference CATA, Honolulu, HI, USA, 4–6 March 2013; pp. 261–266.
2. Kabir, S.R.; Akhtaruzzaman, M.; Haque, R. Performance Analysis of Different Feature Detection Techniques for Modern and Old Buildings. In Proceedings of the 3rd International Conference on Recent Trends and Applications in Computer Science and Information Technology, Tiranë, Albania, 23 November 2018; pp. 120–127.
3. Rebec, K.M.; Deanović, B.; Oostwegel, L. Old Buildings Need New Ideas: Holistic Integration of Conservation-Restoration Process Data Using Heritage Building Information Modelling. *J. Cult. Herit.* **2022**, *55*, 30–42. [CrossRef]
4. Mahinda, M.C.P.; Udhyani, H.P.A.J.; Alahakoon, P.M.K.; Kumara, W.G.C.W.; Hinas, M.N.A.; Thamboo, J.A. Development of An Effective 3D Mapping Technique for Heritage Structures. In Proceedings of the 2021 3rd International Conference on Electrical Engineering (EECon), Colombo, Sri Lanka, 24 September 2021; pp. 92–99. [CrossRef]
5. Tuytelaars, T.; Mikolajczyk, K. Local Invariant Feature Detectors: A Survey. *FNT Comput. Graph. Vis.* **2007**, *3*, 177–280. [CrossRef]
6. Santosh, D.; Achar, S.; Jawahar, C.V. Autonomous Image-Based Exploration for Mobile Robot Navigation. In Proceedings of the 2008 IEEE International Conference on Robotics and Automation, Pasadena, CA, USA, 19–23 May 2008; pp. 2717–2722. [CrossRef]
7. Milford, M.; McKinnon, D.; Warren, M.; Wyeth, G. Feature-based Visual Odometry and Featureless Place Recognition for SLAM in 2.5 D Environments. In Proceedings of the Australasian Conference on Robotics and Automation (ACRA 2011), Melbourne Australia, 7–9 December 2011; pp. 1–8.
8. Sminchisescu, C.; Bo, L.; Ionescu, C.; Kanaujia, A. Feature-Based Pose Estimation. In *Visual Analysis of Humans*; Moeslund, T.B., Hilton, A., Krüger, V., Sigal, L., Eds.; Springer: London, UK, 2011; pp. 225–251. [CrossRef]
9. Hu, Y. Research on a Three-Dimensional Reconstruction Method Based on the Feature Matching Algorithm of a Scale-Invariant Feature Transform. *Math. Comput. Model.* **2011**, *54*, 919–923. [CrossRef]
10. Nixon, M.S.; Aguado, A.S. *Feature Extraction and Image Processing*, 1st ed.; Newnes: Oxford, UK; Boston, MA, USA, 2002.
11. Amiri, M.; Rabiee, H.R. RASIM: A Novel Rotation and Scale Invariant Matching of Local Image Interest Points. *IEEE Trans. Image Process.* **2011**, *20*, 3580–3591. [CrossRef]
12. Weng, D.; Wang, Y.; Gong, M.; Tao, D.; Wei, H.; Huang, D. DERF: Distinctive Efficient Robust Features From the Biological Modeling of the P Ganglion Cells. *IEEE Trans. Image Process.* **2015**, *24*, 2287–2302. [CrossRef]
13. Levine, M.D. Feature Extraction: A Survey. *Proc. IEEE* **1969**, *57*, 1391–1407. [CrossRef]
14. Ha, Y.-S.; Lee, J.; Kim, Y.-T. Performance Evaluation of Feature Matching Techniques for Detecting Reinforced Soil Retaining Wall Displacement. *Remote Sens.* **2022**, *14*, 1697. [CrossRef]
15. Viola, P.; Wells III, W.M. Alignment by maximization of mutual information. *Int. J. Comput. Vis.* **1997**, *24*, 137–154. [CrossRef]
16. Myronenko, A.; Song, X. Intensity-Based Image Registration by Minimizing Residual Complexity. *IEEE Trans. Med. Imaging* **2010**, *29*, 1882–1891. [CrossRef]
17. Liu, X.; Tian, Z.; Ding, M. A Novel Adaptive Weights Proximity Matrix for Image Registration Based on R-SIFT. *AEU-Int. J. Electron. Commun.* **2011**, *65*, 1040–1049. [CrossRef]
18. Leng, C.; Xiao, J.; Li, M.; Zhang, H. Robust Adaptive Principal Component Analysis Based on Intergraph Matrix for Medical Image Registration. *Comput. Intell. Neurosci.* **2015**, *2015*, 829528. [CrossRef] [PubMed]
19. Friedrichs, K.; Münster, S.; Kröber, C.; Bruschke, J. Creating Suitable Tools for Art and Architectural Research with Historic Media Repositories. In *Digital Research and Education in Architectural Heritage*; Münster, S., Friedrichs, K., Niebling, F., Seidel-Grzesińska, A., Eds.; Communications in Computer and Information Science; Springer International Publishing: Cham, Switzerland, 2018; Volume 817, pp. 117–138. [CrossRef]
20. Ali, H.; Whitehead, A. Subset Selection for Landmark Modern and Historic Images. In Proceedings of the 2nd International Conference on Signal and Image Processing, Geneva, Switzerland, 21–22 March 2015; pp. 69–79. [CrossRef]
21. Ali Heider, K.; Whitehead, A. Modern to Historic Image Matching: ORB/SURF an Effective Matching Technique. In Proceedings of the Computers and Their Applications, Las Vegas, NV, USA, 24–26 March 2014.
22. Becker, A.-K.; Vornberger, O. Evaluation of Feature Detectors, Descriptors and Match Filtering Approaches for Historic Repeat Photography. In *Image Analysis*; Felsberg, M., Forssén, P.-E., Sintorn, I.-M., Unger, J., Eds.; Lecture Notes in Computer Science; Springer International Publishing: Cham, Switzerland, 2019; Volume 11482, pp. 374–386. [CrossRef]
23. Anderson-Bell, J.; Schillaci, C.; Lipani, A. Predicting non-residential building fire risk using geospatial information and convolutional neural networks. *Remote Sens. Appl. Soc. Environ.* **2021**, *21*, 100470. [CrossRef]
24. Agarwal, S.; Snavely, N.; Simon, I.; Seitz, S.M.; Szeliski, R. Building Rome in a Day. In Proceedings of the 2009 IEEE 12th International Conference on Computer Vision, Kyoto, Japan, 29 September–2 October 2009; pp. 72–79. [CrossRef]
25. Uttama, P.L.; Delalandre, M.; Ogier, J.M. Segmentation and Retrieval of Ancient Graphic Documents. In *Graphics Recognition. Ten Years Review and Future Perspectives*; Springer: Cham, Switzerland, 2006; pp. 88–98.

26. Ali, H.; Whitehead, A. Feature Matching for Aligning Historical and Modern Images. *Int. J. Comput. Appl.* **2014**, *21*, 188–201.
27. Wolfe, R. Modern to Historical Image Feature Matching. 2015. Available online: http://robbiewolfe.ca/programming/honoursproject/report.pdf (accessed on 3 February 2023).
28. Wu, G.; Wang, Z.; Li, J.; Yu, Z.; Qiao, B. Contour-Based Historical Building Image Matching. In Proceedings of the 2nd International Symposium on Image Computing and Digital Medicine—ISICDM, Chengdu, China, 13–15 October 2018; ACM Press: Chengdu, China, 2018; pp. 32–36. [CrossRef]
29. Hasan, M.S.; Ali, M.; Rahman, M.; Arju, H.A.; Alam, M.M.; Uddin, M.S.; Allayear, S.M. Heritage Building Era Detection Using CNN. *IOP Conf. Ser. Mater. Sci. Eng.* **2019**, *617*, 012016. [CrossRef]
30. Maiwald, F.; Schneider, D.; Henze, F.; Münster, S.; Niebling, F. Feature matching of historical images based on geometry of quadrilaterals. *Int. Arch. Photogramm. Remote Sens. Spatial Inf. Sci.* **2018**, *XLII-2*, 643–650. [CrossRef]
31. Yue, L.; Li, H.; Zheng, X. Distorted Building Image Matching with Automatic Viewpoint Rectification and Fusion. *Sensors* **2019**, *19*, 5205. [CrossRef] [PubMed]
32. Si, L.; Hu, X.; Liu, B. Image Matching Algorithm Based on the Pattern Recognition Genetic Algorithm. *Comput. Intell. Neurosci.* **2022**, *2022*, 7760437. [CrossRef]
33. Edward, J.; Yang, G.-Z. RANSAC with 2D Geometric Cliques for Image Retrieval and Place Recognition. In Proceedings of the CVPR Workshop, Boston, MA, USA, 7–12 June 2015.
34. Avrithis, Y.; Tolias, G. Hough Pyramid Matching: Speeded-Up Geometry Re-Ranking for Large Scale Image Retrieval. *Int. J. Comput Vis.* **2014**, *107*, 1–19. [CrossRef]
35. Lowe, D.G. Distinctive Image Features from Scale-Invariant Keypoints. *Int. J. Comput. Vis.* **2004**, *60*, 91–110. [CrossRef]
36. Rublee, E.; Rabaud, V.; Konolige, K.; Bradski, G. ORB: An Efficient Alternative to SIFT or SURF. In Proceedings of the 2011 International Conference on Computer Vision, Barcelona, Spain, 6–13 November 2011; pp. 2564–2571. [CrossRef]
37. Leutenegger, S.; Chli, M.; Siegwart, R.Y. BRISK: Binary Robust Invariant Scalable Keypoints. In Proceedings of the 2011 International Conference on Computer Vision, Barcelona, Spain, 6–13 November 2011; pp. 2548–2555. [CrossRef]
38. Smith, S.M.; Brady, J.M. Susan-a new approach to low level image processing. *Int. J. Comput. Vis.* **1997**, *23*, 45–78. [CrossRef]
39. Nixon, M.; Aguado, A. *Feature Extraction & Image Processing for Computer Vision*; Elsevier: Amsterdam, The Netherlands, 2012. [CrossRef]
40. Tsafrir, D.; Tsafrir, I.; Ein-Dor, L.; Zuk, O.; Notterman, D.A.; Domany, E. Sorting Points into Neighborhoods (SPIN): Data Analysis and Visualization by Ordering Distance Matrices. *Bioinformatics* **2005**, *21*, 2301–2308. [CrossRef]
41. Harris, C.; Stephens, M. A Combined Corner and Edge Detector. In *Proceedings of the Alvey Vision Conference 1988*; Alvey Vision Club: Manchester, UK, 1988; pp. 23.1–23.6. [CrossRef]
42. Shi, F.; Huang, X.; Duan, Y. Robust Harris-Laplace Detector by Scale Multiplication. In *Advances in Visual Computing*; Bebis, G., Boyle, R., Parvin, B., Koracin, D., Kuno, Y., Wang, J., Wang, J., Wang, J., Pajarola, R., Lindstrom, P., Eds.; Lecture Notes in Computer Science; Springer: Berlin/Heidelberg, Germany, 2009; Volume 5875, pp. 265–274. [CrossRef]
43. Sarangi, S.; Sahidullah, M.; Saha, G. Optimization of Data-Driven Filterbank for Automatic Speaker Verification. *Digit. Signal Process.* **2020**, *104*, 102795. [CrossRef]
44. Mutlag, W.K.; Ali, S.K.; Aydam, Z.M.; Taher, B.H. Feature Extraction Methods: A Review. *J. Phys. Conf. Ser.* **2020**, *1591*, 012028. [CrossRef]
45. Kumar, G.; Bhatia, P.K. A Detailed Review of Feature Extraction in Image Processing Systems. In Proceedings of the 2014 Fourth International Conference on Advanced Computing & Communication Technologies, Rohtak, India, 8–9 February 2014; pp. 5–12. [CrossRef]
46. Wang, X.; Jabri, A.; Efros, A.A. Learning Correspondence from the Cycle-Consistency of Time. *Comput. Vis. Pattern Recognit.* **2019**, 2566–2576. [CrossRef]
47. Muhammad, U.; Tanvir, M.; Khurshid, K. Feature Based Correspondence: A Comparative Study on Image Matching Algorithms. *Int. J. Adv. Comput. Sci. Appl.* **2016**, *7*. [CrossRef]
48. Zhao, C.; Cao, Z.; Yang, J.; Xian, K.; Li, X. Image Feature Correspondence Selection: A Comparative Study and a New Contribution. *IEEE Trans. Image Process.* **2020**, *29*, 3506–3519. [CrossRef]
49. Howe, P.D.; Livingstone, M.S. Binocular Vision and the Correspondence Problem. *J. Vis.* **2005**, *5*, 800. [CrossRef]
50. Torresani, L.; Kolmogorov, V.; Rother, C. Feature Correspondence Via Graph Matching: Models and Global Optimization. In *Computer Vision—ECCV 2008*; Forsyth, D., Torr, P., Zisserman, A., Eds.; Lecture Notes in Computer Science; Springer: Berlin/Heidelberg, Germany, 2008; Volume 5303, pp. 596–609. [CrossRef]
51. Kolmogorov, V.; Zabih, R. Computing Visual Correspondence with Occlusions Using Graph Cuts. In Proceedings of the Eighth IEEE International Conference on Computer Vision. ICCV 2001, Vancouver, BC, Canada, 7–14 July 2001; Volume 2, pp. 508–515. [CrossRef]
52. Kabbai, L.; Abdellaoui, M.; Douik, A. Image Classification by Combining Local and Global Features. *Vis. Comput.* **2019**, *35*, 679–693. [CrossRef]
53. Mikolajczyk, K.; Mikolajczyk, K. Scale & Affine Invariant Interest Point Detectors. *Int. J. Comput. Vis.* **2004**, *60*, 63–86. [CrossRef]
54. Keyvanpour, M.R.; Vahidian, S.; Ramezani, M. HMR-Vid: A Comparative Analytical Survey on Human Motion Recognition in Video Data. *Multimed. Tools Appl.* **2020**, *79*, 31819–31863. [CrossRef]

55. Chen, L.; Rottensteiner, F.; Heipke, C. Feature Detection and Description for Image Matching: From Hand-Crafted Design to Deep Learning. *Geo-Spat. Inf. Sci.* **2021**, *24*, 58–74. [CrossRef]
56. Krig, S. Interest Point Detector and Feature Descriptor Survey. In *Computer Vision Metrics*; Springer International Publishing: Cham, Switzerland, 2016; pp. 187–246. [CrossRef]
57. Hassaballah, M.; Abdelmgeid, A.A.; Alshazly, H.A. Image Features Detection, Description and Matching. In *Image Feature Detectors and Descriptors*; Awad, A.I., Hassaballah, M., Eds.; Studies in Computational Intelligence; Springer International Publishing: Cham, Switzerland, 2016; Volume 630, pp. 11–45. [CrossRef]
58. Leng, C.; Zhang, H.; Li, B.; Cai, G.; Pei, Z.; He, L. Local Feature Descriptor for Image Matching: A Survey. *IEEE Access* **2019**, *7*, 6424–6434. [CrossRef]
59. González-Aguilera, D.; Ruiz de Oña, E.; López-Fernandez, L.; Farella, E.; Stathopoulou, E.K.; Toschi, I.; Remondino, F.; Rodríguez-Gonzálvez, P.; Hernández-López, D.; Fusiello, A.; et al. Photomatch: An open-source multi-view and multi-modal feature matching tool for photogrammetric applications. *Int. Arch. Photogramm. Remote Sens. Spatial Inf. Sci.* **2020**, *XLIII-B5-2020*, 213–219. [CrossRef]
60. Sun, J.; Shen, Z.; Wang, Y.; Bao, H.; Zhou, X. LoFTR: Detector-Free Local Feature Matching with Transformers. *Comput. Vis. Pattern Recognit.* **2021**, 8922–8931. [CrossRef]
61. Zitová, B.; Flusser, J. Image Registration Methods: A Survey. *Image and Vision Computing* **2003**, *21*, 977–1000. [CrossRef]
62. Litjens, G.; Kooi, T.; Bejnordi, B.E.; Setio, A.A.A.; Ciompi, F.; Ghafoorian, M.; van der Laak, J.A.W.M.; van Ginneken, B.; Sánchez, C.I. A Survey on Deep Learning in Medical Image Analysis. *Med. Image Anal.* **2017**, *42*, 60–88. [CrossRef]
63. Flusser, J.; Suk, T. A Moment-Based Approach to Registration of Images with Affine Geometric Distortion. *IEEE Trans. Geosci. Remote Sens.* **1994**, *32*, 382–387. [CrossRef]
64. Goshtasby, A.; Stockman, G.; Page, C. A Region-Based Approach to Digital Image Registration with Subpixel Accuracy. *IEEE Trans. Geosci. Remote Sens.* **1986**, *GE-24*, 390–399. [CrossRef]
65. Hsieh, Y.C.; McKeown, D.M.; Perlant, F.P. Performance Evaluation of Scene Registration and Stereo Matching for Cartographic Feature Extraction. *IEEE Trans. Pattern Anal. Machine Intell.* **1992**, *14*, 214–238. [CrossRef]
66. Hellier, P.; Barillot, C. Coupling Dense and Landmark-Based Approaches for Nonrigid Registration. *IEEE Trans. Med. Imaging* **2003**, *22*, 217–227. [CrossRef]
67. Mikolajczyk, K.; Schmid, C. A Performance Evaluation of Local Descriptors. *IEEE Trans. Pattern Anal. Mach. Intell.* **2005**, *27*, 1615–1630. [CrossRef] [PubMed]
68. Noble, F.K. Comparison of OpenCV's Feature Detectors and Feature Matchers. In Proceedings of the 2016 23rd International Conference on Mechatronics and Machine Vision in Practice (M2VIP), Nanjing, China, 28–30 November 2016; pp. 1–6. [CrossRef]
69. Dhana Lakshmi, M.; Mirunalini, P.; Priyadharsini, R.; Mirnalinee, T.T. Review of Feature Extraction and Matching Methods for Drone Image Stitching. In *Proceedings of the International Conference on ISMAC in Computational Vision and Bio-Engineering 2018 (ISMAC-CVB), Palladam, India, 16–17 May 2018*; Pandian, D., Fernando, X., Baig, Z., Shi, F., Eds.; Lecture Notes in Computational Vision and Biomechanics; Springer International Publishing: Cham, Switzerland, 2019; Volume 30, pp. 595–602. [CrossRef]
70. Spasova, V.G. Experimental evaluation of keypoints detector and descriptor algorithms for indoors person localization. *Annu. J. Electron.* **2014**, *8*, 85–87.
71. Vijayan, V.; Kp, P. FLANN Based Matching with SIFT Descriptors for Drowsy Features Extraction. In Proceedings of the 2019 Fifth International Conference on Image Information Processing (ICIIP), Shimla, India, 15–17 November 2019; pp. 600–605. [CrossRef]
72. Luo, Z.; Zhou, L.; Bai, X.; Chen, H.; Zhang, J.; Yao, Y.; Li, S.; Fang, T.; Quan, L. ASLFeat: Learning Local Features of Accurate Shape and Localization. In Proceedings of the IEEE/CVF Conference on Computer Vision and Pattern Recognition, Seattle, WA, USA, 13–19 June 2020. [CrossRef]
73. Rosten, E.; Drummond, T. Machine Learning for High-Speed Corner Detection. In *Computer Vision—ECCV 2006*; Leonardis, A., Bischof, H., Pinz, A., Eds.; Lecture Notes in Computer Science; Springer: Berlin/Heidelberg, Germany, 2006; Volume 3951, pp. 430–443. [CrossRef]
74. Calonder, M.; Lepetit, V.; Strecha, C.; Fua, P. BRIEF: Binary Robust Independent Elementary Features. In *Computer Vision—ECCV 2010*; Daniilidis, K., Maragos, P., Paragios, N., Eds.; Lecture Notes in Computer Science; Springer: Berlin/Heidelberg, Germany, 2010; Volume 6314, pp. 778–792. [CrossRef]
75. Martin, K.A.C. A BRIEF History of the "Feature Detector". *Cereb Cortex* **1994**, *4*, 1–7. [CrossRef] [PubMed]
76. Tareen, S.A.K.; Saleem, Z. A Comparative Analysis of SIFT, SURF, KAZE, AKAZE, ORB, and BRISK. In Proceedings of the 2018 International Conference on Computing, Mathematics and Engineering Technologies (iCoMET), Sukkur, Pakistan, 3–4 March 2018; pp. 1–10. [CrossRef]
77. Azimi, E.; Behrad, A.; Ghaznavi-Ghoushchi, M.B.; Shanbehzadeh, J. A Fully Pipelined and Parallel Hardware Architecture for Real-Time BRISK Salient Point Extraction. *J. Real-Time Image Proc.* **2019**, *16*, 1859–1879. [CrossRef]
78. Awad, A.I.; Hassaballah, M. (Eds.) Image Feature Detectors and Descriptors: Foundations and Applications. In *Studies in Computational Intelligence*; Springer International Publishing: Cham, Switzerland, 2016; Volume 630. [CrossRef]
79. Chen, J.; Shan, S.; He, C.; Zhao, G.; Pietikäinen, M.; Chen, X.; Gao, W. WLD: A Robust Local Image Descriptor. *IEEE Trans. Pattern Anal. Mach. Intell.* **2010**, *32*, 1705–1720. [CrossRef]
80. Zhang, H.; Wohlfeil, J.; Grießbach, D. Extension and evaluation of the AGAST feature detector. *ISPRS Ann. Photogramm. Remote Sens. Spatial Inf. Sci.* **2016**, *III–4*, 133–137. [CrossRef]

81. Xiong, X.; Choi, B.-J. Comparative Analysis of Detection Algorithms for Corner and Blob Features in Image Processing. *Int. J. Fuzzy Log. Intell. Syst.* **2013**, *13*, 284–290. [CrossRef]
82. Ghafoor, A.; Iqbal, R.N.; Khan, S. Robust Image Matching Algorithm. In Proceedings of the 4th EURASIP Conference focused on Video/Image Processing and Multimedia Communications (IEEE Cat. No.03EX667), Zagreb, Croatia, 2–5 July 2003; Volume 1, pp. 155–160. [CrossRef]
83. Lecun, Y.; Bottou, L.; Bengio, Y.; Haffner, P. Gradient-Based Learning Applied to Document Recognition. *Proc. IEEE* **1998**, *86*, 2278–2324. [CrossRef]
84. Dalal, N.; Triggs, B. Histograms of Oriented Gradients for Human Detection. In Proceedings of the 2005 IEEE Computer Society Conference on Computer Vision and Pattern Recognition (CVPR'05), San Diego, CA, USA, 20–25 June 2005; Volume 1, pp. 886–893. [CrossRef]
85. Jakubovic, A.; Velagic, J. Image Feature Matching and Object Detection Using Brute-Force Matchers. In Proceedings of the 2018 International Symposium ELMAR, Zadar, Croatia, 16–19 September 2018; pp. 83–86. [CrossRef]
86. Norouzi, M.; Fleet, D.J.; Salakhutdinov, R.R. Hamming distance metric learning. In Proceedings of the Neural Information Processing Systems (NeurIPS 2012), Lake Tahoe, NV, USA, 3–8 December 2012; Volume 25, pp. 1061–1069.
87. Lu, Y.; Liu, A.-A.; Su, Y.-T. Detection in Biomedical Images. In *Computer Vision for Microscopy Image Analysis*; Elsevier: Amsterdam, The Netherlands, 2021; pp. 131–157. [CrossRef]

Article

Development of Static and Dynamic Colorimetric Analysis Techniques Using Image Sensors and Novel Image Processing Software for Chemical, Biological and Medical Applications

Woo Sik Yoo [1,2,*], **Jung Gon Kim** [1], **Kitaek Kang** [1] and **Yeongsik Yoo** [3]

1   WaferMasters, Inc., Dublin, CA 94568, USA
2   Institute of Humanities Studies, Kyungpook National University, Daegu 41566, Republic of Korea
3   College of Liberal Arts, Dankook University, Yongin 16890, Republic of Korea
*   Correspondence: woosik.yoo@wafermasters.com

Abstract: Colorimetric sensing techniques for point(s), linear and areal array(s) were developed using image sensors and novel image processing software for chemical, biological and medical applications. Monitoring and recording of colorimetric information on one or more specimens can be carried out by specially designed image processing software. The colorimetric information on real-time monitoring and recorded images or video clips can be analyzed for point(s), line(s) and area(s) of interest for manual and automatic data collection. Ex situ and in situ colorimetric data can be used as signals for process control, process optimization, safety and security alarms, and inputs for machine learning, including artificial intelligence. As an analytical example, video clips of chromatographic experiments using different colored inks on filter papers dipped in water and randomly blinking light-emitting-diode-based decorative lights were used. The colorimetric information on points, lines and areas, with different sizes from the video clips, were extracted and analyzed as a function of time. The video analysis results were both visualized as time-lapse images and RGB (red, green, blue) color/intensity graphs as a function of time. As a demonstration of the developed colorimetric analysis technique, the colorimetric information was expressed as static and time-series combinations of RGB intensity, HSV (hue, saturation and value) and CIE L*a*b* values. Both static and dynamic colorimetric analysis of photographs and/or video files from image sensors were successfully demonstrated using a novel image processing software.

Keywords: pH indicator; litmus paper; color sensing; colorimetric quantification; dynamic analysis; photograph; video; image processing software

Citation: Yoo, W.S.; Kim, J.G.; Kang, K.; Yoo, Y. Development of Static and Dynamic Colorimetric Analysis Techniques Using Image Sensors and Novel Image Processing Software for Chemical, Biological and Medical Applications. *Technologies* **2023**, *11*, 23. https://doi.org/10.3390/technologies11010023

Academic Editors: Gwanggil Jeon and Imran Ahmed

Received: 25 December 2022
Revised: 20 January 2023
Accepted: 25 January 2023
Published: 28 January 2023

1. Introduction

Color is considered a property of objects and substances indicating status and conditions. Colorimetric information, including brightness and contrast, are frequently used in characterizing and describing properties of objects and substances of interest, both in daily life and various scientific fields, such as chemical, biological and medical research and industry [1–7].

One good example is the colorimetric identification of pH levels (power of hydrogen or potential for hydrogen in the range of 1–14) using litmus paper made from absorbent paper [8–10]. It is used to specify how acidic or basic a water-based solution is by comparison with a standard color chart for the specific litmus paper. It is frequently used in checking swimming pool water quality, food safety, properties of unknown solutions, and vaginal acidity levels for feminine health reasons [10–14]. For proper test results, an appropriate waiting time is needed before the pH reading can evaluate the complete chemical reaction. Time-dependent information on the color change, from the time of contact to completion of the reaction is ignored, even though it potentially contains valuable information on reactivity, diffusivity, non-uniformity, and so on. The color of the litmus paper is usually offset between wet and dry conditions and can cause pH reading errors in visual inspection.

In addition to the development of individual functional sensors and image analysis techniques, the development of integrated sensing systems for practical usage and automated data acquisition and analysis of database accumulation is strongly desired for efficient monitoring of time-dependent colorimetric information on the objects and substances under the process in progress. We have developed novel image analysis software with image capturing and video recording functions and introduced a few application examples in various fields, including engineering, cultural heritage evaluation/conservation, archaeology, chemical, biological and medical applications [10,15–23].

Numerous colorimetric assays exist for the measurement of analytes. Applications range from simple pH measurements to more complex assays, such as pesticides and pharmaceutical compounds, which can be used in applications for water testing and health monitoring. The majority of such tests utilize laboratory spectrophotometry. Simple calibration curves and consistent absorbance measurements are used for quantitative determination of the analyte concentration. The quantitative determination is performed by comparing the absorbance at wavelengths specific to the chromophore of interest to the calibration curve. [24–27]. Progress in image analysis techniques and the development of smart integrated systems using them will enhance our understanding of the objects and substances of interest under various conditions. The development of video clip analysis techniques and utilization of commercially available general image sensors, such as digital cameras, USB cameras, smartphones and camcorders, are important for the effective use of colorimetric information in chemical, biological and medical applications.

The colorimetric analysis technique has been extremely useful for digital forensic studies of cultural heritage identification processes [28–30]. The world's oldest metal-type printed book, made in 1239 in Korea, has been identified by comparing six nearly identical books from Korea from the 13th to 16th centuries using this image analysis technique.

In this paper, static and dynamic colorimetric image analysis techniques and systems for chemical, biological and medical applications are proposed and demonstrated using experimental chromatography samples and randomly blinking light-emitting-diode-based decorative lights. The static and dynamic colorimetric information was expressed as RGB (red, green, blue) intensity, HSV (hue, saturation and value) and CIE L*a*b* values as traceable, quantitative colorimetric data. Munsell color indices and hexadecimal color codes can also be extracted from any points or desired partial area of images.

2. Experimental Section

Two sets of experiments were performed to demonstrate static and dynamic colorimetric image analysis techniques developed for chemical, biological and medical applications. Photographs and video clips were captured using smartphones, USB (universal serial bus) cameras and digital microscopes. Individual photographs and video clips were analyzed using a novel image analysis software (PicMan (ver. 22.12) from WaferMasters, Inc., Dublin, CA, USA). For USB cameras and digital microscopes, individual photographs and video clips were recorded through the image analysis software PicMan [10,15–23]. The static image analysis can be performed by snapshot images (photographs) from any image sensors (cameras), any digital image files from reliable/calibrated devices and image files from reliable/trusted source. The dynamic image analysis can be performed using real-time video images or pre-recorded video clips from any image sensors (cameras). The software is constantly updated with newer functions. The authors used the latest version of PicMan at the time of manuscript submission (December 2022).

To avoid ethical issues with this study, experiments using biological specimens are not presented in this paper. Two sets of artificial experimental results are described to demonstrate the concept of static and dynamic colorimetric analysis techniques using image sensors and novel image processing software in chemical, biological and medical applications and so on.

For the first set, highlighter pens with four different colors, white filter papers and a container for water were prepared for chromatography experiments (Figure 1a). Dots and lines were drawn on the filter papers using the highlighter pens before dipping the bottom of the filter paper into the water for water absorption and diffusion for chromatography. Diffusion characteristics of highlighter ink with water absorption were recorded as video clips and the colorimetric information was extracted as a function of position and time.

(a)

(b)

Figure 1. (**a**) Highlighter color pens and white filter papers used for chromatography experiments.; (**b**) decorative lights with randomly blinking LEDs used for dynamic colorimetric image analysis.

For the second set, decorative lights with randomly blinking light-emitting diodes (LEDs) were prepared (Figure 1b). Blinking decorative lights were recorded as video clips for the colorimetric analysis using the image analysis software. This is to simulate the color characterization of bioluminescence.

Quantitative colorimetric information on points, lines and areas of interest, in still images sampled from the video clips, were extracted to demonstrate the traceability and feasibility of statistical analysis of the extracted data. For dynamic analysis, colorimetric information in the video clips was extracted as a function of time and location.

The static analysis was performed at specific point(s), line(s) and area(s) of the image(s) at a given time. The dynamic analysis was performed at specific point(s), line(s) and area(s) of time-series image(s) (i.e., video clips) by either synthesizing time cross-section image(s) or color/intensity information as a function of time.

3. Results

3.1. Chromatography Experiment

White filter papers with color dots and lines in different marked patterns, using highlighter color pens, were prepared. The bottom parts of the marked filter papers were immersed into water to observe ink diffusion patterns and the speed of diffusion. As the water is absorbed by the filer paper, the ink color diffuses and travels with the water. The experimental scenes were recorded as video clips for image analysis using the software (PicMan).

Figure 2 shows a screen captured image of the PicMan software being used for color analysis of three specific points (P1, P2 and P3) and a line segment L1, on a frame extracted from one of the video clips. The video clip used in the figure has eight (8) colored dots marked by highlighter pens. The absorption of water by the filter paper and the diffusion of colored ink dots were recorded with 30 frames per second (fps) for 180 s. The total number of frames for this specific video clip is 5400 (=180 s × 30 fps). In other words, each pixel on the image has 5400 points of time-series, RGB intensity data for a period of 180 s (30 points per second, frame rate). The figure shows an RGB intensity graph along the line segment L1. The RBG intensity range is between 0 and 255 because the intensity of each color channel is assigned to 8-bit (2^8 = 256) levels. The line intensity data can be exported as CSV files for further analysis and customized graphical display. The colors of the lines in the intensity graph represent each color component. The gray line is the average of RGB intensity in the 0~255 range. The solid lines in the line intensity graph represent the intensity of RGB channels in the line segment L1, while the dotted lines represent the line intensity of RGB channels outside of the extended portion of the line segment L1. As seen from the line intensity graph, the color ink on the line segment L1 travels with the water, leaving a bright yellowish color as a trail. Visual inspection can also be used, but it would be very difficult to quantitatively record results and the subjectivity of all records could not be avoided.

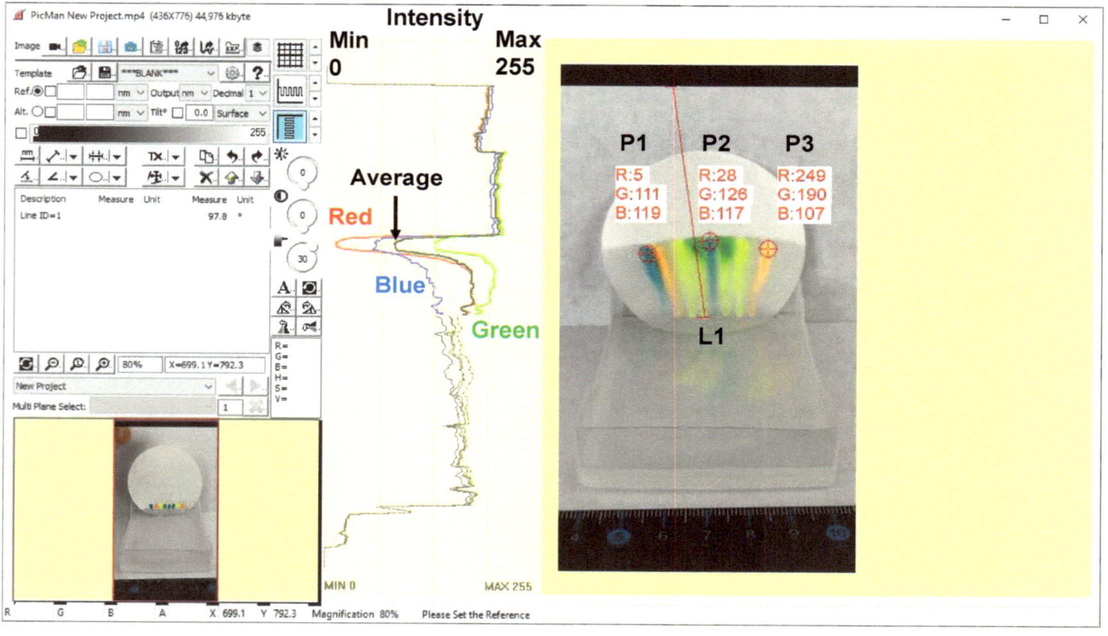

Figure 2. Screen-captured image of the PicMan software being used for color analysis of three specific points (P1, P2 and P3) and a line segment L1 on a frame extracted from one of the video clips.

For image display on a monitor screen, color is reproduced by combinations of RGB intensity. Each pixel has three 8-bit intensity values (3 channel × 8 bit = 24 bit) and can display 2^{24} (=16,777,216) colors. The RGB color system is one of many color systems used in practice. For image display, an RGB color system is used and all colors are uniquely identified as RGB-based hexadecimal color codes. For printing, CMY (cyan, magenta, yellow) and CYMK (cyan, yellow, magenta, key) color models are used. The HSL (hue, relative chroma, intensity) and HSV (hue, saturation, value) color systems are also frequently used. The CIELAB color space often referred to as L*a*b* is a color space defined by the International Commission on Illumination (CIE) in 1976 [31–33]. Color conversion between different color systems is often required for imaging system calibration and comparison with color standards or acquired images of standard color palettes. To accommodate the technical requirements for practical use, the PicMan software added the color conversion function. Color information on point(s), line(s) and area(s) of interest can be converted into the desired format and exported as the CSV file format. Figure 3 shows examples of color conversion of three points (P1, P2 and P3) on the image, as displayed on the screen, between color systems. HSV and L*a*b* color systems can also be used instead of the RGB color scale. The color information can also be exported in the hexadecimal codes and Munsell color system. To use general photographs and video images with RGB scale for calorimetric determination, it is convenient to use the original format from imaging devices without colorimetric conversion into other color systems.

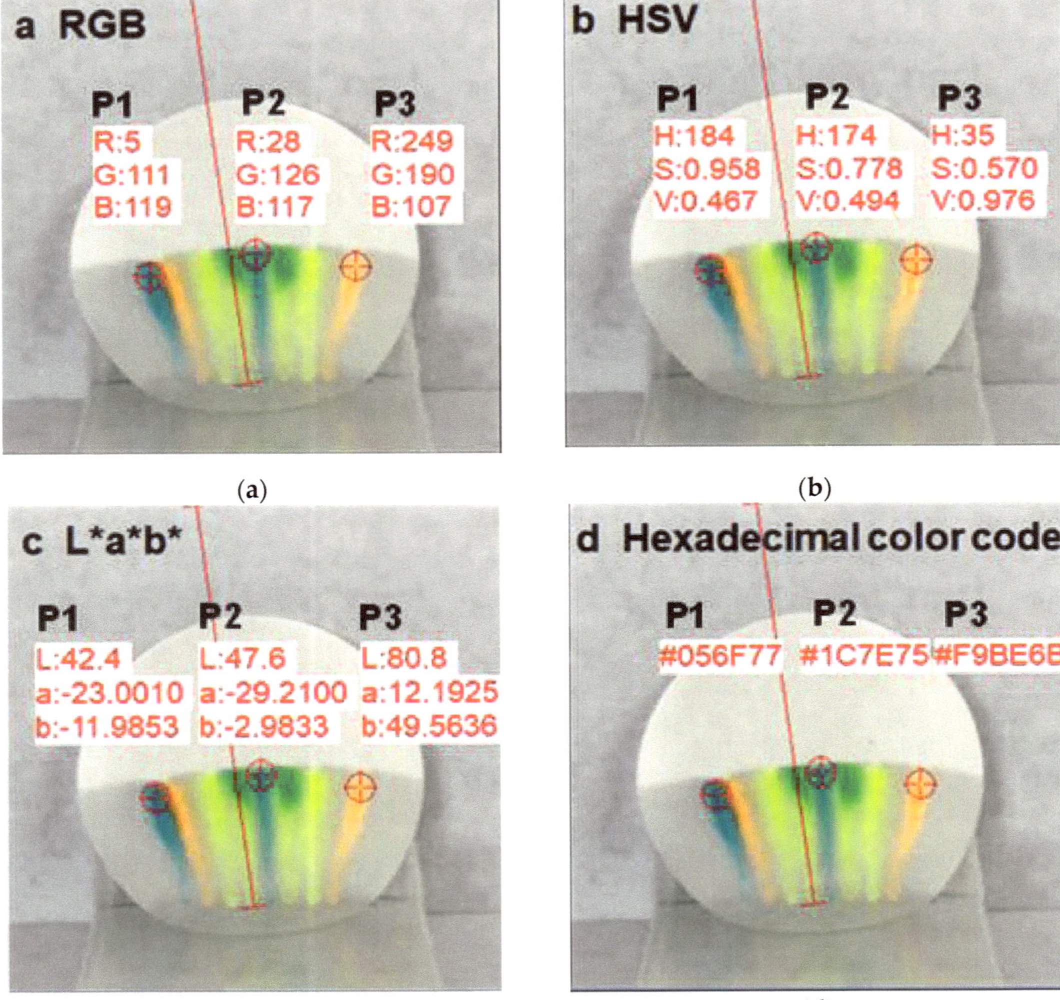

Figure 3. Examples of color conversion of three points (P1, P2 and P3) on the image displayed on the screen between color systems. (**a**) RGB; (**b**) HSV; (**c**) L*a*b*; and (**d**) hexadecimal color code.

Objects and substances which evolve with time require dynamic analysis, thus video image recording techniques are commonly used. However, dynamic image analysis is not easy because it contains a lot of information on a single-frame image and a large number of frames. A single-frame high-resolution image can easily contain more than 10 mega pixels (10 million pixels, 10 MP) of RGB intensity data. Visual inspection by trained personnel, including researchers and operators, is typically employed for this type of task. It is easy to make human errors and form biased opinions. Automatic image analysis techniques must be developed for effective data mining and the prevention of human errors.

Figure 4 shows 45 images sampled every 4 s from a 180-second-long video clip (30 fps × 180 s = 5400 frames). It only shows less than 1% (45 frames/5400 frames = 0.83%) of the total information recorded in the video clip. We are not utilizing 99.17% of the available images. By scanning well-organized time-series images, we can form an opinion on the movement of colored ink from the chromatography experiment. Without proper image analysis software, we can only rely on manual, time consuming measurements to form even a qualitative and subjective opinion. We can only verbally describe how the colored ink was diffusing with water absorption by the filter paper.

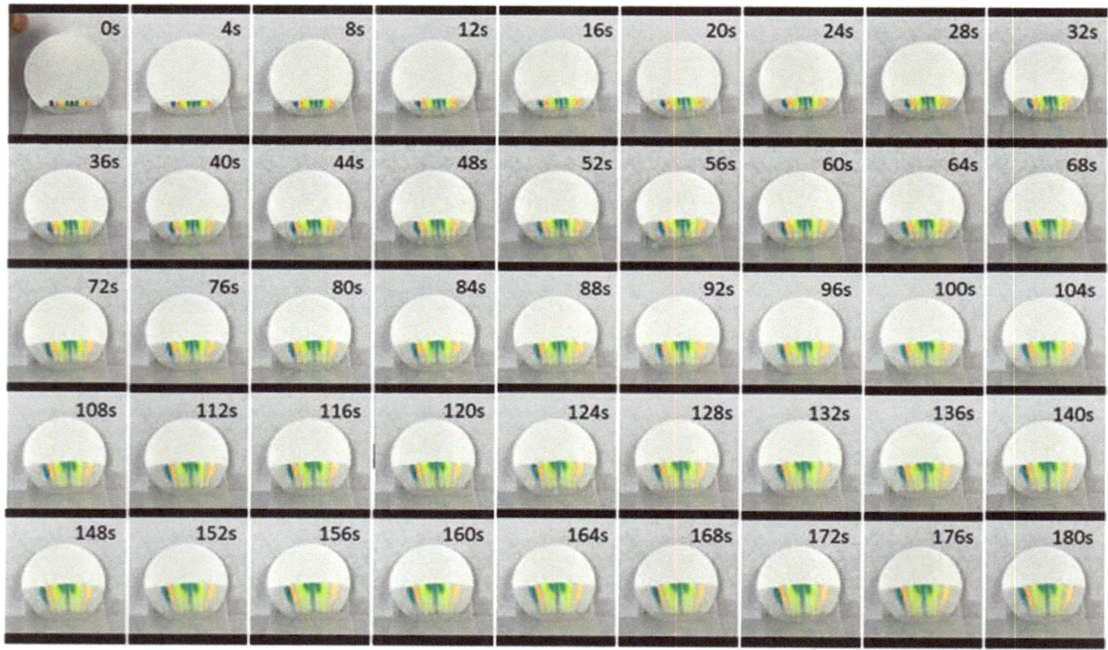

Figure 4. Time series of 45 images of chromatography experiment (sampled every 4 seconds) from a 180 s-long video clip (30 fps × 180 s = 5,400 frames).

Time cross-section operation of PicMan was used to synthesize new images from the start (frame 0) to the end (frame 5400) of the 180-second-long video clips of a chromatography experiment. Figure 5 shows four vertically sloped line segments and four horizontal line segments of interest for information extraction. The synthesized time cross-section images of the eight-line segments from 5400 frames of images are shown in Figure 6. The horizontal axis represents time in seconds. It can be used as a reference for frame number by multiplying the frame rate of 30 fps by the time elapsed. As seen in the figure, ink diffusion speed and color change with time (or water absorption and ink diffusion) are well summarized as a time-synchronized, newly synthesized image. From the upper half of a synthesized image (four vertically sloped line segments), we can easily recognize ink diffusion speed, ink color separation and cumulative diffusion length. From the bottom half of the synthesized image (four horizontal line segments), we can see the time required for the ink to cross the horizontal lines. The synthesized images can also be used for additional color information analysis using PicMan for additional insights. All data can be exported in various formats: CSV files, line graphs, modified or processed images and/or newly synthesized images for efficient and effective use of image-based experimental data sets.

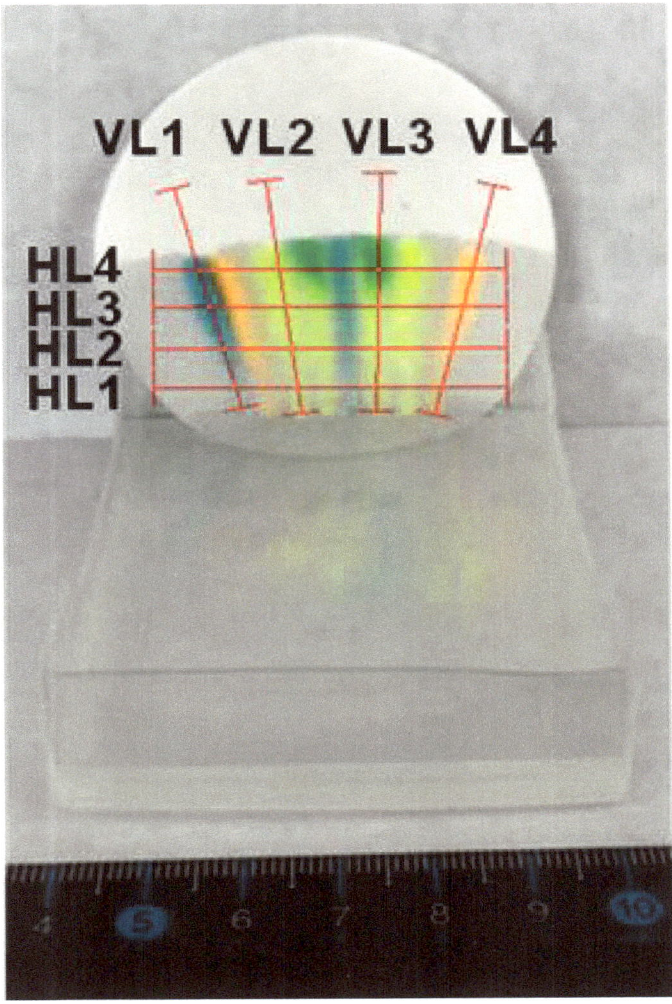

Figure 5. Time cross-section operation of four vertically sloped line segments (VL1~VL4) and four horizontal line segments (HL1~HL4) from entire frames of images for information extraction using PicMan.

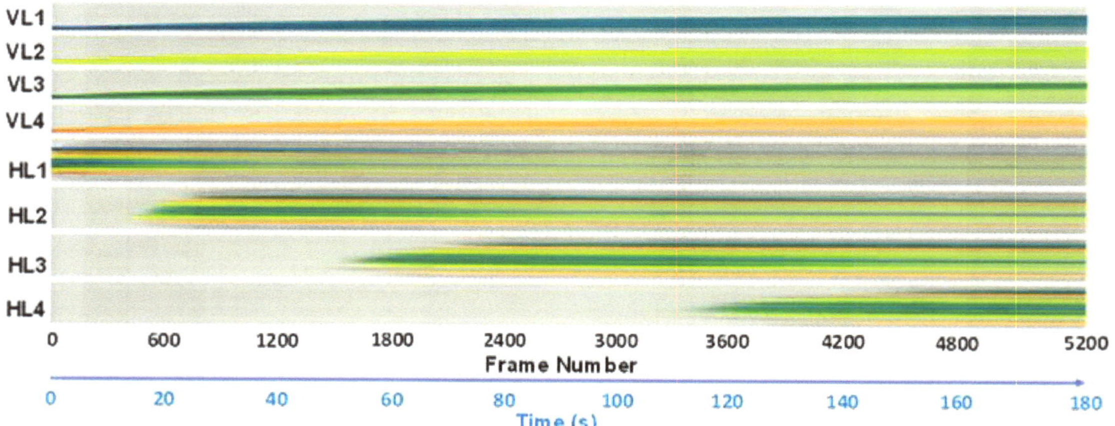

Figure 6. Exported time cross-section image across the eight-line segments (for vertical line segments VL1~VL4 and horizontal line segments HL1~HL4) for 0–180 s (30 fps × 180 s = 5400 frames) using PicMan.

3.2. LED-Based Decorative Light Experiment

Color, brightness, contrast, uniformity and their changes with time and conditions are considered as meaningful signals in image-based data processing for chemical, biological and medical applications. We typically take photographs and record video clips to document what we observe under certain circumstances, either simple observations or controlled experiments. Finding meaningful signals from acquired image(s) and video clip(s) is the most important step. Once the signals are found, the decoding and analysis of the signals into meaningful stories or scenarios begins. It is similar to a detective's investigation process.

To simulate this process, we recorded video clips of three decorative lights which have independently and randomly blinking, three-color, LEDs inside. From one of the recorded video clips, we have extracted color and brightness information on a line segment (L-L') as a function of time (or frames) in Figure 7. It was an 18.5 s long video clip with 30 fps. The total number of frames was 555. All color and brightness information on the line segment L-L', per frame, was extracted and assembled into a new, time cross-section image in the time sequence. Images extracted from the video clip at 4, 8, 12 and 16 s from the recording start were also shown for reference. Three squares (A1, A2 and A3) indicated in the top image allow statistical analysis of color and brightness information as a function of time (or frames) to decode LED blinking sequences or time charts per individual LED in three different decorative lights. White filter papers with color dots and lines in different marked patterns, using highlighter color pens, were prepared. The bottom part of the marked filter papers was immersed into water to observe ink diffusion patterns and the speed of diffusion. As the water is absorbed by the filter paper, the ink color diffuses and travels with the water. The experimental scenes were recorded as video clips for image analysis using the software (PicMan).

From the time cross-section image in Figure 7 (bottom left), we can easily recognize rough blinking intervals and color changes in each of the decorative lights. However, it is difficult to decode the mixture of an individual LED's blinking sequence by visual inspection, due to the filtering effect from the color and black letters of the light housing. For accurate decoding of an individual LED's blinking sequence, we have to select the area of minimum disturbances from the design pattern and letters on the decorative lights. We have selected three areas (A1, A2 and A3 in Figure 7) of 100-pixel squares (=10 pixel × 10 pixel) and extracted average color information on the selected areas per frame. The extracted color information was graphed as RGB intensity as a function of time per decorative light (Figure 8). The blinking sequences of nine LEDs (a set of RGB LEDs per decorative light) were perfectly decoded.

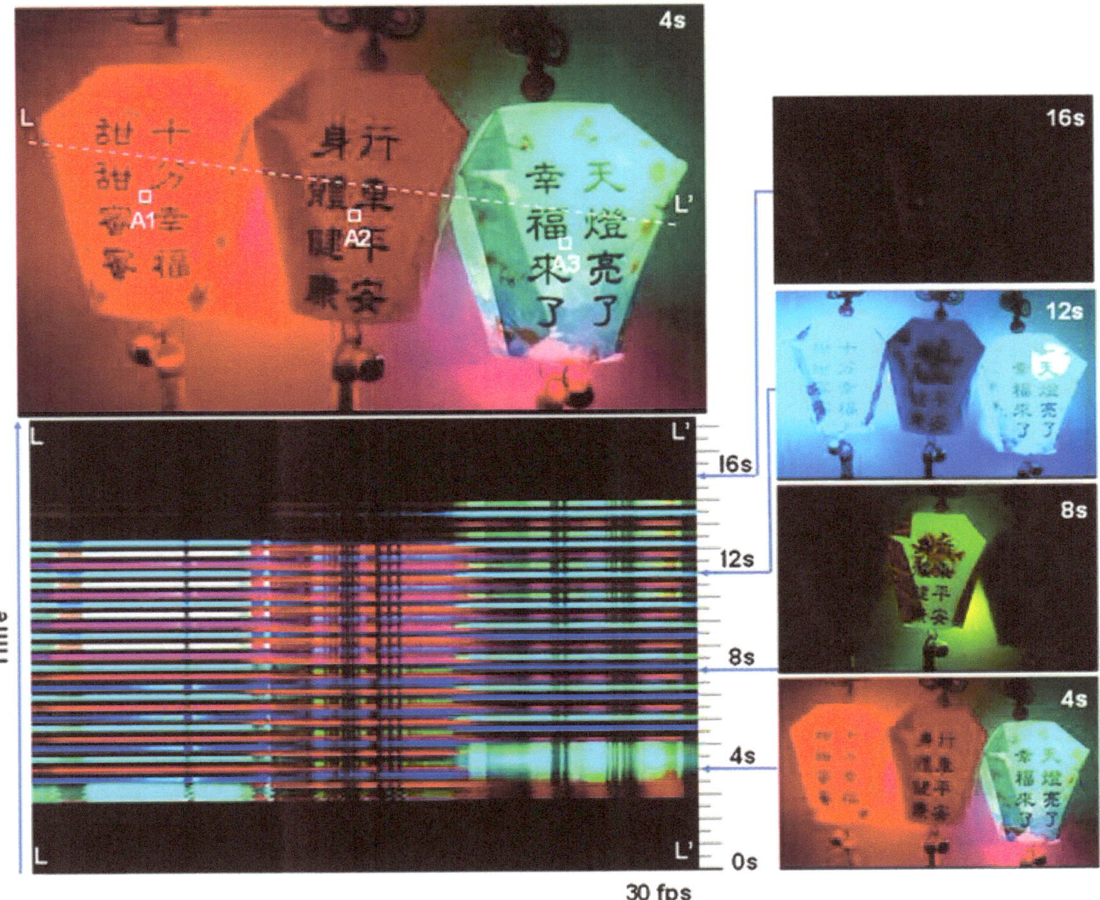

Figure 7. Randomly blinking decorative lights and time cross-section images across the line segment L-L' are shown. Top image: three 100-pixel (10 pixel × 10 pixel) areas of interest for average color analysis.

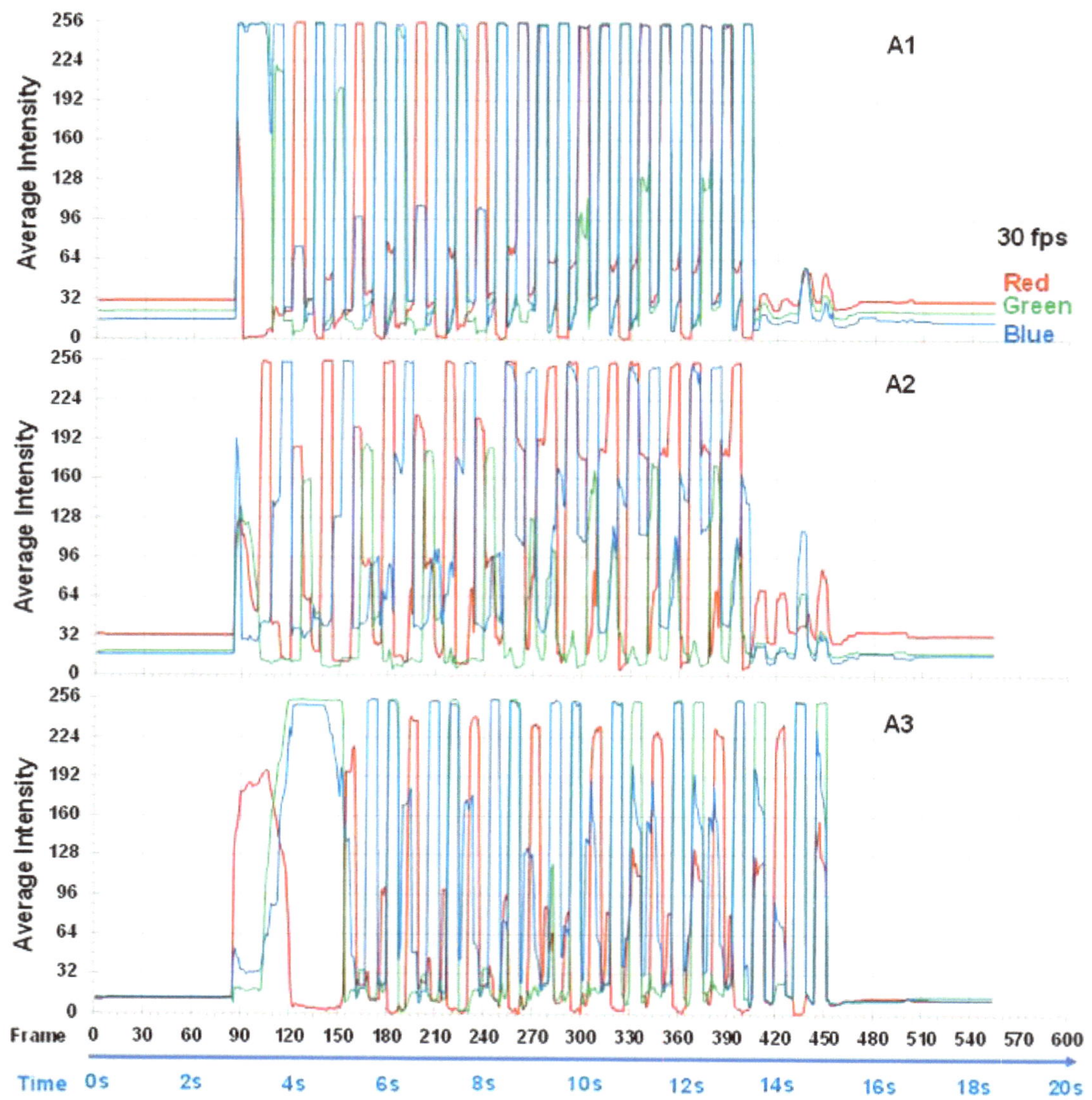

Figure 8. Decoded LED blinking sequence from analysis of a video clip using PicMan (average of 100 pixels (=10 pixel × 10 pixel) at square areas ((**A1**), (**A2**) and (**A3**)).

4. Discussion

With the advances in image sensor technology, application software development and computer performance, we can record a large quantity of high-resolution and high-quality photographs and video clips. However, we can only use a very small portion of the information from recorded images. It is typically far less than a fraction of one percent of the information contained in image files. This is true for all industries and academic fields, including the semiconductor industry, materials science and engineering, chemical, biological and medical fields. It is time to focus on image analysis and investigate how to use the information contained in images effectively and efficiently. Newer concepts for image data processing and extracted data reconstruction must be developed.

We have demonstrated new ways of image analysis, data extraction and image reconstruction techniques using the novel image processing software, PicMan. It is easy to integrate image sensors with PicMan to develop customized integrated image sensing and image analysis systems. Color conversion functions can also be used for system calibration. Of course, all basic functions can either be found or developed by individual researchers. However, it is neither cost-effective nor is it the best possible option for image acquisition and analysis for practical usage. It is important to collaborate with experts to provide engineering solutions with necessary skill sets. We have to remind ourselves how much information in the image files is not utilized or even reviewed and how important it could be. We have to mine meaningful data from all images, as much as possible, before acquiring additional image data. To do this we must improve our data mining skills.

Almost all image-based colorimetric studies are designed for very specific tasks for static analysis and are not available for dynamic analysis or the expansion/modification of applications [1–9,24–27]. The development of image-based, static and dynamic colorimetric analysis software (PicMan) for universal applications can provide significant flexibility in image analysis and information extraction from image(s) and video clip(s) of various formats. Various application examples can be found elsewhere [10,15–23,28–30,34].

5. Conclusions

Concepts of colorimetric sensing techniques for point(s), linear and areal array(s) were introduced along with the use of newly developed image processing software (PicMan) combined with image sensors for greater insight into the fields of chemical, biological and medical applications. As analytical examples, using the image analysis software, video clips of chromatographic experiments using different colored inks on filter papers dipped in water and randomly blinking LED-based decorative lights were used. The colorimetric information on real-time monitoring and recorded images or video clips can be analyzed for the point(s), line segment(s) and area(s) of interest for manual and automatic data collection.

The colorimetric information on point(s), line segment(s) and area(s) with different sizes from the video clips were extracted, analyzed and recreated as synthesized images as a function of time. As a demonstration of the developed colorimetric analysis technique, the colorimetric information was expressed as static and time-series combinations of RGB (red, green, blue) intensity, HSV (hue, saturation and value), CIE L*a*b* values, hexadecimal color codes and Munsell color index. Monitoring and recording of colorimetric information on one or more specimens were demonstrated by decoding blinking time sequences of individual LEDs in three independently operating, randomly blinking LED-based decorative lights.

Ex situ and in situ colorimetric data can be used as signals for process control, process optimization, safety and security alarms, and inputs for machine learning and data mining, including artificial intelligence. The importance of improving image analysis techniques for the efficient and effective use of information carried by images is demonstrated and emphasized.

For ensuring colorimetric data accuracy and repeatability, the qualification and calibration of image sensors must be established. The image resolution, frame rate, shutter speed and image format must be specified for consistent results. Application-specific customized functions for proper colorimetric analysis must be developed based on the requirement for image analysis/processing software.

Author Contributions: All authors equally contributed in this study. Conceptualization, W.S.Y., K.K. and Y.Y.; material preparation, Y.Y.; methodology, Y.Y. and J.G.K.; software, W.S.Y. and K.K.; data acquisition and analysis, Y.Y., J.G.K. and W.S.Y.; writing—original draft preparation, review and editing, W.S.Y. and Y.Y. All authors have read and agreed to the published version of the manuscript.

Funding: This research received no external funding.

Institutional Review Board Statement: Not applicable.

Informed Consent Statement: Not applicable.

Data Availability Statement: Not applicable.

Conflicts of Interest: The authors declare no conflict of interest.

References

1. Šafranko, S.; Živković, P.; Stanković, A.; Medvidović-Kosanović, M.; Széchenyi, A.; Jokić, S. Designing ColorX, Image Processing Software for Colorimetric Determination of Concentration, To Facilitate Students' Investigation of Analytical Chemistry Concepts Using Digital Imaging Technology. *J. Chem. Educ.* **2019**, *96*, 1928–1937. [CrossRef]
2. Vijayasankaran, N.; Varma, S.; Yang, Y.; Mun, M.; Arevalo, S.; Gawlitzek, M.; Swartz, T.; Lim, A.; Li, F.; Zhang, B.; et al. Effect of cell culture medium components on color of formulated monoclonal antibody drug substance. *Biotechnol. Prog.* **2013**, *29*, 1270–1277. [CrossRef] [PubMed]
3. Woolf, M.S.; Dignan, L.M.; Scott, A.T.; Landers, J.P. Digital postprocessing and image segmentation for objective analysis of colorimetric reactions. *Nat. Protoc.* **2021**, *16*, 218–238. [CrossRef] [PubMed]
4. Barros, J.A.V.A.; de Oliveira, F.M.; Santos, G.D.O.; Wisniewski, C.; Luccas, P.O. Digital image analysis for the colorimetric determination of aluminum, total iron, nitrite and soluble phosphorus in waters. *Anal. Lett.* **2017**, *50*, 414–430. [CrossRef]
5. Dische, Z. Qualitative and quantitative colorimetric determination of heptoses. *J. Biol. Chem.* **1953**, *204*, 983–998. [CrossRef] [PubMed]
6. Nimeroff, I. Colorimetry, National Bureau of Standards Monograph 104. 1968. Available online: https://nvlpubs.nist.gov/nistpubs/Legacy/MONO/nbsmonograph104.pdf (accessed on 24 December 2022).
7. Böck, F.C.; Helfer, G.A.; da Costa, A.B.; Dessuy, M.B.; Flôres Ferrão, M. PhotoMetrix and colorimetric image analysis using smartphones. *J. Chemom.* **2020**, *34*, e3251. [CrossRef]
8. Narimani, R.; Azizi, M.; Esmaeili, M.; Rasta, S.H.; Khosroshahi, H.T. Genome analysis of cellulose and hemicellulose degrading Micromonospora sp. CP22. *3 Biotech* **2020**, *10*, 416. [CrossRef]
9. Alberti, G.; Zanoni, C.; Magnaghi, L.R.; Biesuz, R. Disposable and Low-Cost Colorimetric Sensors for Environmental Analysis. *Int. J. Environ. Res. Public Health* **2020**, *17*, 8331. [CrossRef]
10. Yoo, Y.; Yoo, W.S. Turning Image Sensors into Position and Time Sensitive Quantitative Colorimetric Data Sources with the Aid of Novel Image Processing/Analysis Software. *Sensors* **2020**, *20*, 6418. [CrossRef]
11. Hemalatha, R.; Ramalaxmi, B.A.; Swetha, E.; Balakrishna, N.; Mastromarino, P. Evaluation of vaginal pH for detection of bacterial vaginosis. *Indian J. Med. Res.* **2013**, *138*, 354–359.
12. O'Hanlon, D.E.; Moench, T.R.; Cone, R.A. Vaginal pH and Microbicidal Lactic Acid When Lactobacilli Dominate the Microbiota. *PLoS ONE* **2013**, *8*, e80074. [CrossRef] [PubMed]
13. Vaginal pH Balance: Symptoms, Remedies, and Tests. Available online: https://www.medicalnewstoday.com/articles/322537 (accessed on 24 December 2022).
14. Maintaining Vaginal Health. Available online: https://health.cornell.edu/sites/health/files/pdf-library/Maintaining-Vaginal-Health.pdf (accessed on 24 December 2022).
15. Kim, G.; Kim, J.G.; Kang, K.; Yoo, W.S. Image-Based Quantitative Analysis of Foxing Stains on Old Printed Paper Documents. *Heritage* **2019**, *2*, 2665–2677. [CrossRef]
16. Yoo, W.S.; Kim, J.G.; Kang, K.; Yoo, Y. Extraction of Colour Information from Digital Images Towards Cultural Heritage Characterisation Applications. *SPAFA J.* **2021**, *5*, 1–14. [CrossRef]
17. Yoo, W.S.; Kang, K.; Kim, J.G.; Yoo, Y. Extraction of Color Information and Visualization of Color Differences between Digital Images through Pixel-by-Pixel Color-Difference Mapping. *Heritage* **2022**, *5*, 3923–3945. [CrossRef]
18. Yoo, Y.; Yoo, W.S. Digital Image Comparisons for Investigating Aging Effects and Artificial Modifications Using Image Analysis Software. *J. Conserv. Sci.* **2021**, *37*, 1–12. [CrossRef]
19. Kim, J.G.; Yoo, W.S.; Jang, Y.S.; Lee, W.J.; Yeo, I.G. Identification of Polytype and Estimation of Carrier Concentration of Silicon Carbide Wafers by Analysis of Apparent Color using Image Processing Software. *ECS J. Solid State Sci. Technol.* **2022**, *11*, 064003. [CrossRef]
20. Yoo, W.S.; Kang, K.; Kim, J.G.; Jung, Y.-H. Development of Image Analysis Software for Archaeological Applications. *Adv. Southeast Asian Archaeol.* **2019**, *2*, 402–411.

21. Yoo, W.S.; Han, H.S.; Kim, J.G.; Kang, K.; Jeon, H.-S.; Moon, J.-Y.; Park, H. Development of a tablet PC-based portable device for colorimetric determination of assays including COVID-19 and other pathogenic microorganisms. *RSC Adv.* **2020**, *10*, 32946–32952. [CrossRef] [PubMed]
22. Wakamoto, K.; Otsuka, T.; Nakahara, K.; Namazu, T. Degradation Mechanism of Pressure-Assisted Sintered Silver by Thermal Shock Test. *Energies* **2021**, *14*, 5532. [CrossRef]
23. Chua, L.; Quan, S.Z.; Yan, G.; Yoo, W.S. Investigating the Colour Difference of Old and New Blue Japanese Glass Pigments for Artistic Use. *J. Conserv. Sci.* **2022**, *38*, 1–13. [CrossRef]
24. Ellman, G.L.; Diane, K.; Valentino, C.; Robert, A., Jr.; Featherstone, M. A new and rapid colorimetric determination of acetylcholinesterase activity. *Biochem. Pharmacol.* **1961**, *7*, 88–90. [CrossRef]
25. Abels, K.; Salvo-Halloran, E.M.; White, D.; Ali, M.; Agarwal, N.R.; Leung, V.; Ali, M.; Sidawi, M.; Capretta, A.; Brennan, J.D.; et al. Quantitative Point-of-Care Colorimetric Assay Modeling Using a Handheld Colorimeter. *ACS Omega* **2021**, *6*, 22439–22446. [CrossRef] [PubMed]
26. Che Sulaiman, I.S.; Chieng, B.W.; Osman, M.J.; Ong, K.K.; Rashid, J.I.A.; Wan Yunus, W.M.Z.; Noor, S.A.M.; Kasim, N.A.M.; Halim, N.A.; Mohamad, A. A review on colorimetric methods for determination of organophosphate pesticides using gold and silver nanoparticles. *Microchim. Acta* **2020**, *187*, 131.
27. Shah, M.M.; Ren, W.; Irudayaraj, J.; Sajini, A.A.; Ali, M.I.; Ahmad, B. Colorimetric Detection of Organophosphate Pesticides Based on Acetylcholinesterase and Cysteamine Capped Gold Nanoparticles as Nanozyme. *Sensors* **2021**, *21*, 8050. [CrossRef] [PubMed]
28. Yoo, W.S. The World's Oldest Book Printed by Movable Metal Type in Korea in 1239: The Song of Enlightenment. *Heritage* **2022**, *5*, 1089–1119. [CrossRef]
29. Yoo, W.S. How Was the World's Oldest Metal-Type-Printed Book (The Song of Enlightenment, Korea, 1239) Misidentified for Nearly 50 Years? *Heritage* **2022**, *5*, 1779–1804. [CrossRef]
30. Yoo, W.S. Direct Evidence of Metal Type Printing in The Song of Enlightenment, Korea, 1239. *Heritage* **2022**, *5*, 3329–3358.
31. Color Model. Available online: https://en.wikipedia.org/wiki/Color_model (accessed on 24 December 2022).
32. Color Space. Available online: https://en.wikipedia.org/wiki/Color_space (accessed on 24 December 2022).
33. Color Conversion. Available online: https://en.wikipedia.org/wiki/HSL_and_HSV (accessed on 24 December 2022).
34. PicManTV. Available online: https://www.youtube.com/@picman-TV (accessed on 18 January 2023).

technologies

Article

Embedded System Performance Analysis for Implementing a Portable Drowsiness Detection System for Drivers

Minjeong Kim [1],* **and Jimin Koo** [2],*

[1] Monta Vista High School, Cupertino, CA 95014, USA
[2] Cupertino High School, Cupertino, CA 95014, USA
* Correspondence: minjeongsunnykim@gmail.com (M.K.); jimin.skoo@gmail.com (J.K.)

Abstract: Drowsiness on the road is a widespread problem with fatal consequences; thus, a multitude of systems and techniques have been proposed. Among existing methods, Ghoddoosian et al. utilized temporal blinking patterns to detect early signs of drowsiness, but their algorithm was tested only on a powerful desktop computer, which is not practical to apply in a moving vehicle setting. In this paper, we propose an efficient platform to run Ghoddoosian's algorithm, detail the performance tests we ran to determine this platform, and explain our threshold optimization logic. After considering the Jetson Nano and Beelink (Mini PC), we concluded that the Mini PC is most efficient and practical to run our embedded system in a vehicle. To determine this, we ran communication speed tests and evaluated total processing times for inference operations. Based on our experiments, the average total processing time to run the drowsiness detection model was 94.27 ms for the Jetson Nano and 22.73 ms for the Beelink (Mini PC). Considering the portability and power efficiency of each device, along with the processing time results, the Beelink (Mini PC) was determined to be most suitable. Additionally, we propose a threshold optimization algorithm, which determines whether the driver is drowsy, or alert based on the trade-off between the sensitivity and specificity of the drowsiness detection model. Our study will serve as a crucial next step for drowsiness detection research and its application in vehicles. Through our experiments, we have determined a favorable platform that can run drowsiness detection algorithms in real-time and can be used as a foundation to further advance drowsiness detection research. In doing so, we have bridged the gap between an existing embedded system and its actual implementation in vehicles to bring drowsiness technology a step closer to prevalent real-life implementation.

Keywords: drowsiness detection; embedded systems; WebRTC; AioRTC; facial detection; blink detection

Citation: Kim, M.; Koo, J. Embedded System Performance Analysis for Implementing a Portable Drowsiness Detection System for Drivers. *Technologies* **2023**, *11*, 8. https://doi.org/10.3390/technologies11010008

Academic Editors: Valeri Mladenov, Gwanggil Jeon and Imran Ahmed

Received: 16 November 2022
Revised: 25 November 2022
Accepted: 27 December 2022
Published: 30 December 2022

1. Introduction

1.1. Background

Drowsiness on the road is a widespread problem with fatal consequences. Each year, 1.35 million people are killed on roadways around the world, according to the World Health Organization's Global Status Report on Road Safety in 2018 [1]. 328,000 drowsy driving-related crashes occur annually in the United States alone, according to a study by the AAA Foundation for Traffic Safety [2]. Furthermore, NHTSA estimates suggest that fatigue-related crashes cost society $109 billion each year, on top of property damage [2]. Studies from the National Sleep Foundation revealed that 50% of U.S. adult drivers admit to consistently driving while drowsy, and 40% admit to falling asleep behind the wheel at least once in their driving careers [3]. Thus, drowsy driving is a ubiquitous problem that claims heavy social and economic tolls. It must be promptly addressed to save irreplaceable human lives and economic damage.

1.2. Prior Work

Because drowsy driving is a fatal problem with significant consequences, various approaches have been proposed in the field. Some solutions to detect drowsy driving involve the use of biological signals such as electroencephalogram (EEG) data [4–10], optical brain wave signals through functional Near InfraRed Spectroscopy (fNIRS) [11], heart rate variability [4,12], etc. [13]; For example, Arefnezhad et al.'s [8] method measures PERcentage of eyelid CLOSure (PERCLOS), a widely used drowsiness detection metric, using EEG electrode data and a Bayesian filtering solution. In their work, Theta and Delta power bands on the EEG spectrum were identified to be correlated with PERCLOS values. Such methods aim to detect physiological changes in the drivers' biosignals that indicate drowsiness. However, these approaches are not practical for application because the driver has to wear bulky, intricate instruments on their head for extended periods of time, which may be too complex for lay people to use and can impede driving.

Some existing technologies have also detected drowsy driving using driving performance data such as vehicle turning curvature, accelerations, line weaving behaviors, etc. [14,15] For example, built-in systems like the Mercedes Attention Assist system assesses personal driving styles to detect irregularities considering 70 parameters and external factors [16]. Similarly, Castignani et al. [17] proposed the use of existing smartphone sensors to develop a driver profile and detect "risky driving events", taking aspects like route topology and weather into consideration. However, such external factors are more indirect methods for drowsiness detection, as the correlation between driving patterns and drowsiness has not been adequately established. Other studies on drowsiness detection span a wide range of systems and metrics, such as head angle, explicit drowsy signals (yawning, nodding off), blink count and patterns [18–21]. For example, Xu et al. [22] and others proposed systems based on PERCLOS [23,24] and approximated blink statistics, which demonstrated a drowsy behavior-detecting accuracy of over 90%. However, the work focused on detecting simple eye closure rather than predicting drowsiness levels.

Most eye-tracking based drowsiness detection algorithms were based on finding the PERCLOS, as mentioned above, which indicates the percentage of time during which the eye is closed over a duration of time. For example, Feng You et al. [25] proposed a program that detects drowsiness by checking whether eyes are closed using eye aspect ratios calculated from facial landmarks. The program uses a deep-cascaded convolutional neural network to detect the face, then a Dlib facial landmark detector to analyze these facial landmarks. A support vector machine (SVM) was developed to classify if the eyes are open or closed in each frame. If the PERCLOS is greater than a designated threshold, the driver is labeled as drowsy. Most of the research focused on the development of the eye close-open detection model rather than the relationship between eye aspect ratio and drowsiness level. Feng You et al. emphasizes that unlike PERCLOS-80, which predicts that eyes are closed if more than 80% of the pupil is covered by the eyelid, their proposed method measures eye aspect ratios. Using this metric allows the program to account for differences in each user's facial features. Similarly, Manishi et al. [26] proposed a method that uses a Viola-Jones algorithm to detect the eyes and mouth, from which a convolutional neural network is used to detect if the eyes and mouth are open or closed. Closed eyes or wide-open mouths indicate drowsiness. Using the NTHU Drowsy Driver Detection Dataset, this model takes into consideration the different situations that the user may use the program, such as in the dark and with or without glasses.

However, we believe that when there is a detectable difference in the percent of eyelid closure, the driver is already in a dangerous situation. Ideally, drowsiness detection algorithms should "predict" drowsiness in earlier stages by analyzing more subtle signs. Thus, we continued to search for a drowsiness detection algorithm that could detect drowsiness levels while the driver still has enough time to take precautionary steps.

We found that Ghoddoosian et al. [27] proposed a drowsiness detection method that could detect the early signs of drowsiness. Their study also includes a dataset with which we trained the model and tested it ourselves. The proposed LSTM model uses a distinct sequence of blinking patterns such as the speed of eyes closing and opening and the duration, amplitude, and frequency of blinks to predict the early signs of drowsiness. The method detects faces using Dlib's face detection model [28], then uses Kazemi and

Sullivan's [29] model to detect facial landmarks. Soukupova and Cech [30]'s model was used to detect blinks. The LSTM model uses data on a sequence of blinks to classify a person's state as alert, low-vigilant or drowsy. The LSTM model was trained on a dataset of 60 different individuals of a wide range of ages and ethnicities who provided videos of themselves in different states of drowsiness. This algorithm appeared the most promising because it could detect early signs of drowsiness, which is why we decided to use it as a base model for our research. However, this method was developed on a powerful desktop computer, and a platform implemented in embedded systems is needed to run the algorithm in vehicles. Embedded systems use CPUs or GPUs that are less performant than desktop CPUs.

At the same time, while there are existing solutions that implement drowsiness detection algorithms in embedded systems, we have found that the programs are too simple and do not consider the complexity of drowsiness expression or require intrusive methods of data collection. For example, Lunbo Xu et al. [22] proposed a smartphone-assisted drowsiness detection system. In their study, Xu et al. validated that all processes of the algorithm run in less than 100 ms/frame, which supports the smartphone's framerate of 10 fps. Xu's research focuses mainly on the development of the eye close-open detection model, and the reported accuracy of the drowsiness detection program measures only the program's ability to determine whether an eye is closed or open. If the PERCLOS exceeds 0.25, the labels the user as drowsy. Similarly, Eddie E. Galarza et al. [31] proposed a drowsiness detection program that runs on an Android-based smartphone. The drowsiness detection program, however, detects various possible behaviors of drowsiness: head swaying, blinking, and looking to the left or right. The reported accuracy of the proposed algorithm considers only the average of the accuracies of detecting each of these behaviors, rather than the state of being drowsiness. Further research exists on drowsiness detecting embedded systems that connect to a separate device. For example, in his review paper on drowsiness detection systems, Anis. Rafid et al. [32], presented several embedded systems including smartphones. Most of them were connected to external devices that had to be worn, such as an EEG System [10] or other devices that collected physiological data. These devices were tested in lab settings, and as mentioned above, the practicality of using these devices in a moving vehicle setting is questionable. Some also used camera's inertia sensors such as accelerometer or gyroscope to detect changes in driver's driving techniques [21,33]. However, as mentioned previously, the correlation between such driving patterns and drowsiness has not been adequately established.

Table 1 summarizes the methods of prior research. Prior solutions for drowsiness detection on can be categorized by the input signal, such as bio signals, driving patterns, and facial landmarks. The solutions involving facial-landmark detection can be further categorized by studies that detects simple eye closure or PERCLOS, and the one that analyzes various parameters of blinks. We believe that predictiveness of drowsiness detection, or its ability to detect the initial stages of drowsiness through subtle expressions, is the highest in the blink-analyzing method, which has not yet been implemented in an embedded system formfactor.

The ideal drowsiness detection system would be a portable embedded system that can predict drowsiness using the method suggested by [27]. However, we were not able to find such a system yet. To implement such a system, research is needed to understand the capabilities of embedded systems and the design requirements. For example, the computational power to run the algorithm, electrical power consumption, camera resolution, and communication speed are a few of the requirements. We have found existing research that defines the quantitative metrics necessary to run a cyber security application involving a neural network in an embedded system [34,35]. However, up to the author's knowledge, we could not find such research for a drowsiness system which can run powerful drowsiness prediction neural networks.

Table 1. Comparison of prior work in terms of predictiveness, direct sign of drowsiness from eyes, and embedded system formfactor. The prior works were categorized by input signals: biological signals, driving patterns, facial landmarks.

Category of Input Signals to Detect Drowsiness		References	Predictiveness	Direct Sign of Drowsiness from Eyes	Embedded System Formfactor
Biological Signals		[4–13]	Not clear	No	No
Driving Patterns		[14–17,33]	Not clear	No	Yes
Facial Landmarks	Simple closed eye detection based	[18]	Too late	Yes	Yes
	PERCLOS based	[22–24]	Too late	Yes	Yes
	Blink pattern based	[27–30]	Yes	Yes	No

1.3. Our Approach

We propose a portable drowsiness detection system consists of a phone and a mini server which can be paired and placed in a vehicle. The phone serves as the camera and touch-display, and the mini server which is plugged into the vehicle's auxiliary power outlet runs the drowsiness detection model in real-time.

Our system uses Wi-Fi to connect the phone and the embedded server locally without connecting to the internet. If our system ran by connecting to the internet, it would have a greater delay (time lag) regardless of the three embedded systems, because there would be more routers involved between the client (phone) and the server. However, even if we used the internet, the overall round-trip delay would be in the hundreds of milliseconds, which would not be problematic for predicting drowsiness. This is because the throughput performance, rather than the delay, needs to be more enforced in following the 33 ms limit. We carefully decided not to use the internet because of privacy concerns, as we believe that users would not want their videos or drowsiness statistics to be accessible through any external server. Furthermore, the cost of LTE communication to the internet from the vehicle can be costly, which makes it less favorable for users.

Because an average vehicle's power outlet (also known as cigarette outlet) only puts out about 120 watts of power, we had to find a device for the server that runs with around or less than 120 watts of power.

We also propose a threshold finder and drowsiness value voting algorithm. We put our research efforts into the following two parts: developing a portable embedded system for performing drowsiness detection and developing an algorithm that finds the threshold for each model and votes for the final drowsiness value. For the drowsiness detection model, we used the same drowsiness detection model as proposed by Ghoddoosian et al. [27]. Our proposed approach not only considers the temporal aspect of drowsiness detection but also allows the model to be utilized in a real-time vehicle setting.

1.3.1. Portable Embedded System

In this paper, we also propose the hardware/software implementation of a portable drowsiness detection system that can be used in the vehicle. In the user's vehicle, a powerful Mini PC server is set up, and the phone connects to the server and run the drowsiness detection web app. This approach also ensures privacy since the user's data stays in the mini server instead of being shared on the internet. In order to develop our system, quantitative analysis of the communication speed and of the performance of the system in a portable setup needs to be performed. The analysis allows us to decide which of the candidate hardware should be used for our proposed system. We tested the candidate systems and selected the device that can both accommodate the complex AI performed for drowsiness detection and be conveniently placed in a vehicle. As shown in Figure 1, the client can be placed near the front of the vehicle where the camera of the client can stably detect the driver's face. The server can be placed anywhere in the vehicle. For example, it can be placed in the trunk and use the 2 V power outlet as its source of power.

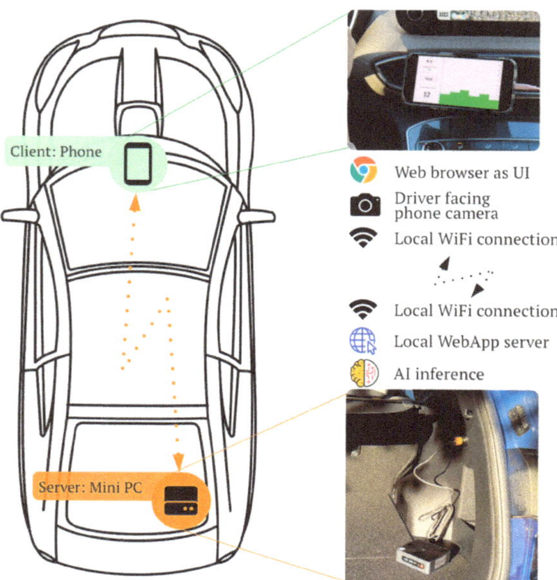

Figure 1. Vehicle installation example of our proposed solution. The client is a phone that can be attached near the dashboard area. The front facing camera should see drivers face to detect blinks. The server can be positioned anywhere in the vehicle as long as 12 V power can be provided. The client and server is communicating wirelessly through Wi-Fi connection.

1.3.2. Threshold Optimization Algorithm and Voting Algorithm

Furthermore, we propose a method for optimizing the threshold for our model, which divides blink sequences into the following categories: alert or drowsy. The drowsy category includes both the "low vigilant" and "drowsy" categories from Ghoddoosian et al.'s model. We hypothesize that on the road, our drowsiness detection model requires a trade-off between sensitivity and specificity. Sensitivity refers to the ability of the model to accurately detect drowsiness when the driver is drowsy, while specificity refers to the ability of the model to correctly detect alertness when the driver is not drowsy. We decided that the sensitivity of our model is much more important than specificity, as poor specificity may cause inconvenience by falsely alerting the driver, but poor sensitivity can lead to life-threatening situations. Currently, most AI-related papers aim to optimize parameters to maximize the F1 score of their models, which is the percentage of the sum of true positive and true negative predictions. However, by changing some parameters, we can control the trade-off between the false positives (FPs) and false negatives (FNs) and emphasize sensitivity over specificity. Additionally, we propose a voting algorithm, which finds the weighted average of various models' predictions based on the true negative and true positive rates of each of the models.

1.4. Review of the Research Process and Outline of This Paper

Figure 2 describes a detailed process for the research we have conducted. First, by analyzing prior research and investigating web app technology, we designed an embedded-system architecture that can predict drowsiness levels in real-time in vehicles. We implemented our drowsiness detection algorithm in a Desktop PC before testing candidate embedded systems to check the functionality of each.

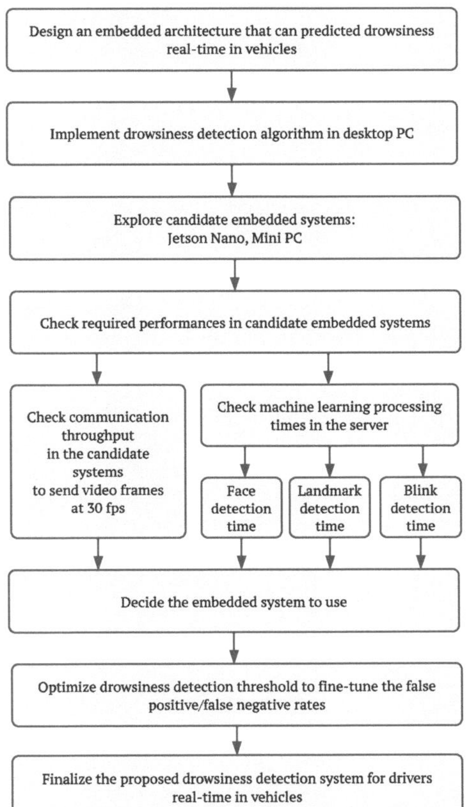

Figure 2. Overall research process to develop proposed drowsiness detection embedded system in vehicles.

We explored possible embedded system candidates that can run our system and developed methods to check the performance of candidate embedded systems. Specifically, we checked the communication throughput to ensure that our system can run without accumulating lag, and also checked processing speeds to process the video streams for detecting drowsiness without lagging. Critical processes were identified: face detection, landmark detection, and blink detection. These will be explained in detail in the sections that follow. From these results, we were able to choose the most practical device for our embedded system. Then, we developed a threshold optimization method that classifies the drowsiness level into certain drowsiness states. The threshold optimization is performed based on the false positive/false negative rates of the models. All of our findings provide the necessary information to finalize the proposed drowsiness detection system for drivers in vehicles.

The remainder of the paper is organized as follows. Section 2 describes our hardware and software setups that explain the algorithms we used to develop and analyze our candidate embedded systems' performance. Section 3 describes how we ran these procedures to validate our system's performance and our Threshold Optimization Algorithm. Section 4 explains and analyzes the specific results from the collected data, and lastly, Section 5 concludes our study.

2. Materials

In this section, we describe the hardware and software configurations of our proposed embedded system that runs the drowsiness detection algorithm in real time, including the drowsiness detection algorithms, threshold optimization method, and the real-time voting algorithm to calculate the final drowsiness value.

2.1. Hardware Setup

As described earlier, we are proposing a drowsiness detection system that uses two parts to detect drowsiness: a client and a server. The client can be any mobile phone that has a camera and can connect to Wi-Fi. The phone should also have a web browser that can load our web app. The server needs to be powerful enough to run the web server and drowsiness detection algorithm, which means that it needs to have adequate communication performance and processing power.

To measure the performance of the communication and server's processing time, we used a laptop as our client because it provides better tools and an ideal SW environment for collecting data from the web app client. The laptop's default camera (supporting 720 p image resolution and 30 fps maximum framerate) was used as the client camera to take video and stream to the server. Since most phones also support fast Wi-Fi connection and have high resolution cameras, the client side can easily be replaced by a mobile phone.

We chose to investigate a Jetson Nano Developer Kit and a Beelink Mini PC as potential embedded systems based on their performance, power efficiency, and formfactor. The specifications for each of the devices used are detailed in Figure 3.

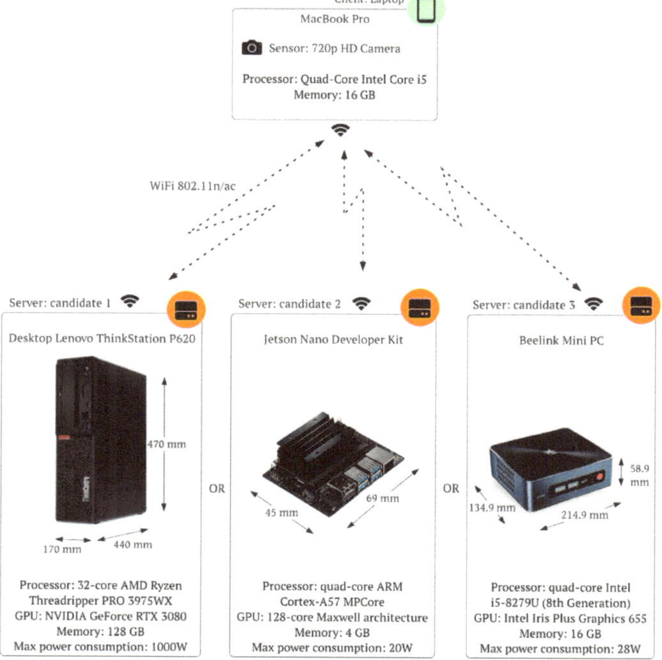

Figure 3. HW Setups for experiments. MacBook Pro was used as a client instead of a phone and three devices are listed as servers: Desktop PC, Jetson Nano development kit, Mini PC. One of the candidate servers was powered up to run the drowsiness detection system. The dimension and specification of the candidate servers are described below each device.

We selected the Jetson Nano as one of our candidates because of its GPU-accelerated performance and efficient power consumption. The Nano contains a quad-core Cortex-A57 processor, which makes it similar to other embedded systems. However, it also has 128 Maxwell architecture-based GPU cores, which accelerate the AI interference process. While there are many other embedded systems such as Raspberry Pis, Arduinos, or BeagleBoards, these devices do not have powerful GPUs or neural network accelerators to perform machine learning processes efficiently. Suzen et al. [36] compared the performances of NVidia's TX, Jetson Nano, and Raspberry Pi 4 Model B (the newest Raspberry Pi model) by running a Deep-CNN that classifies clothing images into 13 categories. They found that the Jetson Nano could run inference operations more than five times faster than a Raspberry Pi. The Jetson Nano also only uses 10 W of power, which is less than the 120 W output of a typical vehicle's power outlet.

Our other candidate was the Beelink Mini PC, which we selected because of its high-performance CPU and GPU, which are similar to those of a typical laptop performance. The Mini PC has an 8th generation Intel i5-8279U process and typical Intel GPU. The power consumption can be up to 28 W, which is still less than the 120 W output of a typical vehicle's power outlet.

Before implementing the embedded server system, we used a desktop PC with AMD Ryzen Threadripper PRO processor. Although the PC is not in an embedded system formfactor, we used it to develop our system to train the model and compile the web-server code. The PC has a 32-core AMD Ryzen Threadripper processor with a powerful NVDIA GeForce RTX 3080 GPU. After developing the code in this machine, we transferred the package into a Docker image and ran it on the Jetson Nano and Mini PC.

When running the experiment, we connected the client and server via Wi-Fi 802.11 n/ac connection. Both devices were in the same room and only one of the servers was turned on to run the web server. The server had a separate local monitor and keyboard/mouse to control the web server status.

When running in vehicles, the phone can be held by an aftermarket phone holder and connected to a USB port in a car. The server can be connected to the 12 V power outlet via a 12 V to 19 V DC converter, and installed in any secure place that does not shift significantly when the car is in motion.

2.2. Software

The overall architecture and block diagrams of our web app's software are shown in Figure 4. The software consists mainly of two parts: the client side and the server side, which we will explain below.

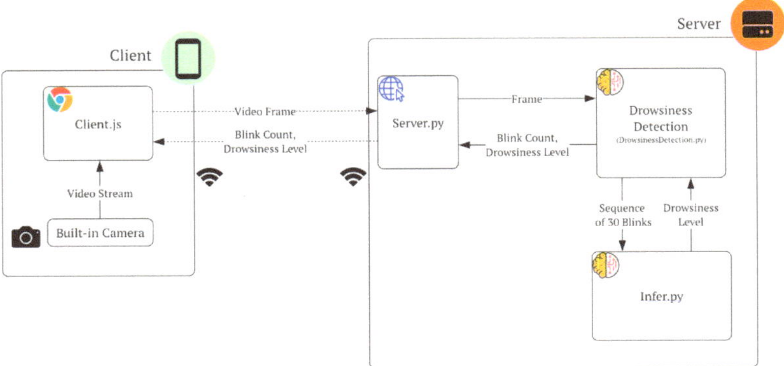

Figure 4. Proposed drowsiness detection SW architecture that runs the web app interface. Major block diagrams are shown for the client and server side.

2.2.1. Client

Our system was built to be compatible with phones: the phone functions form the "client" part, utilizing the camera and screen to communicate with users, while the server functions form the computational device. When the client connects to the server, it automatically downloads the files needed to run the web app, which include index.html, client.js, and base.css. These files are provided in the AioRTC server example. Index.html provides the input interface to the user and displays the elements that users interact with as in Figure 5, while client.js includes the majority of the functionality of the web app, such as video display, WebRTC communication, display of the drowsiness detection level, and history of the drowsiness levels in real-time. Client.js receive a video stream from the built-in camera, from which is sends frames to the server, as shown in the Client side of Figure 4. The CSS file contains the styling elements, which include color schemes and the formatting of the main page. Currently, we are using a simple method of alerting drivers by displaying their drowsiness value and changing the color of a bar graph to red when the user's drowsiness level passes a certain threshold.

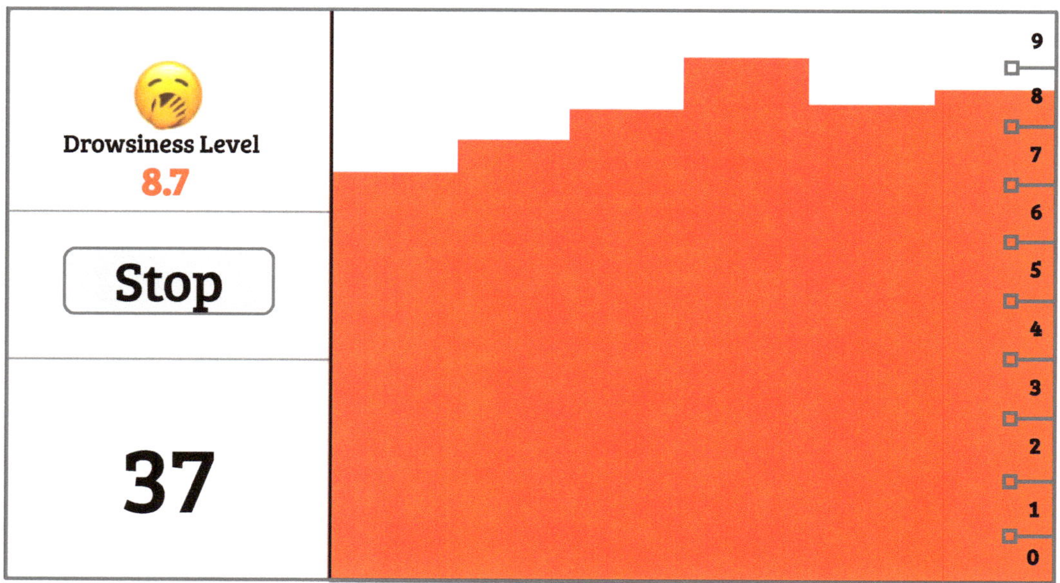

Figure 5. Running example of the user interface. The emoji on the top left describes the current drowsiness status and the bar chart displays on the right the historical drowsiness value over time. On the bottom left, the number of detected blinks is shown so that users can know whether the algorithm is running without error.

2.2.2. Server

For our server, we utilized AioRTC (https://github.com/aiortc/aiortc (accessed on 25 April 2021)), which is an open source library for Web Real-Time Communication (WebRTC) and Object Real-Time Communication (ORTC) in Python. We chose to use AioRTC because its implementation is simple and readable compared to WebRTC. In addition, our code handles a local server, meaning there is required communication between the client and the backend (models, server.py). AioRTC was most suitable because it can send data from the 'client' to the 'server' without having to download data locally before analyzing it, ultimately reducing lag. When the device begins detecting drowsiness upon the driver's request, the client (Client.js) sends video frames to the Server.py. We modified server.py to include a section that sends and receives data from the drowsiness prediction server, as shown in the Server side of Figure 4. First, the server sends a fixed 2D array of the video frame to the drowsiness prediction file, DrowsinessDetection.py. DrowsinessDetection.py uses the RNN drowsiness detection model to predict and return drowsiness values along with the total number of collected blinks to server.py. The client receives the information and displays it on the HTML page. We explain DrowsinessDetection.py in more detail in Section 2.3.1.

2.3. Drowsiness Detection Model

In the following sections, we summarize the algorithms used and developed by Ghoddoosian et al. for the drowsiness detection model, as well as our implementation that enables our proposed program to monitor the driver in real-time in the server client setup. We utilized Ghoddoosian et al.'s blink detection and feature extraction algorithms and offline training process making minimal changes to the code. For the live monitoring step, we modified the blink feature extraction code to include a function that sends data and receives a drowsiness value. This process is further explained in Section 2.3.3. Furthermore, we developed a Threshold optimization algorithm and Voting Algorithm, which both serve to calculate the final real-time drowsiness prediction.

2.3.1. Blink Detection and Blink Feature Extraction

For our blink detection and blink feature extraction steps, we used the models proposed in Ghoddoosian et al.'s paper. We cloned their source code published in GitHub and combined it with the AioRTC webserver example. Their blink detection process can be divided into three steps: face detection, facial landmark detection, and blink detection. First, the program uses Dlib's pre-trained face detector, which is based on the standard Histogram of Oriented Gradients + Linear SVM method for object detection [28]. Ghoddoosian et al. then used Kazemi and Sullivan's [28] facial landmark detector because it was trained with an "in-the-wild dataset" which included videos filmed in various conditions (illuminations, facial expressions, head positions, rotations, etc.), thus making the model robust to a variety of environmental conditions [27]. For the blink detection step, Ghoddoosian et al. used Soukupova and Cech's blink detection model [30] to perform the first blink detection step. Once a blink is detected, the blink retrieval algorithm uses the eye aspect ratios of the eyes to extract four features of the blink: amplitude, velocity, frequency, and duration. For further explanation of these values and their formulas, we refer readers to [27]. Each of these features is then normalized for each individual according to one-third of the blink features extracted from their alert video. This step is significant for this model as all the data are trained together, so differences across each individual's blinking pattern must be accounted for.

2.3.2. Offline Training and Threshold Optimization Algorithm

Regarding the drowsiness detection model, we used the model proposed by Ghoddoosian et al. The structure of the model is shown in Figure 6 from [27]. Ghoddoosian et al. introduced an HM-LSTM network to incorporate the temporal element of drowsiness detection, since the Hidden Markov Model (HMM) from [27] indicates that expressions of drowsiness follow a temporal pattern. The model was trained on a dataset of 180 10 min recordings, totaling to 30 h of RGB videos. 60 participants were asked to film themselves in three different drowsiness states: alert, low-vigilance, and drowsy [27]. Of the 60 participants, 51 were men and 9 were women, some with facial hair and some with glasses, and of five different ethnicities. Participants were given instructions on how to film their videos; specifically, the phone should have been about an arm's length away from the user and in the location/angle representing that of an actual vehicle setting (suggested placing the phone on computer screen).

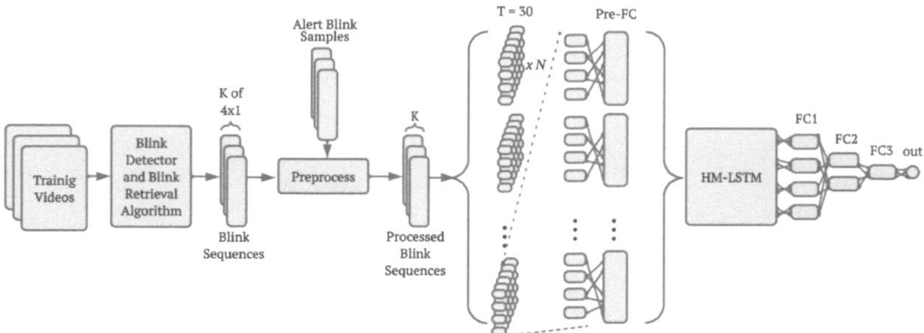

Figure 6. Drowsiness detection algorithm suggested by Ghoddoosian et al.

This dataset was divided into five separate folders, referred to as "folds," each consisting of 36 videos (3 videos each from 12 participants). During the cross-validation step, a new model was created and trained with videos from four of these folders and tested with those of the remaining folder.

After the blinks of each of these videos are analyzed according to the algorithm explained in the previous section, the values are normalized and passed in as input for the HM-LSTM model. For details on the dataset with which the model was trained and the features of this HM-LSTM network, we refer readers to [27].

In Ghoddoosian et al.'s paper, training.py trains the model by classifying each blink sequence's predicted drowsiness value, a number from 0.0 to 10.0, into three categories: alert, low-vigilance, and drowsy. Then, it compares the predicted category of blinks to the label of these blinks. The blinks are classified according to the ranges listed below.

- Alert: 0.0 predicted value < 3.3
- Low vigilance: 3.3 predicted value 6.6
- Drowsy: 6.6 < predicted value 10.0

For our proposed solution for drowsiness detection for drivers, we added an additional step after the model training. First, we group the low-vigilance and drowsy categories together since we want to alert the driver once they begin expressing any signs of drowsiness. Additionally, we do not fix the threshold to one value (i.e., 3.33) as Ghoddoosian's training algorithm does. Instead, we developed a customized threshold value which is determined by the ratio of false positive to false negative values of each model's confusion matrix. False negative (FN) rates indicate the fraction of predictions that incorrectly predicted a blink sequence as drowsy, while false positive (FP) rates indicate the fraction of predictions that incorrectly predicted a blink sequence as not drowsy. Both cases negatively impact the user experience, but we believe that decreasing the false negative rate is more important than lowering the false positive rate because safety is more important than convenience. We intentionally alter the threshold to perform the trade-off between the values. These rates change according to the threshold since the threshold determines whether a blink is predicted as drowsy or not drowsy. The process of finding the optimal threshold value will be explained in Methods Section 3.3. To test our optimization algorithm, we used the same dataset as the one used by Ghoddoosian et al.

2.3.3. Online Monitoring and Voting Algorithm

The HM-LSTM model proposed in [27] is trained and runs only off-line after all data has been collected. 10 min videos from 60 people were used to train and test the model. However, an online monitoring algorithm needs to be developed to predict the user's drowsiness level in real-time.

Figure 7 shows the sequence of our online monitoring algorithm. We implemented it in two main files: DrowsinessDetection.py and Infer.py. We developed DrowsinessDetection.py based on the blinkvideo.py, which Ghoddoosian et al. uses to extract blink features for the model input. Instead of saving the blink features as a text file, however, DrowsinessDetection.py calls a function in Infer.py using a list of these values as the parameter. Infer.py then uses TensorFlow to run a session for each of the drowsiness detection models and get the drowsiness value. DrowsinessDetection.py returns the drowsiness value and number of blinks collected back to the AioRTC server file, as mentioned in Section 2.2.2.

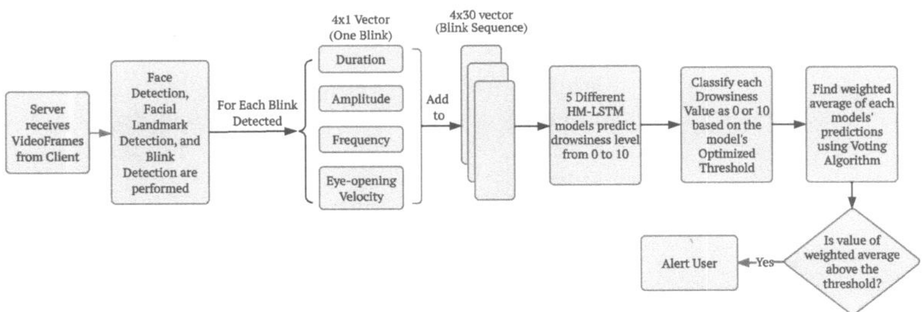

Figure 7. Real-time drowsiness detection and alerting algorithm running in server.

To predict the final drowsiness level, we developed a voting algorithm that find the weighted average of the predicted drowsiness values from each of the 5 models. Notably, Ghoddoosian et al. validated the drowsiness detection algorithm using a leave-one-out cross-validation step, which results in five different drowsiness detection models trained with varying combinations of datasets. Specifically, the training dataset was split into five different groups; the first model was trained using groups 2, 3, 4 and 5 and tested using group 1, the second model was trained using groups 1, 3, 4, 5 and tested using group 2, etc. For further explanation of these models, we refer readers to [27]. Each model outputs a drowsiness prediction value, which is categorized according to the thresholds found by the Threshold Optimization Algorithm. Alert blink sequences are assigned the value 0.0, and drowsy blink sequences are assigned the value 10.0. To obtain the final drowsiness value from the values predicted by the five models, our Voting Algorithm which finds the final value from their weighted average. Each model is assigned a weight value V_i, which is the sum of the true negative rate of the model and double the true positive rate for model i. We assigned a greater coefficient to the true positive rate than the true negative rate because we value sensitivity over specificity as explained in the introduction.

$$V_i = 2TP_i + TN_i \tag{1}$$

We then take the sum, S, of all V_i across models. S will be used to normalize the weighted values, as shown below.

$$S = \sum_{i=1}^{N} V_i \tag{2}$$

Finally, we define the prediction value, P, as the final predicted drowsiness value as shown below.

$$P = \sum_{i=1}^{N} \frac{V_i}{S} b_i \tag{3}$$

b_i is the binary value indicating whether the driver is drowsy or alert (1 if drowsy, 0 otherwise). If the value P surpasses 0.5, we predict the driver is drowsy, and the device alerts the user. The value P is displayed on the client's screen in the form of a bar chart, as illustrated in Figure 5.

3. Methods

We proposed a portable embedded solution for a drowsiness detection system that can be mounted on a vehicle. In order to validate our solution, we decided to check the following characteristics: communication throughput between the client and the server, processing time to detect the face, facial landmarks, and blinks, and the algorithm for finding optimized threshold value based on the false positive and false negative rates of the drowsiness detection models. In the following sections, we describe the detailed measurement procedures to collect the necessary data and the analysis we performed for each of the items above.

3.1. Communication Speed

Our proposed system requires fast enough communication speed for the drowsiness system to work properly. If the speed is not fast enough to transmit video frame rate from client to server, the client will start to accumulate unsent video frames causing delays and eventually a memory overflow and system crash.

Communication speed can be explained in terms of delay and throughput. The delay between the client and server determines the amount of time it takes for a frame sent from the client to be accessed by the server. The throughput is defined as the amount of data that can be sent in a given time unit. In our case, we used frames per second (fps) as the unit. In Figure 8, the delay is described as a video frame that is sent from the client to the server and sent back to the client. The throughput is expressed in terms of time intervals (t1, t2, t3, t4), which is the inverse of the throughput. In our implementation, frames must be sent to the server at a rate equal to the camera's frame rate in order for the system to function in real-time without accumulating lag. On the other hand, the communication delay, which is the round-trip travel time of sending a frame from the server and receiving it, has a negligible effect on the efficiency of the proposed system since it does not cause lag to accumulate. An increase in the delay time would simply cause a delay in getting the drowsiness value displayed on the client's screen. Since delay time is typically in the order of a few tens of milliseconds for Wi-Fi, its effect will be negligible. Thus, we only analyzed the throughput of our system.

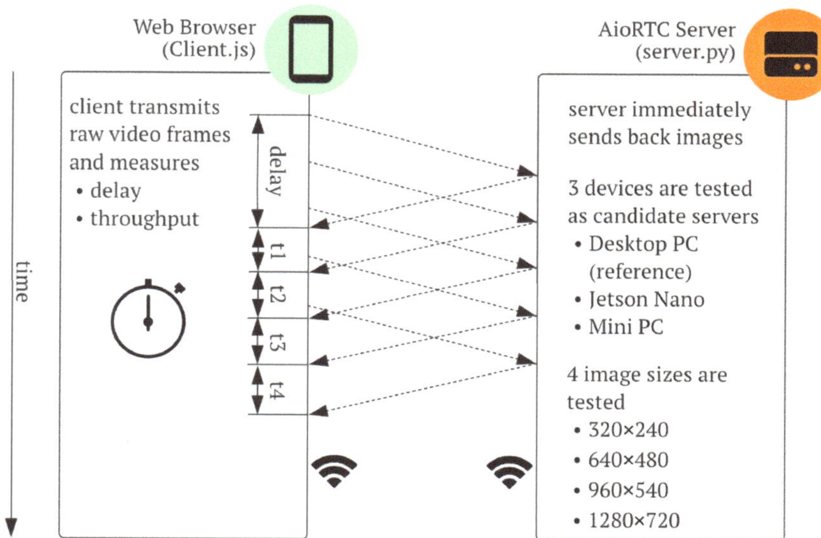

Figure 8. Method for measuring communication speed between client and server. The delay time is defined as the time for a video frame to send to server and received back to client. The through is defined as the number frames that can be transferred per one second. We inserted a code to check the arrival time of the frames and measured the time difference between frames to measure the throughput.

To evaluate the communication throughput of our proposed system, we set up the AioRTC server to send back the original video frames and modified the client-side code (client.js) to record the time when the client receives the returned frame from the server, which is described as t1,t2, t3, t4. We collected 500 data points in the client web browser and downloaded the collected data for processing later. Both the delay and throughput were measured, but only the throughput was analyzed in the results section.

Furthermore, in order to see whether the communication throughput supports different image resolutions, we sent four different image resolutions 320×240, 640×680, 960×540, and 1280×720. If the communication throughput is not high enough, certain video resolutions may not be supported and cause accumulating lag over time.

We measured the communication throughput of the three candidate devices for servers separately.

3.2. Processing Time

In order to implement a drowsiness detection system in an embedded system inside a vehicle, we needed to make sure the server is capable of processing the algorithm in real-time. For a video with a frame rate of 30 fps, the processing needs to be completed within a 33 ms period. To calculate the processing time, we added a few lines of code that print out timestamps in our full software, which includes the AioRTC server and drowsiness detection algorithm. Only some steps of the drowsiness detection algorithms are run for every frame; those steps are time-sensitive, because if a total processing time per frame exceeds 33 ms, the unprocessed frames will start to accumulate on the server side. The processing steps that need to be run for every frame are face detection (Dlib), landmark detection (68-point landmark, OpenCV), and blink detection (SVM classifier explained in [28]). The setup to measure the processing time is shown in Figure 9.

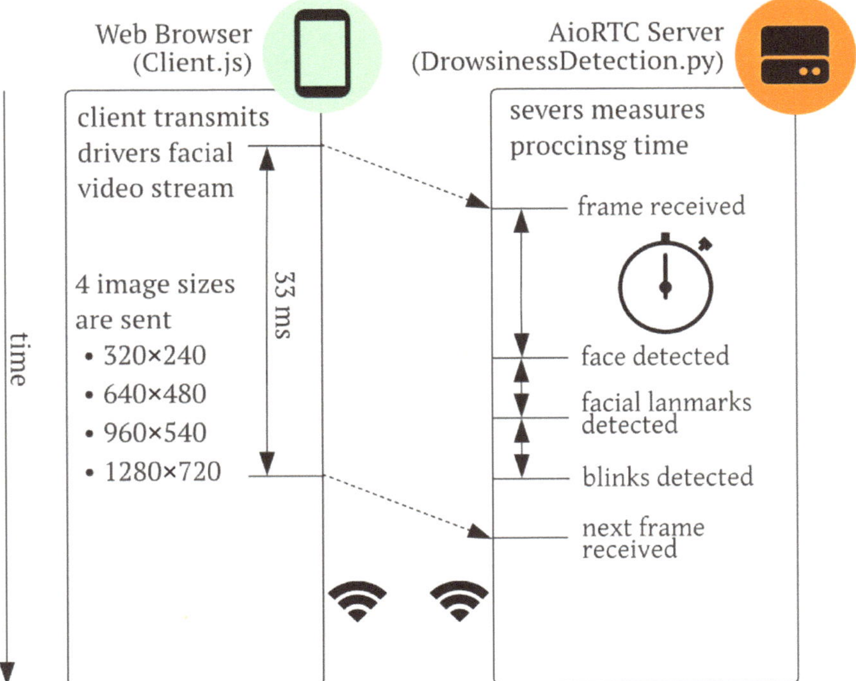

Figure 9. The timestamps which were used to analyze the processing time of drowsiness detection algorithm: the time required to detect face, detect facial landmarks, detect blinks are logged.

There are also steps in the drowsiness detection algorithm that are less time-sensitive. For example, the drowsiness prediction step using the HM-LSTM model [27] only needs to be performed after each blink. As a person can typically blink up to 20 times a minute, up to a few seconds is allowed for this step giving room for a relatively long delay time for inference. Thus, we did not measure the inference time for the HM-LSTM model.

We measured the processing time by printing timestamps at the following events in the server code, DrowsinesDetection.py:

- Right after the frame is received from the client
- Right after the face is detected (face detection)
- Right after 68 facial landmarks are detected (landmark detection)
- Right after blink detection is performed, which checks if a blink is occurring in the frame (blink detection)

By finding the differences between these logged timestamps, we calculated the net processing time for each step. The same process was done for each of our candidate platforms at four different frame resolutions:

- Devices: Desktop PC, Jetson Nano, Mini PC
- Resolutions: 320×240, 640×680, 960×540, and 1280×720

For each measurement, 450 data points were collected, and the mean and standard deviation values were calculated.

3.3. Threshold Optimization Algorithm

In order to find the optimal threshold, we first need to see how different thresholds affects the FN, FP, and 2*FN + FP values. Furthermore, we wanted to compare how those values change when we use the optimal threshold versus the default threshold. Ideally, the optimized threshold would produce a decreased FN rate at the expense of an increased FP rate, which would indicate that the algorithm will correctly detect drowsiness more frequently while giving false alarms more frequently as well. Since Ghoddoosian et al. [27] divided their dataset into five folds, resulting in five different drowsiness detection models created in the cross-validation step, we calculated the FN, FP, and 2*FN + FP values and tested the Threshold Optimization Algorithm for each of the models. The following process was performed for each model.

We propose a method to sweep through the various thresholds and find the threshold that produces the minimum 2FN + FP value, as shown in Algorithm 1. Since Ghoddoosian et al., used a threshold value of 3.33 to divide the drowsy and low vigilant blinks, we sweep our threshold for dividing drowsy and not drowsy blinks from 3.33 to 10 in increments of 0.33 (21 total thresholds). To evaluate the FP and FN rate of each threshold, we created a function that calculates the values of the confusion matrix for the given threshold value, model, and dataset as input. From the confusion matrix, we get the FP and FN metrics, and calculate the value of 2FN + FP. We plotted each of these values in a line graph to observe the general trend of these values according to the varying threshold. We used the Pandas method idxmin() to find the threshold producing the minimum 2FN + FP value. After finding the optimal threshold value and the pertaining FN, FP, and 2*FN + FP values, we compared them with the same values produced by the default threshold (6.67).

Algorithm 1: Threshold Optimization Algorithm. The algorithm finds the optimal threshold value which minimizes the value of 2FN + FP

1:	**Input:**
2:	O: array of loaded models
3:	B: array of loaded blink sequences
4:	L: array of loaded blink sequences' labels (0: not drowsy, 10: drowsy)
5:	**Output:** T: array of optimized threshold value for each model
6:	$M \leftarrow length(O)$
7:	$N \leftarrow length(B)$
8:	**for** $m = 1, 2, \ldots, M$ **do**
9:	initialize map V
10:	**for** $t = 3.3', 3.6', \ldots, 10$ **do** $\triangleright$iterate through thresholds
11:	initialize array C of length N
12:	**for** $i = 1, 2, \ldots, N$ **do** $\triangleright$iterate through blink seq
13:	P = output of model $O[m]$ on blink sequence $B[i]$
14:	**if** $P \geq threshold$ **then** $\triangleright P$ is a value from 0 to 10
15:	$C[i] \leftarrow 10$ $\triangleright$classify blink sequence as drowsy
16:	**Else**
17:	$C[i] \leftarrow 0$ $\triangleright$classify blink sequence as not drowsy
18:	**end if**
19:	**end for**
20:	Calculate confusion matrix by comparing C and L
21:	Calculate FN and FP from confusion matrix
22:	$V(t) \leftarrow 2FN + FP$
23:	**end for**
24:	$T[m] = \underset{t}{\operatorname{argmin}} V(t)$ $\triangleright$find optimal threshold for model $O[m]$
25:	**end for**

4. Results and Discussion

We successfully implemented the proposed client/server platform to predict drowsiness for drivers in vehicles. The implemented code functioned desirably, providing the drowsiness level and classifying the drowsiness state. The platform serves as a web app from a local server so that any client device, such as a phone or laptop, can launch the web app. However, since the server needs to be powerful enough to run AI models that detect drowsiness levels from the raw camera images, we developed the methods to validate the performance of candidate server devices by measuring the processing time. Additionally, since the communication speed needs to be fast enough to transmit the video frame data and processing speed, we developed the methods to measure each of the quantities above. In the following section, we present the results for each candidate device in this section.

The platform we proposed can be used by other researchers who want to develop their own drowsiness prediction algorithms, as they can install and test their systems in their vehicles and check the communication speed and processing time as we have shown. In Section 4.1, we describe how we checked the communication speed, and in Section 4.2, we describe how we measured the processing time of our candidate servers. Detailed interpretations of the results are shown.

We successfully calculated the effect of changing the thresholds on the FN and FP rates of our models and optimized the threshold based on the value of 2FN + FP. We compared the effectiveness of the optimized threshold values by comparing the FP, FN values calculated from the optimized threshold with those of the default threshold. This approach can also be used for other applications where a trade-off between specificity and sensitivity needs to be made. Section 4.3 explains the details of the calculated results and the analysis.

4.1. Communication Speed

As shown in Figure 10, the median time difference between frames being received by the client side was consistently close to 33.3 ms for all three cases: the computers, Jetson Nano, and Beelink Mini PC for all image sizes. As explained in Section 3.1, these values indicate that the client side (MacBook Pro for experimental setup) was able to send the frames to the server side and receive them back with no observable lag. Time differences varied from a minimum of about 29 ms to a maximum of up to 35 ms, displaying a range of around 6 ms. Table 2 shows the average and standard deviation of all time differences across all devices and resolutions. The average of all time differences between frames is 33.30 ms. As for the standard deviations, the average is 4.154 ms across all devices and resolutions. This standard deviation results from fluctuations of transfer time and frame processing time rather than frame capturing time variance in the camera.

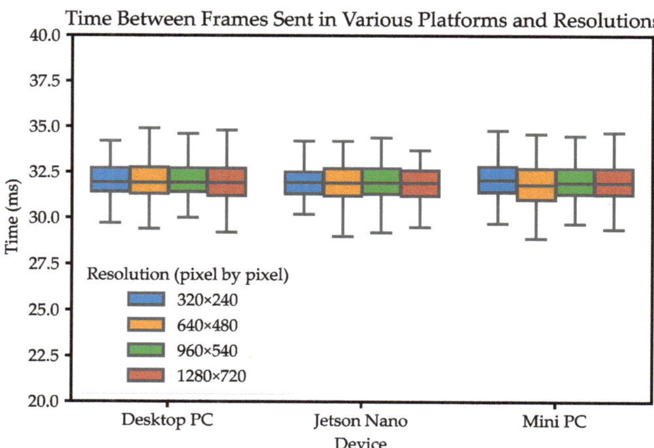

Figure 10. The time intervals between frames when a video stream is sent from client to server and then received back. Regardless of video frame resolutions or server platforms, the time intervals were around 33 ms which corresponds to 30 fps. Small variations (±4.154 ms) existed from the mean value between frames.

Table 2. The time intervals between frames when a video stream is sent from client to server and then received back. Ideally the value of interval should match the frame rate of the video. The server was changed between desktop PC, Jetson Nano, and Mini PC.

Video Resolution	Desktop PC (ms)	Jetson Nano (ms)	Mini PC (ms)
320×240	33.319 ± 3.892	33.326 ± 4.197	33.333 ± 3.930
640×480	33.255 ± 4.121	33.186 ± 5.255	33.320 ± 4.462
960×540	33.303 ± 3.722	33.299 ± 4.108	33.315 ± 3.932
1280×720	33.322 ± 3.984	33.303 ± 3.933	33.316 ± 3.867

In regards to the communication speed, there were no apparent issues with the throughput or delay across all platforms with a camera running at 30 fps. The communication speed is affected by the Wi-Fi performance of the devices in the client and server, wireless link quality, and the video frame processing time in client and server. Given that the wireless link will be inside of a vehicle and modern cellular phones that support high speed Wi-Fi such as 802.11 G/N/AC, we do not expect a shortage of speed for the current setting.

Calculating the required communication throughput from the raw image size and frame rate, we could see that sending videoframes with resolution of 1280×720 at 30 fps requires a relatively higher range of Wi-Fi speed. For example, multiplying 1280×720 (pixels) $\times$ 24 bit (RGB each 8 bit) $\times$ 30 fps leads to a link speed of 632 Mbps. 1280×720 was the maximum resolution and 30 fps was the maximum frame rate that we could use with the built-in MacBook Pro camera. However, when we send the videoframe through WebRTC we can choose to send frames in compressed format either VP8 or H264, which would still allow us to use the same Wi-Fi throughput for higher frame rate and resolution for future uses. Certain blink features, such as eye-opening velocity, can be analyzed more accurately with higher frame rates.

4.2. Processing Time

Table 3 shows the various processing times of detecting the face, facial landmarks, and blinks at various video resolutions in the candidate servers. Across all video resolutions, the Jetson Nano typically took more than 4 times more time than the Desktop PC or Mini PC. The same data is shown in Figure 11, where the average processing time for each operation on each device is displayed as well. In general, the average total processing time for the Windows 10 computer (Desktop PC) across all resolutions was 18.54 ms with an average standard deviation of 0.85 ms. The average total processing time for the Jetson Nano was 94.27 ms with an average standard deviation of 3.02 ms. Lastly, the average total processing time for the Beelink (Mini PC) was 22.73 ms with an average standard deviation of 1.56 ms.

Table 3. Inference time duration of face detection, landmark detection, blink detection, and total at various video resolutions. Desktop PC and Mini PC show similar performance while at Jetson Nano the processing time is longer. Units are in milliseconds.

Inference	Video Resolution	Desktop PC (ms)	Jetson Nano (ms)	Mini PC (ms)
Face Detection	320×240	17.177 ± 0.816	89.799 ± 1.064	19.031 ± 1.368
	640×480	17.752 ± 0.925	96.453 ± 2.048	21.935 ± 1.928
	960×540	13.165 ± 0.762	73.182 ± 2.282	16.078 ± 1.169
	1280×720	13.082 ± 0.571	73.585 ± 2.288	16.218 ± 1.152
Landmark Detection	320×240	2.543 ± 0.121	8.533 ± 0.348	3.410 ± 0.332
	640×480	2.570 ± 0.283	8.633 ± 0.590	3.346 ± 0.651
	960×540	2.391 ± 0.116	8.503 ± 0.314	3.375 ± 0.369
	1280×720	2.319 ± 0.320	8.528 ± 0.221	3.362 ± 0.383
Blink Detection	320×240	0.835 ± 0.063	2.645 ± 5.164	1.037 ± 0.124
	640×480	0.803 ± 0.065	2.408 ± 0.106	1.083 ± 0.170
	960×540	0.790 ± 0.316	2.398 ± 0.108	1.019 ± 0.128
	1280×720	0.745 ± 0.037	2.401 ± 0.069	1.035 ± 0.125
Total	320×240	20.554 ± 0.876	100.997 ± 5.317	23.478 ± 1.533
	640×480	21.126 ± 1.042	107.524 ± 2.121	26.364 ± 2.018
	960×540	16.345 ± 0.817	84.083 ± 2.344	20.471 ± 1.306
	1280×720	16.146 ± 0.671	84.514 ± 2.305	20.615 ± 1.366

As demonstrated by the data mentioned above, the Jetson Nano Computer took significantly longer to run the inference operations at each frame compared to other platforms. Even with the fastest operation time in Jetson Nano, in the case of 950×540 resolution, it took 84 ms to process one frame, which far exceeds the desirable 33 ms value for the 30 fps camera video frame rate. Thus, we choose not to use Jetson Nano for our drowsiness detection system because an operation time longer than its frame rate will accumulate as lag over the time, making it impossible to alert drivers in real time. The Windows 10 computer performed inference operations in the shortest amount of time, but the size of the computer and the required power to run the computer makes it impractical for use in vehicles. In the case of the Mini PC, the operation time meets the frame rate requirement. Even the longest operation time took 26.364 ms, which was less than 33 ms. The mobile size of the Mini PC and its efficient power consumption (28 W at maximum) makes it possible to apply this system in vehicles. Therefore, we concluded that the Mini PC is best suited to run our inference operations, which include Face Detection, Landmark Detection, and Blink Detection. Notably, predicting drowsiness values using the analyzed blinks is also a significant step of the drowsiness detection algorithm, but since this step is not performed at every frame and only when a new blink is detected, we did not measure the processing time.

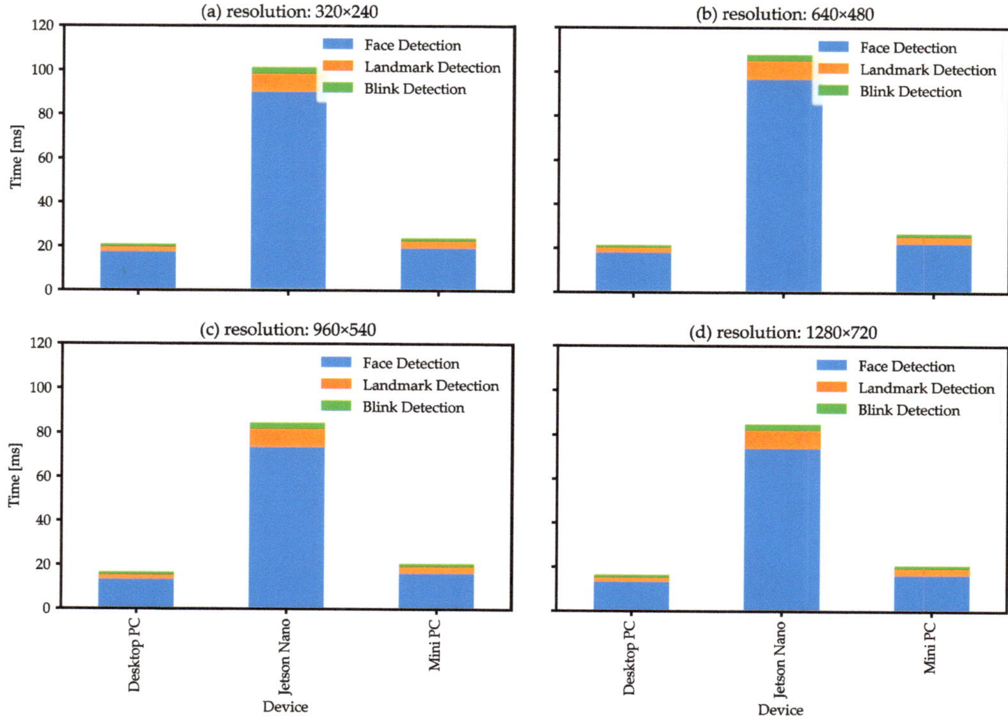

Figure 11. Inference time duration of face detection, landmark detection, blink detection, and total at various video resolutions. Across all video resolutions the Jetson Nano took four times more processing time than Desktop PC or Mini PC. The processing time in Desktop PC and Mini PC was smaller than 33 ms meeting the 30 fps frame rate requirement.

In reality, there are 4 processes in drowsiness detection: face, landmark, blink, drowsy. The first three operations must happen at every frame, while the last one only needs to happen every blink, which happens in much larger time intervals, which is less than 20 time the minute. Thus, we did not conduct the processing time analysis for the drowsiness detection algorithm itself.

We found that the face detection step is the most demanding operation compared to the landmark detection and blink detection steps. The face detection operation took more than 80% of the total processing time for all resolutions across all devices. The face detection processing time also varied depending on the resolution, not necessarily proportional to the image size. For example, in Figure 11, 960×540 and 1280×720 resolutions took smaller time than 320×240 or 640×480 resolutions. This may be due to differences in the image scaling process that the program performs before inputting the image to the face detection model. Depending on the specific image size, the scaling operation could simply involve increasing the pixel intensity, while for their sizes may involve complex fraction multiplications. For landmark and blink detection, we did not observe changes due to image resolution.

At a frame rate of 30 fps, the processing can be done in time on both a Mini PC and desktop PC. However, if we increase the frame rate to 60 fps in future, the processing needs to happen within 16.6 ms, and only a Desktop PC with image resolutions of 960×540 or 1280×720 would barely meet this requirement. In such cases, we think extra hardware such as the Google Coral stick in combination with the Mini PC could be used to improve the inference speed of facial detection.

4.3. Threshold Optimization

To display the relationship between threshold values and false negative (FN) and false positive (FP) rates, we calculated the false negative and false positive rates of each model for each threshold value from 3.33 to 10.0 in increments of 0.33 and graphed as shown in Figure 12.

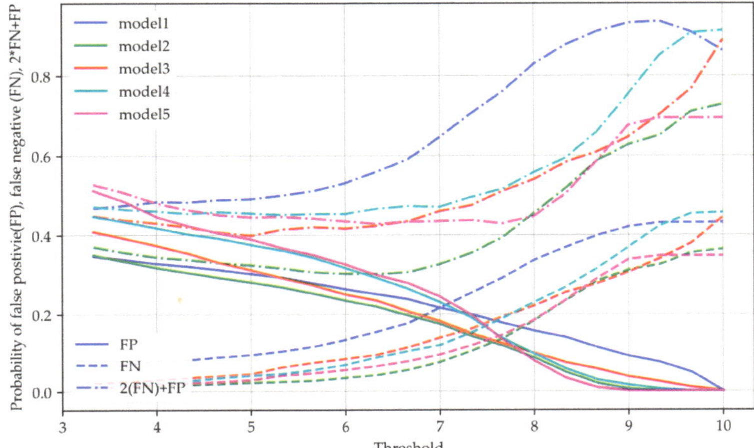

Figure 12. The false positive (FP), false negative (FN), and cost function of 2FN + FP values are shown across a range of threshold values, 3.3 to 10. Each color represents different models (1–5) generated from the cross-validation process as elaborated in Section 2.3.3. Optimal threshold can be determined by finding a threshold that minimizes 2FN + FP. This approach solves the trade-off problem between false positive and false negative rates.

As shown in Figure 12, the FP rate tends to decrease as the threshold increases because a higher threshold leads to a greater confidence of drowsiness detection. The solid curves, which represent FP curves, display an FP rate between 0.3 and 0.5 when the threshold is 3.33 but decrease to 0.1 when the threshold is approximately 0.8. On the contrary, the FN rate tends to increase as the threshold increases because there are more instances falsely detected as negative when the threshold gets higher. The dashed curves, which represent the FN rates, display an FN rate close to 0 when the threshold is 3.3 but increases to around 0.4 when the threshold is 10. The two different tendencies create a trade-off situation when choosing a threshold value. Thus, we created another metric that weighs the sensitivity and specificity differently. For example, 2FN + FP will penalize FN twice more than FP. The dotted-dashed curve, which represents the 2FN + FP, displays slight dips at relatively low thresholds before increasing at a fast rate. Except for model 1, the 2FN + FP value tends to stay between 0.3 and 0.5 until the threshold is 7, then rises rapidly from 0.5 to 0.9 as the threshold increases. Since we wanted to find a threshold that minimizes 2FN + FP, where minimizing the FN value is more important, the threshold will be between 3.3 and 7. Flat lines throughout the curves indicate that differences in the threshold do not affect the accuracy of the models at those regions.

Model 1 displays consistently poorer performance than the other models. This can be seen from the higher values of blue solid line, blue dashed line, and blue dotted-dashed line. We predict that incorrect labels of some of the videos specifically in the test set of Model 5 could have caused this decrease in performance. Notably, this significant dip in performance will be taken into account when calculating the final drowsiness value, which is the weighted average of all of the drowsiness classifications. We further describe this step in Section 2.3.3.

Table 4 summarizes the minimum values of 2FN + FP, FP, and FN for each model. In addition, we compared the values at the optimized threshold to those at the default threshold (6.67). With the optimal thresholds, the minimum values of FN were smaller, and values of FP were larger compared to those with default threshold. In the table, Model 2 and Model 5's optimal thresholds are both 6.33, which is very close to default 6.6, and the values of FP and FN are similar for both thresholds. However, for Model 3 and Model 4 which have different optimal thresholds than the default threshold, the FN rates are 63.3%, 48.1% and lower at the expense of increased FP rates of 47.6% and 38.5% for model 3 and model 4, respectively. Thus, the threshold optimization algorithm will reduce the number of false negative cases at the expense of increasing the number of false positive cases. Thus, the drivers can almost always be alerted of their drowsiness, despite occasionally being annoyed by false alarms.

Table 4. Two sets of 2FN + FP, FP, and FN values are listed: the first set shows these values for the optimal threshold values and the second show these values for the default threshold value (6.67). When the optimal threshold differs from the default threshold (as in models 3 and 4), the FN value is minimized at the expense of an increased FP value.

Model Number	2FN + FP at Optimal Threshold	FP at Optimal Threshold	FN at Optimal Threshold	Optimal Threshold	2FN + FP at Default Threshold	FP at Default Threshold	FN at Default Threshold
1	0.47	0.35	0.061	3.33	0.59	0.24	0.177
2	0.30	0.22	0.041	6.33	0.31	0.20	0.054
3	0.40	0.31	0.043	5.00	0.43	0.21	0.114
4	0.45	0.36	0.055	5.33	0.47	0.26	0.104
5	0.43	0.30	0.064	6.33	0.43	0.28	0.077

We chose a weight of 2 for FN arbitrarily to penalize FP more than FN. However, a more precise weight can be assigned through additional research on drivers' experiences and how they feel about the false alarms versus missed alerts when drowsy. This effort as a future work will require both quantitative and qualitative analysis for improving both the driving experience and safety.

4.4. Future Work

The dataset we used for our drowsiness detection algorithm was the one used by Ghoddoosian et al., which is composed of 10 min recordings made by 60 people for each of the alert, low-vigilant, and drowsy states. We believe that since drowsiness behaviors differ from person to person, the decoding accuracy of the model can be greatly improved by personalizing the model for each user (training the model with a dataset consisting solely of the user's videos).

Furthermore, for our Threshold Optimization and Voting algorithm, we assigned a weight of two for the TF and FN values, respectively. We arbitrarily chose the value of two simply for the purpose of our experiment and think further research can be conducted to find the optimal ratio of the weight of TP to TN or FN to FP values in finding the optimal threshold.

5. Conclusions

In this paper, we proposed and developed an embedded system that allows a neural network-based drowsiness detection model to run in real-time in vehicles and discovered that the most practical system setups of a Beelink Mini PC as the server and a phone as the web app client. The drowsiness detection algorithm can run smoothly with no accumulating lag time. For the Beelink Mini PC, the total processing time for all drowsiness detection processes is 22.73 ms, which is less than the maximum 33 ms value required to process frames of a 30 fps camera video stream. We have also shown that communication throughput between the client and server was adequate to send video images between the two. Both the measured average and the median time interval between frames were around 33.33 ms. Thus, we have implemented a system that can run drowsiness prediction effectively via a client and server platform. We also developed an algorithm that calculates an optimal threshold value considering the trade-off between the safety and the convenience of the user. In some models, the algorithm reduced the false negative rates by 63.3% at the expense of increasing the false positive rate by 47.6%.

In the future, a personalized drowsiness model (trained based on a dataset of a single person) can be developed to improve the accuracy in comparison to the existing model used in this paper, which was trained with a dataset of 60 participants. Our Threshold Optimization Algorithm can be improved further by exploring different cost functions which optimize the ratio between the FN rate and FP rate. Overall, we believe that our drowsiness prediction platform can be used by other researchers aiming to create a real-time embedded system implementation of drowsiness prediction in vehicles.

Author Contributions: Conceptualization, J.K. and M.K.; methodology, J.K.; software, M.K.; validation, J.K. and M.K.; formal analysis, J.K.; investigation, J.K. and M.K.; resources, M.K.; data curation, J.K. and M.K.; writing—original draft preparation, J.K. and M.K.; writing—review and editing, J.K. and M.K.; visualization, J.K.; supervision, M.K.; project administration, J.K. and M.K.; funding acquisition, J.K. and M.K. All authors have read and agreed to the published version of the manuscript.

Funding: This research received no external funding. The APC was funded by Minjeong Kim and Jimin Koo.

Institutional Review Board Statement: Not applicable.

Informed Consent Statement: Not applicable.

Data Availability Statement: 3rd Party Data. Restrictions apply to the availability of these data. Data was obtained from Ghoddoosian et al. and are available from the authors [27] with the permission of Ghoddoosian et al.

Acknowledgments: We would like to thank Jared Lera and the Applied Computing Foundation for providing guidance throughout our project and for suggesting Ghoddoosian's model as a foundation for our drowsiness detection system. Additionally, we would like to thank Vassilis Athitsos from the University of Texas at Arlington for guiding us on the structure and submission process of this paper.

Conflicts of Interest: The authors declare no conflict of interest.

References

1. Global Status Report on Road Safety 2018. Available online: https://www.who.int/publications-detail-redirect/9789241565684 (accessed on 17 October 2022).
2. Fatigued Driving-National Safety Council. Available online: https://www.nsc.org/road/safety-topics/fatigued-driver (accessed on 11 September 2022).
3. About Half of Americans Admit to Driving While Drowsy. Available online: https://pittsburgh.legalexaminer.com/transportation/about-half-of-americans-admit-to-driving-while-drowsy/ (accessed on 11 September 2022).
4. Fujiwara, K.; Abe, E.; Kamata, K.; Nakayama, C.; Suzuki, Y.; Yamakawa, T.; Hiraoka, T.; Kano, M.; Sumi, Y.; Masuda, F.; et al. Heart Rate Variability-Based Driver Drowsiness Detection and Its Validation With EEG. *IEEE Trans. Biomed. Eng.* **2019**, *66*, 1769–1778. [CrossRef]
5. Borghini, G.; Astolfi, L.; Vecchiato, G.; Mattia, D.; Babiloni, F. Measuring Neurophysiological Signals in Aircraft Pilots and Car Drivers for the Assessment of Mental Workload, Fatigue and Drowsiness. *Neurosci. Biobehav. Rev.* **2014**, *44*, 58–75. [CrossRef]
6. Kartsch, V.J.; Benatti, S.; Schiavone, P.D.; Rossi, D.; Benini, L. A Sensor Fusion Approach for Drowsiness Detection in Wearable Ultra-Low-Power Systems. *Inf. Fusion* **2018**, *43*, 66–76. [CrossRef]
7. Yeo, M.V.M.; Li, X.; Shen, K.; Wilder-Smith, E.P.V. Can SVM Be Used for Automatic EEG Detection of Drowsiness during Car Driving? *Saf. Sci.* **2009**, *47*, 115–124. [CrossRef]
8. Arefnezhad, S.; Hamet, J.; Eichberger, A.; Frühwirth, M.; Ischebeck, A.; Koglbauer, I.V.; Moser, M.; Yousefi, A. Driver Drowsiness Estimation Using EEG Signals with a Dynamical Encoder–Decoder Modeling Framework. *Sci. Rep.* **2022**, *12*, 2650. [CrossRef]

9. Sikander, G.; Anwar, S. Driver Fatigue Detection Systems: A Review. *IEEE Trans. Intell. Transp. Syst.* **2019**, *20*, 2339–2352. [CrossRef]
10. Lin, C.-T.; Chuang, C.-H.; Huang, C.-S.; Tsai, S.-F.; Lu, S.-W.; Chen, Y.-H.; Ko, L.-W. Wireless and Wearable EEG System for Evaluating Driver Vigilance. *IEEE Trans. Biomed. Circuits Syst.* **2014**, *8*, 165–176. [CrossRef]
11. Lin, C.-T.; Chang, C.-J.; Lin, B.-S.; Hung, S.-H.; Chao, C.-F.; Wang, I.-J. A Real-Time Wireless Brain–Computer Interface System for Drowsiness Detection. *IEEE Trans. Biomed. Circuits Syst.* **2010**, *4*, 214–222. [CrossRef]
12. Vicente, J.; Laguna, P.; Bartra, A.; Bailón, R. Drowsiness Detection Using Heart Rate Variability. *Med. Biol. Eng. Comput.* **2016**, *54*, 927–937. [CrossRef]
13. Lee, B.-G.; Chung, W.-Y. Driver Alertness Monitoring Using Fusion of Facial Features and Bio-Signals. *IEEE Sens. J.* **2012**, *12*, 2416–2422. [CrossRef]
14. Jung, S.-J.; Shin, H.-S.; Chung, W.-Y. Driver Fatigue and Drowsiness Monitoring System with Embedded Electrocardiogram Sensor on Steering Wheel. *IET Intell. Transp. Syst.* **2014**, *8*, 43–50. [CrossRef]
15. Kundinger, T.; Sofra, N.; Riener, A. Assessment of the Potential of Wrist-Worn Wearable Sensors for Driver Drowsiness Detection. *Sensors* **2020**, *20*, 1029. [CrossRef]
16. What Is ATTENTION ASSIST®? | Mercedes-Benz Safety Features | Fletcher Jones Motorcars. Available online: https://www.fjmercedes.com/mercedes-benz-attention-assist/ (accessed on 12 November 2022).
17. Castignani, G.; Derrmann, T.; Frank, R.; Engel, T. Driver Behavior Profiling Using Smartphones: A Low-Cost Platform for Driver Monitoring. *IEEE Intell. Transp. Syst. Mag.* **2015**, *7*, 91–102. [CrossRef]
18. Satish, K.; Lalitesh, A.; Bhargavi, K.; Prem, M.S.; Anjali, T. Driver Drowsiness Detection. In Proceedings of the 2020 International Conference on Communication and Signal Processing (ICCSP), Melmaruvathur, India, 28–30 July 2020; pp. 380–384.
19. Jabbar, R.; Shinoy, M.; Kharbeche, M.; Al-Khalifa, K.; Krichen, M.; Barkaoui, K. Driver Drowsiness Detection Model Using Convolutional Neural Networks Techniques for Android Application. In Proceedings of the 2020 IEEE International Conference on Informatics, IoT, and Enabling Technologies (ICIoT), Doha, Qatar, 2–5 February 2020; pp. 237–242.
20. Kolpe, P.; Kadam, P.; Mashayak, U. Drowsiness Detection and Warning System Using Python. In Proceedings of the 2nd International Conference on Communication & Information Processing (ICCIP), Tokyo, Japan, 27–29 November 2020.
21. Mohammad, F.; Mahadas, K.; Hung, G.K. Drowsy Driver Mobile Application: Development of a Novel Scleral-Area Detection Method. *Comput. Biol. Med.* **2017**, *89*, 76–83. [CrossRef]
22. Xu, L.; Li, S.; Bian, K.; Zhao, T.; Yan, W. Sober-Drive: A Smartphone-Assisted Drowsy Driving Detection System. In Proceedings of the 2014 International Conference on Computing, Networking and Communications (ICNC), Honolulu, HI, USA, 3–6 February 2014; pp. 398–402.
23. García, I.; Bronte, S.; Bergasa, L.M.; Almazán, J.; Yebes, J. Vision-Based Drowsiness Detector for Real Driving Conditions. In Proceedings of the 2012 IEEE Intelligent Vehicles Symposium, Alcala de Henares, Spain, 3–7 June 2012; pp. 618–623.
24. Mandal, B.; Li, L.; Wang, G.S.; Lin, J. Towards Detection of Bus Driver Fatigue Based on Robust Visual Analysis of Eye State. *IEEE Trans. Intell. Transp. Syst.* **2017**, *18*, 545–557. [CrossRef]
25. You, F.; Li, X.; Gong, Y.; Wang, H.; Li, H. A Real-Time Driving Drowsiness Detection Algorithm With Individual Differences Consideration. *IEEE Access* **2019**, *7*, 179396–179408. [CrossRef]
26. Manishi; Kumari, N. Development of an Enhanced Drowsiness Detection Technique for Car Driver. *Xian Dianzi Keji Daxue Xuebao J. Xidian Univ.* **2020**, *14*, 3181–3186. [CrossRef]
27. Ghoddoosian, R.; Galib, M.; Athitsos, V. A Realistic Dataset and Baseline Temporal Model for Early Drowsiness Detection. *arXiv* **2019**, arXiv:190407312.
28. Dalal, N.; Triggs, B. Histograms of Oriented Gradients for Human Detection. In Proceedings of the 2005 IEEE Computer Society Conference on Computer Vision and Pattern Recognition (CVPR'05), San Diego, CA, USA, 20–25 June 2005; Volume 1, pp. 886–893.
29. Kazemi, V.; Sullivan, J. One Millisecond Face Alignment with an Ensemble of Regression Trees. In Proceedings of the 2014 IEEE Conference on Computer Vision and Pattern Recognition, Columbus, OH, USA, 23–28 June 2014; pp. 1867–1874.
30. Soukupová, T.; Cech, J. Real-Time Eye Blink Detection Using Facial Landmarks. In Proceedings of the 21st Computer Vision Winter Workshop, Rimske Toplice, Slovenia, 3–5 February 2016.
31. Galarza, E.E.; Egas, F.D.; Silva, F.M.; Velasco, P.M.; Galarza, E.D. Real Time Driver Drowsiness Detection Based on Driver's Face Image Behavior Using a System of Human Computer Interaction Implemented in a Smartphone. In *Proceedings of the International Conference on Information Technology & Systems (ICITS 2018), Libertad, Ecuador, 10–12 January 2018*; Rocha, Á., Guarda, T., Eds.; Advances in Intelligent Systems and Computing; Springer International Publishing: Berlin/Heidelberg, Germany, 2018; Volume 721, pp. 563–572. ISBN 978-3-319-73449-1.
32. Rafid, A.-U.-I.; Niloy, A.; Islam Chowdhury, A.; Sharmin, N. A Brief Review on Different Driver's Drowsiness Detection Techniques. *Int. J. Image Graph. Signal Process.* **2020**, *12*, 41–50. [CrossRef]
33. Bergasa, L.M.; Almería, D.; Almazán, J.; Yebes, J.J.; Arroyo, R. DriveSafe: An App for Alerting Inattentive Drivers and Scoring Driving Behaviors. In Proceedings of the 2014 IEEE Intelligent Vehicles Symposium Proceedings, Dearborn, MI, USA, 8–11 June 2014; pp. 240–245.
34. Shafiq, M.; Tian, Z.; Bashir, A.K.; Du, X.; Guizani, M. CorrAUC: A Malicious Bot-IoT Traffic Detection Method in IoT Network Using Machine-Learning Techniques. *IEEE Internet Things J.* **2021**, *8*, 3242–3254. [CrossRef]

35. Shafiq, M.; Tian, Z.; Bashir, A.K.; Du, X.; Guizani, M. IoT Malicious Traffic Identification Using Wrapper-Based Feature Selection Mechanisms. *Comput. Secur.* **2020**, *94*, 101863. [CrossRef]
36. Süzen, A.A.; Duman, B.; Şen, B. Benchmark Analysis of Jetson TX2, Jetson Nano and Raspberry PI Using Deep-CNN. In Proceedings of the 2020 International Congress on Human-Computer Interaction, Optimization and Robotic Applications (HORA), Ankara, Turkey, 26–27 June 2020; pp. 1–5.

technologies

Article

Evaluation of Machine Learning Algorithms for Classification of EEG Signals

Francisco Javier Ramírez-Arias [1,2], Enrique Efren García-Guerrero [1], Esteban Tlelo-Cuautle [3], Juan Miguel Colores-Vargas [2], Eloisa García-Canseco [4], Oscar Roberto López-Bonilla [1], Gilberto Manuel Galindo-Aldana [5] and Everardo Inzunza-González [1,*]

[1] Facultad de Ingeniería, Arquitectura y Diseño, Universidad Autónoma de Baja California, Carretera Transpeninsular Ensenada-Tijuana No. 3917, Ensenada 22860, Mexico; francisco.javier.ramirez.arias@uabc.edu.mx (F.J.R.-A.); eegarcia@uabc.edu.mx (E.E.G.-G.); olopez@uabc.edu.mx (O.R.L.-B.)

[2] Facultad de Ciencias de la Ingeniería y Tecnología, Universidad Autónoma de Baja California, Blvd. Universitario No. 1000, Valle de las Palmas, Tijuana 21500, Mexico; miguel.colores@uabc.edu.mx

[3] Departamento de Electrónica, Instituto Nacional de Astrofísica, Óptica y Electrónica, Luis Enrique Erro No. 1, Santa María Tonanzintla, Puebla 72840, Mexico; etlelo@inaoep.mx

[4] Facultad de Ciencias, Universidad Autónoma de Baja California, Carretera Transpeninsular Ensenada-Tijuana No. 3917, Ensenada 22860, Mexico; eloisa.garcia@uabc.edu.mx

[5] Facultad de Ingeniería y Negocios, Guadalupe Victoria, Universidad Autónoma de Baja California, Carretera Estatal No. 3, Gutiérrez, Mexicali 21720, Mexico; gilberto.galindo.aldana@uabc.edu.mx

* Correspondence: einzunza@uabc.edu.mx; Tel.: +52-646-152-8244

Citation: Ramírez-Arias, F.J.; García-Guerrero, E.E.; Tlelo-Cuautle, E.; Colores-Vargas, J.M.; García-Canseco, E.; López-Bonilla, O.R.; Galindo-Aldana, G.M.; Inzunza-González, E. Evaluation of Machine Learning Algorithms for Classification of EEG Signals. *Technologies* **2022**, *10*, 79. https://doi.org/10.3390/technologies10040079

Academic Editors: Gwanggil Jeon and Imran Ahmed

Received: 13 May 2022
Accepted: 24 June 2022
Published: 30 June 2022

Publisher's Note: MDPI stays neutral with regard to jurisdictional claims in published maps and institutional affiliations.

Abstract: In brain–computer interfaces (BCIs), it is crucial to process brain signals to improve the accuracy of the classification of motor movements. Machine learning (ML) algorithms such as artificial neural networks (ANNs), linear discriminant analysis (LDA), decision tree (D.T.), K-nearest neighbor (KNN), naive Bayes (N.B.), and support vector machine (SVM) have made significant progress in classification issues. This paper aims to present a signal processing analysis of electroencephalographic (EEG) signals among different feature extraction techniques to train selected classification algorithms to classify signals related to motor movements. The motor movements considered are related to the left hand, right hand, both fists, feet, and relaxation, making this a multiclass problem. In this study, nine ML algorithms were trained with a dataset created by the feature extraction of EEG signals.The EEG signals of 30 Physionet subjects were used to create a dataset related to movement. We used electrodes C3, C1, CZ, C2, and C4 according to the standard 10-10 placement. Then, we extracted the epochs of the EEG signals and applied tone, amplitude levels, and statistical techniques to obtain the set of features. LabVIEW™2015 version custom applications were used for reading the EEG signals; for channel selection, noise filtering, band selection, and feature extraction operations; and for creating the dataset. MATLAB 2021a was used for training, testing, and evaluating the performance metrics of the ML algorithms. In this study, the model of Medium-ANN achieved the best performance, with an AUC average of 0.9998, Cohen's Kappa coefficient of 0.9552, a Matthews correlation coefficient of 0.9819, and a loss of 0.0147. These findings suggest the applicability of our approach to different scenarios, such as implementing robotic prostheses, where the use of superficial features is an acceptable option when resources are limited, as in embedded systems or edge computing devices.

Keywords: EEG; BCI; feature extraction; artificial intelligence; machine learning; deep learning; artificial neural network; mental commands; signal classification; pattern recognition

1. Introduction

The central nervous system is composed of the spinal cord and the brain; the human brain resides in the skull and is considered an essential part of the central nervous system [1]. The human brain is composed of 100 billion neurons on average [2], the joint action of which is responsible for thoughts, actions, and emotional states. The brain is divided into the right and left hemispheres, where the right hemisphere is in charge of regulating the muscular activity of the left side of the human body, while the left hemisphere regulates the activities of the right side [3]. In order to measure and record the brain's activities, different neuroimaging techniques are used, including magnetoencephalography (MEG) [4], electrocorticography (ECoG) [5], intracortical neuronal recording, functional magnetic resonance (fMRI) [6], near-infrared spectroscopy (NIRS) [7], and electroencephalography (EEG), the latter being the neurophysiological technique [8] most widely accepted by the scientific community and the private sector in the development of research in fields such as neuroscience, robotics, home automation, the Internet of Things, education, etc. [9].

Electroencephalography (EEG) is a non-invasive procedure for measuring the electrical activity generated within the brain as a result of different mental processes [10]. The electrical signals are acquired through electrodes placed on the scalp's surface; thus, waves with different amplitudes and frequencies that refer to a person's mental state are obtained [11]. The frequency ranges span from 0 Hz to 100 Hz. Based on these ranges, the signals are classified as follows: delta, which ranges from 0 Hz to 4 Hz; theta, which contains signals from 4 Hz to 7 Hz; alpha, where the information range is between 8 Hz and 12 Hz; beta, where the range is between 12 Hz and 30 Hz; and gamma, with a range that covers from 30 Hz to 100 Hz [12,13]. Different ranges of signals are essential for identifying different clinical problems, such as schizophrenia [14], Alzheimer's, insomnia, epileptic disorders, brain tumors, and different injuries and infections related to the central nervous system. Furthermore, classification of motor impairment in neural disorders by means of EEG signals processing has been a successful method for identifying central nervous system roots of motor disabilities [15]. Compared with other methods, this neuroimaging technique offers advantages such as portability, temporal resolution, safety, cost, small time constants, simple equipment, and effectiveness [16].

The EEG neuroimaging method is the preferred method for developing brain–computer interfaces (BCIs), both in the academic community and the private sector. Historically, BCI has been clinically applied for understanding motor impairment, both in verbal communication [17] and limb movement [18], as well as cognitive impairment [19], and offers a great advantage over electromyography pattern recognition [20] due to the lack of neuromuscular signals under amputation conditions. BCIs are direct communication and control channels between users' brains and computers where muscle activity is not involved [21,22]. They are currently considered a powerful communication technology as they do not involve muscular routes to complete tasks such as communication, commands, and actions. The basis of these systems is the computer, whose central role is the analysis of EEG signals [23,24]. BCIs are classified as exogenous and endogenous. Exogenous BCIs require external conditions or stimuli so that the brain can generate a particular response based on the stimulus. Endogenous BCIs do not require external stimulation; however, they require some training on the user's part so that they can regulate brain rhythms [24]. Despite the differences mentioned, most BCI models contain the following elements: signal acquisition, information preprocessing, feature extraction, and classification [16,25]. The acquisition of signals is carried out by employing electrodes placed on the scalp's surface [26], through which analog signals are obtained and then digitized by means of analog–digital converters. The next step is the preprocessing of the signals, whereby the following are removed: noise induced by the electrical line; the background noise of the brain; various artifacts that the EEG signals present as a result of some muscular activity such as eye movement, facial muscle activity, etc. [27]. Feature extraction is one of the crucial steps due to its impact on the performance of classification algorithms [28]. Some of the obtained features are in the domains of time and frequency [28], i.e., mean, median, variance, maximum, and minimum, among others [29,30]. The feature extraction process produces a vector containing the most relevant features of the EEG signals, used as input for classification algorithms. The next step is classification, which is carried out by different algorithms, including LDA, SVM [31], KNN [32], D.T. [33], N.B., and ANN [34].

Currently, there are different fields of science, engineering, and research that evaluate and make use of BCIs to develop applications that present solutions to complex problems [35,36]. These have been possible due to advances in high-density electronics, data acquisition systems that allow high-quality EEG signals to be acquired, intelligent systems that use machine and deep learning algorithms, and neural networks that allow pattern recognition and signal classification to be performed with high precision. In [25], the authors explain that BCIs can be used in the following six application scenarios: replace, restore, augment, enhance, supplement, and research tools. The authors of [37] commented that current and future BCI application areas are device control, user status monitoring, assessment, training and education, gaming and entertainment, cognitive enhancement, safety, and security. Intelligent systems commonly incorporate machine learning (ML) approaches [38–40]. ML refers to a system able to learn from training data from certain activities so that the analytical model generation process is automated, and associated tasks can be completed or supplemented [41,42]. Deep learning (DL) is a paradigm within ML based on the use of artificial neural networks (ANNs) [41]. Commonly, ML algorithms focus on classifying EEG signals related to the motor and imaginary movements of hands and feet to carry out control actions, as presented in [43–46]. DL is useful in areas with vast and high-dimensional data; therefore, deep neural networks outperform ML algorithms for most text, images, video, voice, and audio processing techniques [47]. Nevertheless, for low-dimensional data input, especially with insufficient training data, ML algorithms may still achieve superior results [48], which are even more interpretable than deep neural network results [49]. The authors of [50] used power, mean, and energy as features to classify EEG signals related to the right and left hands through artificial neural networks (ANNs) and support vector machine (SVM). In [51], the authors used SVM to control the direction of a wheelchair by extracting the mean, energy, maximum value, minimum value, and dominant frequency characteristics of the EEG signals. In [52], the authors used the fast Fourier transform and principal component analysis as characteristics of the EEG signals to feed the SVM classifier to control a robotic arm. The authors of [53] reported the use of EEG signals to control an exoskeleton and the use of SVM, LDA, and NN for their respective classification. Studies such as the one presented in [54] have used pretrained neural network models to classify EEG signals through time–frequency characteristics. Recent studies have focused on the proper selection of EEG signal characteristics and its effect on the accuracy of ML and DL algorithms, as presented in [30]. ML and DL techniques are widely accepted and help to develop specific tasks within different applications [55–61]. Moreover, they are increasingly used to obtain EEG data for pattern analysis, classification of group membership, and BCIs [29,62–67]. However, there are still open research problems, such as the real-time processing of EEG signal classification and the optimization of ML algorithms for implementation on embedded systems or edge computing devices. Hence, research on and development of reliable, efficient, and robust systems for EEG signal classification, among others, should be pursued [16,68]. The complexity of human movements for the manipulation of tools is very high and diverse; for an adult human brain that has automated different movements, it does not represent a major effort, however, for ML it requires the management of precise information inputs that allow programming and execution of free movement. Previous studies offer multiple classes of motor imagery limb movements based on EEG spectral and time domain descriptors [69]; in this sense, there continues to be a need in machine learning to increase the reliability and accuracy of EEG signals used for programming human-like movements.

For the reasons stated above, the aim of this paper is to evaluate nine ML algorithms for the classification of EEG signals. The purpose is to find which ML model presents the best performance metrics for the identification of movement patterns in EEG signals for the control of a mechatronic system, in this case, a robotic hand prosthesis. The selected dataset consists of more than 1500 EEG recordings of 1–2 min in length from 109 subjects and is publicly available in [70]. In this study, we randomly selected 30 subjects to train, validate, and test the proposed method. The ultimate aim is to facilitate the development of robotic limb prosthetics, which is possible because ML algorithms can recognize patterns in EEG signals with complex dynamics. The hypothesis is that ML algorithms perform better in tasks of signal classification than standard methods. The novelty of this study is to provide a methodology for the classification of EEG signals by training several ML algorithms and employing processing, analysis, and feature extraction techniques in the time domain of various lapses of EEG signals related to motor tasks, which can be translated into commands for the control of mechanisms or mechatronic systems such as wheelchairs, robotic prostheses, and mobile robots.

The rest of this paper is organized as follows: Section 2 presents the materials used for the development of the proposed method; additionally, the description of the dataset used for this paper is presented. Section 3 presents the performance metrics obtained from the proposed ML models and the discussion of the main findings obtained in this study. Section 5 presents the proposed usage scenario in a real-world application. Conclusions and future work are described in Section 6.

2. Materials and Methods

2.1. Hardware and Software

The hardware used for the implementation of the proposed method had the following specifications: Microsoft Windows 10 Pro operating system, system model OptiPlex 3070, system type $\times$64-based PC, Processor Intel Core i5-9500 at 3.00 GHz, six Cores, six logical processors, memory (RAM) of 16.0 GB DDR4 2666 MHz (2×8 GB), and NVIDIA GeForce GT 1030 GDDR5 2 GB PCI-Express $\times$16. The software used for reading the EEG signals, electrode selection, signal segmentation, preprocessing, analysis, feature extraction, and preparation of the dataset was LabVIEW 2015. Furthermore, the following libraries, which are part of the development environment of LabVIEW, were used: Biomedical Toolkit and Signal Express. The MATLAB 2021a version was used for training and testing the different ML algorithms, which are part of the Statistics, Machine Learning and Deep Learning Toolbox.

2.2. Machine Learning Algorithm Training

In this paper, we selected nine ML algorithms to evaluate their performance in the classification of EEG signals related to the motor movements of right hand, left hand, both fists, feet, and relaxation. The nine selected algorithms are naive Bayes (N.B.), k-nearest neighbors (KNN), decision tree (D.T.), support vector machine (SVM), linear discriminant analysis (LDA), Narrow-ANN, Medium-ANN, Wide-ANN, and Bilayered-ANN. These ML algorithms are part of the statistical and machine learning toolbox of MATLAB, which has various tools that can be used for both the pre- and post-processing of data.

Figure 1 shows the block diagram to train, test, and evaluate the selected ML algorithms. First, the dataset is loaded; the chosen dataset is constituted of more than 1500 EEG recordings from 109 subjects that become between 1 and 2 minutes long and can be found in [70]. In this study, 30 people were randomly chosen to train, test, and validate the proposed method. Subsequently, the data are normalized between 0 and 1 to obtain better results. Next, we randomly split the dataset into 80% for training and 20% for testing. Then, the ML model is trained. The next step is to obtain the performance metrics of the ML models (for example, using the confusion matrix), i.e., the performance metrics to evaluate the ML algorithms, such as the area under the curve (AUC) and accuracy, among others.

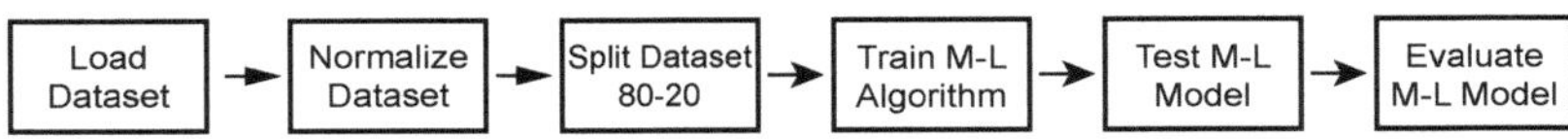

Figure 1. Block diagram for training, testing, and evaluating the ML algorithms.

A typical system for EEG signal classification is conceptually divided into signal acquisition, preprocessing, feature extraction, and classification [23,71]. The EEG signals are acquired by electrodes located on the scalp's surface that transfer information on the electrical neuronal activity to the data acquisition system. In preprocessing, line noise and muscle artifacts are removed from EEG signals. Feature extraction uses several digital signal processing techniques to obtain feature vectors. These vectors are used to train the ML or DL algorithms to classify the EEG signals. The result of the algorithms is a specific class, as illustrated in Figure 2. The following subsections describe the procedure in detail.

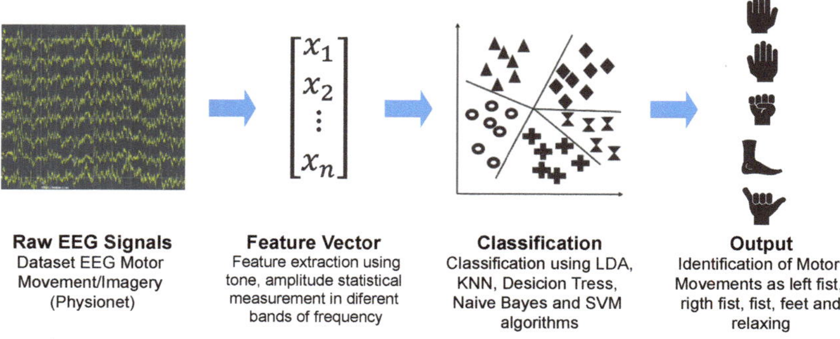

Figure 2. Proposed method for classifying EEG signals.

2.3. Input Data

The dataset used for EEG signal classification was developed by Schalk and colleagues at Nervous System Disorders Laboratory and is publicly available on Physionet [70]. The data consist of more than 1500 EEG recordings of 1–2 min in length from 109 subjects. Patients performed 14 tasks (experiments) while 64 electrodes acquired and recorded the EEG signals through the BCI2000 system [72]. The data are in EDF+ format [73], and they contain 64 EEG signals, each displayed at a rate of 160 samples per second, and an annotation channel, which refers to the actions performed during the task. Table 1 shows the protocol of the Schalk agreement experiment. The diagram of the position of the electrodes used to record the data is the standard 10-10 placement. The dataset consists of 109 folders, and each folder contains 28 files, where 14 of these have the *.edf* extension, and the other 14 have the *.edf.event* extension. The files that contain the EEG signals are those that contain the *.edf* extension. The *.edf.event* files refer to the events during the development of the different tasks. Although the original set of recorded data consists of continuous multichannel data, and the number of users that comprise it is extensive, we only used the EEG signals of 30 randomly selected subjects, and the tasks that are related to the real movements that take place in tasks 3, 5, 7, 9, 11, and 13. In tasks 3, 7, and 11, real movements related to the right and left fists and relaxation are carried out, while in tasks 5, 9, and 13, real movements of both fists and both feet are carried out. Table 1 summarizes the dataset used in the proposed approach.

Table 1. Tasks presented in the dataset to train the ML algorithms for EEG signal classification.

Task	Real Movement	Imaginary Movement	To	T1	T2	Duration
1	Open Eyes	-	Relaxing	-	-	1 min
2	Close Eyes	-	Relaxing	-	-	1 min
3	Fist	-	Relaxing	Left	Right	2 min
4	-	Fist	Relaxing	Left	Right	2 min
5	Fist/Feet	-	Relaxing	Fist	Feet	2 min
6	-	Fist/Feet	Relaxing	Fist	Feet	2 min
7	Fist	-	Relaxing	Left	Right	2 min
8	-	Fist	Relaxing	Left	Right	2 min
9	Fist/Feet	-	Relaxing	Fist	Feet	2 min
10	-	Fist/Feet	Relaxing	Fist	Feet	2 min
11	Fist	-	Relaxing	Left	Right	2 min
12	-	Fist	Relaxing	Left	Right	2 min
13	Fist/Feet	-	Relaxing	Fist	Feet	2 min
14	-	Fist/Feet	Relaxing	Fist	Feet	2 min

2.4. Proposed Method for EEG Signal Processing

Figure 3 depicts the proposed method for EEG signal processing, described in detail in the following subsections.

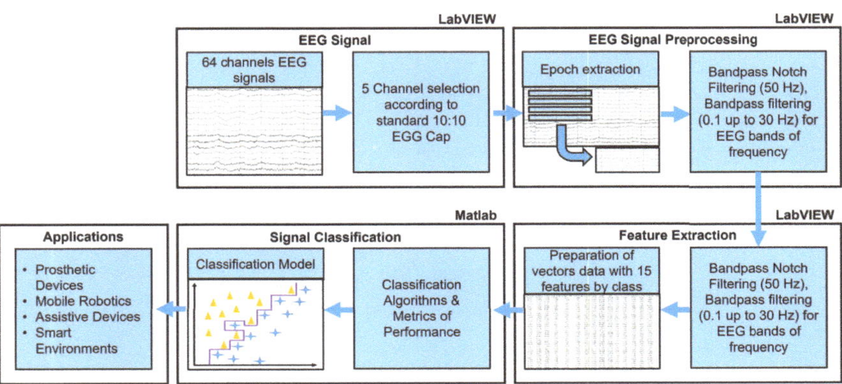

Figure 3. Proposed method for EEG signal classification.

2.5. EEG Signal Acquisition and Channel Selection

The LabVIEW software 2015 version was employed as the development platform, while the Biomedical Toolkit was used to import the EEG signals, due to the signals being in EDF format. The selected electrodes are shown in Figure 4b. These electrodes present neuronal activity correlated to the execution of the left- and right-hand movements (contained in electrodes C3, C4, and CZ [74,75]) and the neuronal activity related to the movement of both feet (contained in electrodes C1 and C2 [76]); because the different EEG channels tend to represent redundant information, as mentioned in [77], electrodes C3, C1, CZ, C2, and C4 were selected in our study. The selected electrodes were located around the center of the skull, within the motor cortex area; their characteristic is that these electrodes are the least affected by different artifacts [78], which allows the reliable extraction of features to be obtained.

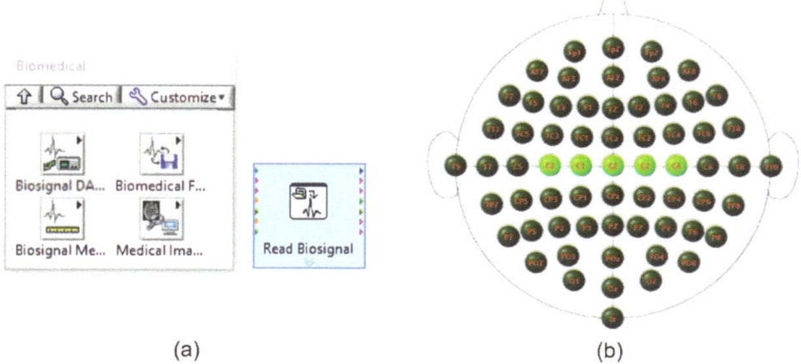

(a) (b)

Figure 4. Selected electrodes for EEG signal classification. (**a**) Biomedical toolkit and (**b**) electrodes selected.

2.6. Preprocessing

The EEG signals used, with a sampled frequency of 160 Hz, are available online [70]. Bandpass filters were required to select only the frequencies of interest and eliminate line noise and some other interferences. For this study, we processed the EEG signals through an IIR bandpass filter, with third-order Butterworth topology from 0.1 to 50 Hz. After this, a 50 Hz notch filter was applied to the signals to eliminate noise from the signal power line. Figure 5 shows the original readings of the electrodes used before and after applying the different filters related to the signal preprocessing operations.

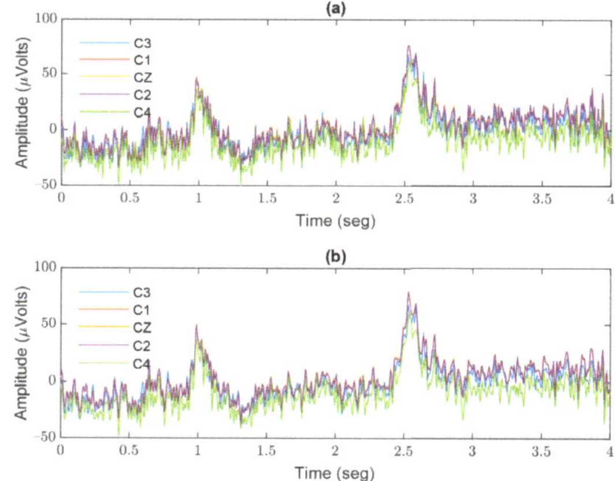

Figure 5. EEG signals acquired from electrodes C3, C1, CZ, C2, and C4. (**a**) Original EEG signal and (**b**) filtered EEG signal.

2.7. EEG Band Separation

Within EEG signal analysis, it is common to separate a signal into different frequency bands, including Delta (1–4 Hz), Theta (4–8 Hz), Alpha (8–12 Hz), Beta (12–30 Hz), and Gamma (30–50 Hz). As shown in Table 2, third-order bandpass Butterworth IIR filters with different cut-off frequencies were used to carry out this separation.

Table 2. Cut-off frequencies of bandpass filters for band extraction of EEG signals.

Band of EEG Signal	Low Cut-Off Frequency	High Cut-Off Frequency
Delta	0.1 Hz	3.99 Hz
Theta	4.0 Hz	7.99 Hz
Alpha	8.0 Hz	11.99 Hz
Beta	12.0 Hz	29.99 Hz
Gamma	30.0 Hz	49.99 Hz

2.8. Feature Extraction

The features of the EEG rhythm can be obtained by using several digital signal processing techniques. These features were used for training the nine ML algorithms. These analysis techniques included measurements of tone, amplitude, and level, as well as statistical analyses. Table 3 shows the type of measurements and features obtained when these techniques were applied to the EGG signal epochs.

Table 3. Features of the EEG signal used to train the ML algorithms.

	Features of the Channels for the Different Electrode Positions															
Band	**C3**	**C3**	**C3**	**C1**	**C1**	**C1**	**Cz**	**Cz**	**Cz**	**C2**	**C2**	**C2**	**C4**	**C4**	**C4**	**Class**
Delta	Amplitude	Frequency	Phase	Peak to Peak	Neg.Peak	Pos.Peak	Median	Mode	Mean	RMS	S.D.	Summation	Variance	Kurtosis	Skewness	Relaxing
Theta	Amplitude	Frequency	Phase	Peak to Peak	Neg.Peak	Pos.Peak	Median	Mode	Mean	RMS	S.D.	Summation	Variance	Kurtosis	Skewness	Left Hand
Alpha	Amplitude	Frequency	Phase	Peak to Peak	Neg.Peak	Pos.Peak	Median	Mode	Mean	RMS	S.D.	Summation	Variance	Kurtosis	Skewness	Right Hand
Beta	Amplitude	Frequency	Phase	Peak to Peak	Neg.Peak	Pos.Peak	Median	Mode	Mean	RMS	S.D.	Summation	Variance	Kurtosis	Skewness	Fist
Gamma	Amplitude	Frequency	Phase	Peak to Peak	Neg.Peak	Pos.Peak	Median	Mode	Mean	RMS	S.D.	Summation	Variance	Kurtosis	Skewness	Feet

Signal Analysis

- Tone measurements. The tone measurements carried out in the EEG signal epochs were the following: amplitude, frequency, and phase.
- Level measurements. The level measurements implemented in the EEG signal epochs were the following: peak-to-peak, negative peak, and positive peak.
- Statistical features. The statistical measurements applied to the different signal epochs were the following:
  - Median [30,79]:

$$Median = \begin{cases} \frac{(N+1)}{2}, & \text{when } N \text{ is odd} \\ \frac{N}{2} + \frac{(N+1)}{2}, & \text{when } N \text{ is ever} \end{cases} \quad (1)$$

- Mode is the number that occurs most frequently in the set;
- Mean [80]:

$$\widetilde{x} = \frac{1}{N} \sum_{i=1}^{N} x_i \tag{2}$$

- Root mean square (RMS) [80]:

$$RMS = \sqrt{\frac{1}{N} \sum_{i=1}^{N} x_i^2} \tag{3}$$

- Standard deviation [80]:

$$S = \sqrt{\frac{1}{N} \sum_{i=1}^{N} (x_i - \widetilde{x})^2} \tag{4}$$

- Summation:

$$\sum_{i=1}^{N} x_i \tag{5}$$

- Variance [80]:

$$S^2 = \frac{1}{N} \sum_{i=1}^{N} (x_i - \widetilde{x})^2 \tag{6}$$

where $\widetilde{x}$ is the mean;
- Kurtosis [80]:

$$Kurtosis = \sum_{i=1}^{N} \frac{(x_i - \widetilde{x})^4}{(N-1)s^4} \tag{7}$$

- Skewness [80]:

$$Skewness = \sum_{i=1}^{N} \frac{(x_i - \widetilde{x})^3}{(N-1)s^3} \tag{8}$$

2.9. Dataset Preparation

The data vectors consist of 15 features, 3 features for each electrode; the electrodes correspond to positions C3, C1, CZ, C2, and C4, which are related to motor movements, and these belong to one of the five classes of "relaxation", "Right hand", "Left hand", and "Fist and Feet". The dataset has 2792 samples, where 558 samples correspond to the "Relaxation" class, 567 to the right hand, 555 to the left hand, 561 to both fists, and 547 to the feet. On average, there are 557 samples per class, which preserves the balance among the classes. Figure A1 in Appendix A shows a fragment of the dataset created by processing EEG signals when different users performed different motor tasks. Figure A2 in Appendix B depicts the graphic user interface (GUI) of the software (App) developed for the feature extraction process. The proposed App allows features to be extracted in different frequency bands, where each frequency band corresponds to a different class. Table 3 shows the features obtained for training the different ML algorithms. Each line represents a vector of features consisting of five electrodes. Three different measurements were made for each electrode, which resulted in a vector with 15 different characteristics used for the training and testing of the ML and DL models. It can be observed that the feature vector is labeled with its respective class. For each of the five classes, 15 different features were obtained in five different frequency bands to improve the classification accuracy of the ML algorithms [81,82]. The proposed dataset can be downloaded at the link from Supplementary Materials.

3. Results

To evaluate the performance of the ML algorithms, we used the following scoring metrics: *Accuracy, Error, Recall, Specificity, Precision,* and *F1-Score.* The performance evaluation of the proposed ML models was initiated by calculating *Sensitivity, Specificity, Precision,* and *Accuracy* [83,84]. *Sensitivity,* also known as *Recall* [84], measures the proportion of positives that are correctly identified as such; it can be calculated by (9). Similarly, *Specificity* measures the proportion of negatives that are correctly identified as such [84]; it can be calculated by (10). *Precision* is the proportion of true positives among the positive predictions [84]; it can be calculated by (11). *Accuracy* can be calculated using (12):

$$Recall = \frac{TruePositives}{FalseNegative + TruePositives}, \tag{9}$$

$$Specificity = \frac{TrueNegatives}{FalsePositives + TrueNegatives}, \tag{10}$$

$$Precision = \frac{TruePositives}{TruePositives + FalsePositives}, \tag{11}$$

$$Accuracy = \frac{TruePositives + TrueNegatives}{TruePositives + FalsePositives + TrueNegatives + FalseNegatives}. \tag{12}$$

F1-Score is a method for combining *Precision* and *Recall* into a single measure that includes both [85]. Neither *Accuracy* nor *Recall* can analyze the complete situation on their own. We might have outstanding *Precision* but poor *Recall,* or vice versa, poor *Precision* but good *Recall.* With *F1-Score,* one can represent both concerns with a single score [86]. Once *Accuracy* and *Recall* for a binary or multiclass classification task have been computed, the two scores may be combined to calculate the *F1-Score* metric; it can be calculated by (13):

$$F1 - Score = \frac{2 * Precision * Recall}{Precision + Recall}. \tag{13}$$

Equations (9)–(13) are valid for binary classification and multiclass issues; however, when used for multiclass problems, they must be calculated for each class and then averaged to obtain each metric per model.

Table 4 shows the average scores obtained in each performance metrics by the nine ML algorithms selected in this study. The first parameter analyzed was accuracy, where the LDA model presented an accuracy score of 0.9229; D.T. obtained 0.9803; KNN obtained 0.8996; N.B. obtained 0.9373; SVM obtained 0.9803; Narrow-ANN, Medium-ANN, and Bilayered-ANN obtained 0.9857; finally, Wide-ANN obtained 0.9821. The Narrow-ANN, Medium-ANN, and Bilayered-ANN models obtained the best accuracy score (0.9857). Regarding the error metric, we can see that the LDA, D.T., N.B., SVM, Narrow-ANN, Medium-ANN, Wide-ANN, and Bilayered-ANN algorithms achieved a score less than 0.1, while the KNN model obtained an Error greater than 0.1; therefore, the models with the lowest error were Narrow-ANN, Medium-ANN, and Bilayered-ANN (0.0143). Considering the recall parameter, we observed that the Narrow-ANN algorithm presented the highest score of 0.9863, while the KNN algorithm obtained the lowest score of 0.9037. Regarding the specificity metric, all the algorithms achieved a score greater than 0.9; the ML models with the best results were the Narrow-ANN, Medium-ANN, and Bilayered-ANN models, all scoring 0.9964. Regarding the precision metric, the Bilayered-ANN algorithm is the one that presented the best result, with 0.9859, while the KNN algorithm presented the lowest score, with 0.9099. Regarding the F1-score parameter, the LDA, D.T., N.B., SVM, Narrow-ANN, Medium-ANN, Wide-ANN, and Bilayered-ANN algorithms achieved scores greater than 0.91, while the KNN model obtained a score below 0.91. The algorithm that presented the best F1-score result was Narrow-ANN, with 0.9859.

Table 4. The average score parameters of the EEG classification algorithms.

	Average Scoring Parameters					
ML Algorithm	**Accuracy**	**Error**	**Recall**	**Specificity**	**Precision**	**F1-Score**
LDA	0.9229	0.0771	0.9219	0.9807	0.9332	0.9228
D.T.	0.9803	0.0197	0.9777	0.9951	0.9792	0.9783
KNN	0.8996	0.1004	0.9037	0.9747	0.9099	0.9047
N.B.	0.9373	0.0627	0.9384	0.9844	0.9382	0.9378
SVM	0.9803	0.0197	0.9789	0.9950	0.9827	0.9803
Narrow-ANN	0.9857	0.0143	0.9863	0.9964	0.9857	0.9859
Medium-ANN	0.9857	0.0143	0.9854	0.9964	0.9856	0.9855
Wide-ANN	0.9821	0.0179	0.9834	0.9955	0.9824	0.9828
Bilayered-ANN	0.9857	0.0143	0.9854	0.9964	0.9859	0.9856

Table 5 presents the performance metrics achieved by each ML algorithm. The metrics used to evaluate the performance of the ML algorithms were the area under the average curve (AUC average), Cohen's Kappa coefficient [87], Matthews correlation coefficient [88], and model loss. Concerning the AUC average metric, all algorithms achieved a score greater than 0.90, where the top three ML models were the SVM, Medium-ANN, and Bilayered-ANN models, which obtained the highest scores (AUC scores). Regarding Cohen's Kappa coefficient, a score above 0.8 indicates exemplary commitment, while zero or less indicates poor commitment. The LDA and KNN algorithms obtained Cohen's Kappa coefficients less than 0.80 but greater than zero, while the D.T., N.B., SVM, Narrow-ANN, Medium-ANN, Wide-ANN, and Bilayered-ANN algorithms achieved Cohen's Kappa coefficients of 0.9384, 0.8040, 0.9384, 0.9552, 0.9552, 0.9440, and 0.9552, respectively, where the Narrow-ANN, Medium-ANN, and Bilayered-ANN algorithms achieved the highest scores. In addition, we used the Matthews correlation coefficient, which has been widely used as a performance metric for ML algorithms since 2000. The best scores obtained were presented by the D.T, N.B., SVM, Narrow-ANN, Medium-ANN, Wide-ANN, and Bilayered-ANN models (0.9736, 0.9225, 0.9757, 0.9824, 0.9819, 0.9783, and 0.9820, respectively), with Narrow-ANN obtaining the best score, while the KNN algorithm achieved the lowest score of 0.8810. The ML model with the lowest loss was Narrow-ANN, with 0.0136, followed by the Medium-ANN and Bilayered-ANN models, both with 0.0147, while the ML algorithm with the highest loss was KNN.

Table 5. Performance metrics of the nine ML algorithms trained for EEG signal classification.

	Performance Metrics			
ML Algorithm	**AUC Average**	**Cohen's Kappa Coefficient**	**Matthews Correlation Coefficient**	**Loss**
LDA	0.9889	0.7592	0.9072	0.0787
D.T.	0.9873	0.9384	0.9736	0.0229
KNN	0.9392	0.6864	0.8810	0.0961
N.B.	0.9935	0.8040	0.9225	0.0616
SVM	0.9988	0.9384	0.9757	0.0217
Narrow-ANN	0.9982	0.9552	0.9824	0.0136
Medium-ANN	0.9998	0.9552	0.9819	0.0147
Wide-ANN	0.9984	0.9440	0.9783	0.0165
Bilayered-ANN	0.9988	0.9552	0.9820	0.0147

Figure 6 shows the ROC curves of the top four ML algorithms trained for the classification of EEG signals related to the state of relaxation, right hand, left hand, both hands, and both feet. These algorithms are LDA, SVM, D.T., and N.B. The algorithm that presented the best performance metrics was SVM, with an AUC average of 0.9988. The ROC curves showed a compromise between sensitivity and specificity. The SVM algorithm was the closest to the upper-left corner of the ROC space, while the D.T. model was closer to the 45-degree diagonal. Classifiers that obtain curves closer to the upper-left corner indicate better performance, while classifiers with ROC curves closer to the 45-degree diagonal of the ROC space are less accurate.

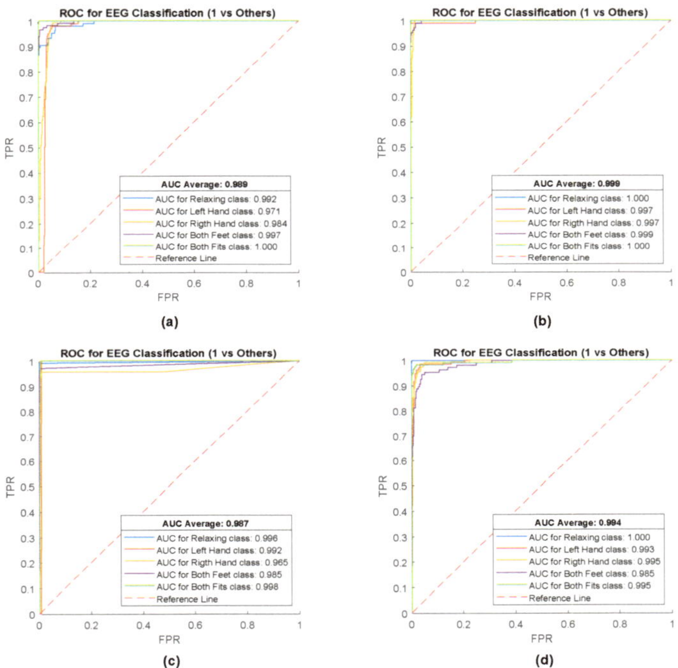

Figure 6. The receiver operating characteristic (ROC) curves of the top four ML algorithms trained for EEG signals classification, related to the movements of hands and feet: (**a**) ROC curves of the LDA algorithm, (**b**) ROC curves of the SVM algorithm, (**c**) ROC curves of the D.T. algorithm, and (**d**) ROC curves of the N.B. algorithm.

Figure 7 shows the ROC curves of the top four DL algorithms (neural networks) trained for the classification of EEG signals. These algorithms are Narrow-ANN, Medium-ANN, Wide-ANN, and Bilayered-ANN. The algorithm that presented the best performance metrics was Medium-ANN, with an AUC average of 0.9998; it was the closest to the upper-left corner of the ROC space.

In machine learning, the presumably best model is chosen from a collection of model candidates obtained by evaluating various model types, hyperparameters, or feature subsets, among others. In this paper, it is proposed to use ConfusionVis, a model-agnostic technique for evaluating and comparing multiclass classifiers based on their confusion matrices [56]. Figure 8 depicts the ConfusionVis achieved for the nine ML models chosen for EEG signal classification. Figure 8a shows the average accuracy score per ML model, where it can be observed that Narrow-ANN had the best accuracy score. Figure 8b illustrates the confusion matrix similarity results, where it can be seen that the D.T., SVM, Narrow-ANN, and Medium-ANN models obtained the best similarity. Figure 8c depicts the error by class scores, where it can be observed that Medium-NN and Narrow-ANN achieved the lowest error score in most classes of movements classified from the EEG signals. Figure 8d shows the error by model scores, where it can also be seen that Medium-ANN obtained the lowest error score, followed by Bilayered-NN, Narrow-ANN, and decision tree (D.T.).

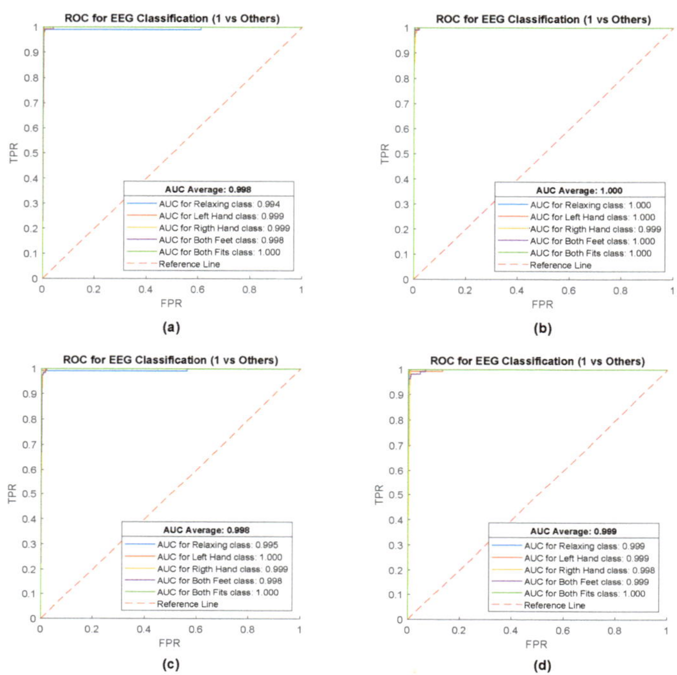

Figure 7. The receiver operating characteristic (ROC) curves of the four ANN algorithms trained for EEG signals classification, related to the movements of hands and feet: (**a**) ROC curves of the Narrow-ANN algorithm, (**b**) ROC curves of the Medium-ANN algorithm, (**c**) ROC curves of the Wide-ANN algorithm, and (**d**) ROC curves of the Bilayered-ANN algorithm.

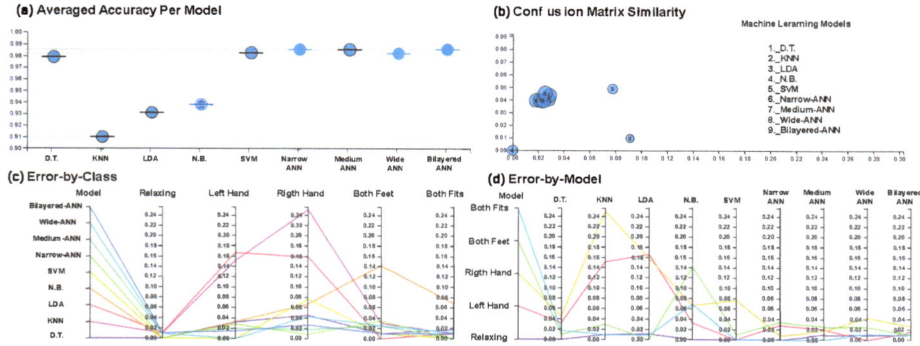

Figure 8. ConfusionVis [56]: Comparative evaluation of the multiclass classifiers based on confusion matrices. (**a**) Averaged accuracy per model, (**b**) confusion matrix similarity, (**c**) error by class, and (**d**) error by model.

Figure 9 shows the training time of the nine ML algorithms tested, with N.B., LDA, and KNN having the shortest training time. However, the results shown in Tables 4 and 5 show that these algorithms had the lowest performance metrics, with the exception of D.T. In contrast, the SVM, Narrow-ANN, Medium-ANN, Wide-ANN, and Bilayered-ANN algorithms had the most considerable training times of 0.13546, 0.37135, 0.16956, 0.36255, and 0.45722 s, respectively, with the Bilayered-ANN algorithm having the longest training time. However, these algorithms had the best performance metrics, as shown in Tables 4 and 5 and Figures 7 and 8. Therefore, the data science engineer or researcher must perform a cost–benefit analysis regarding accuracy and processing time. In most circumstances, engineers favor accuracy over training time, because training is only performed a few times and only the trained ML model is employed. For this reason, in this study, it is more convenient to select the Narrow-ANN model.

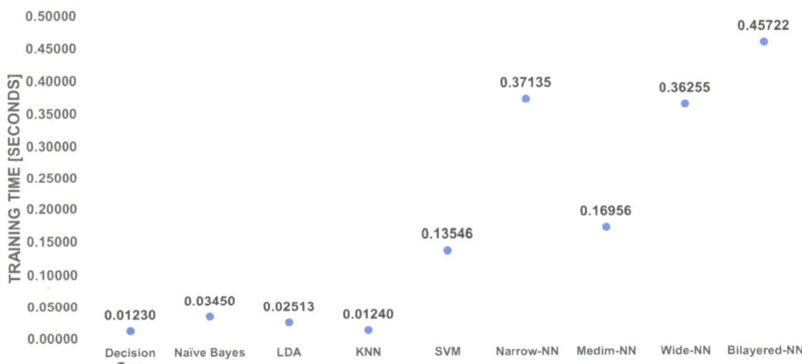

Figure 9. Training time of the nine ML models.

4. Discussion

In this study, we observed that the different features used were helpful for the classification of EEG signals, as proposed in our hypothesis. The presented features are based on the time domain: amplitude, frequency, phase, peak–peak value, negative peak, positive peak, median, mode, average, mean square error value, standard deviation, summation, variance, kurtosis, and skewness. We consider that they are good features for classifying EEG signals related to movements. Using these features, the ML model that achieved the best performance was Medium-ANN, with average area under the curve of 0.9998, Cohen's Kappa coefficient of 0.9552, Matthew correlation coefficient of 0.9819, and loss of 0.0147.

We observed that the performance metrics obtained from the nine machine learning algorithms were good. Using standard features in different frequency bands and related to a particular class allowed machine learning and deep learning algorithms to obtain excellent performance metrics, as shown in Tables 4 and 5 and Figures 6–8; this is because the proposed frequency bands and features improved the separability of the data, making the classification algorithms substantially better.

Regardless, the data science engineer/scientist is in charge of carrying out the corresponding analysis in terms of costs–benefits and precision concerning the information processing time. In most cases, ML models with better precision are chosen, and training time is usually sacrificed. Since the training of the ML algorithms is performed once, only the trained model is used for the assigned task. The Medium-ANN algorithm was selected for this reason and because its performance metrics were the best. Therefore, feature extraction is worth mentioning among the processes that improve relevant information acquisition and ensure better performance metrics when training EEG signal classification algorithms, as shown in different studies. Our results are consistent with other spectrogram methods implemented for identifying EEG patterns in persons with motor impairment using similar brain sources that were analyzed in this study [15]. Many human behavior fields are still a challenge for BCIs; findings from this study may provide complementary data for other studies reporting findings from central nervous system damage with residuals of motor impairment of upper limb movement [18]. In addition to limb paralysis, limb loss represents an obstacle to quality of life for which the results of this study offer a comprehensive and reliable technique for extracting electrical brain sources for human movement programming. As in other research [20], results of the present study provide consistent and accurate information for future controlling inputs for the adaptation of prosthesis. As reported elsewhere [69], we conclude that it is necessary to increase movement classes in EEG features extraction for providing mechatronic systems controlled by means of BCI, suitable and reliable patterns corresponding for target movements.

5. Proposed Usage Scenario

The ML algorithms proposed in this research study could be implemented in high-performance embedded systems or edge computing devices as verified in previous studies [59,89]. These act as the central control system, which is in charge of communicating with the BCI to acquire EEG signals. Likewise, the control system is in charge of carrying out the digital processing of the EEG signals, the extraction of features, the classification, and the translation (decoding and execution) of the control commands. The mechatronic control system would have a trained ML model which would allow a user with some motor disability to perform some motor activities, such as opening and closing the right fist, left fist, or both fists through the classification of EEG signals.

Figure 10 depicts a conceptual diagram of the prospective mechatronic control system. We could consider this model the first step in developing intelligent prostheses that integrate the system's several components. The future characteristics to be developed are lower cost, size, portability, low power consumption, and reliable communication with the BCI.

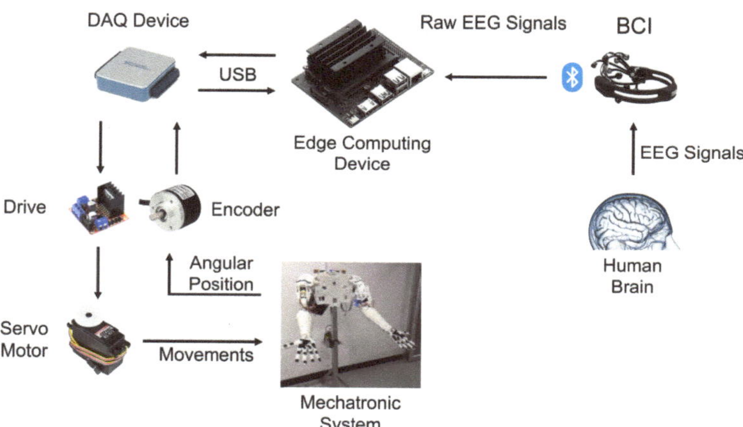

Figure 10. Suggested usage scenario for an application of mechatronic control system.

Limitations of the Study

One of the drawbacks of this research study is the need for a BCI; users should have short hair, as the BCI must be comfortable and enjoyable to them. Furthermore, the electrodes must be maintained in saline solution. Successful implementation also relies on the BCI battery life. Finally, if the emotional state of the participants is altered, accurate measurements cannot be acquired.

6. Conclusions

In this study, a methodology for classifying motor movements by processing the EEG signals of 30 users is presented. The classification of the EEG signals was related to left hand, right hand, both fists, feet, and relaxation movements. As a result of EEG signal processing, a customized dataset was created and used to train the ML algorithms. The dataset was obtained by reading the EEG signal files in EDF+ format, extracting the different segments of the EEG signals, filtering the signals, extracting the features, and labeling their corresponding classes. The customized dataset was created to train and evaluate the performance metrics of different ML algorithms in the classification of EEG signals related to motor movements. The model of Medium-ANN achieved the best performance metrics, with an AUC average of 0.9998, Cohen's Kappa coefficient of 0.9552, Matthews correlation coefficient of 0.9819, and loss of 0.0147. These findings enable the approach to be applied to different scenarios, such as robotic prosthesis implementation, where the utilization of physical qualities is an acceptable alternative when hardware resources are restricted, or in embedded systems or edge computing devices, which have the advantages of low cost, small size, portability, low power consumption, and reliable communication with the BCI.

Furthermore, with the proposed method, we estimate that quantifiable information about motor movement can be obtained through the feature extraction and performance metrics of ML algorithms. We also believe that the proposed method could allow us to generate different datasets that could be used for future studies, as the proposed software was developed and customized to analyze EEG signals.

Supplementary Materials: The following supporting information can be downloaded at: https://www.mdpi.com/article/10.3390/technologies10040079/s1.

Author Contributions: Conceptualization, E.E.G.-G. and E.I.-G.; Data curation, J.M.C.-V., E.G.-C. and O.R.L.-B.; Formal analysis, E.E.G.-G., E.T.-C. and E.G.-C.; Investigation, F.J.R.-A., E.T.-C., G.M.G.-A. and E.I.-G.; Methodology, F.J.R.-A., G.M.G.-A. and E.I.-G.; Project administration, O.R.L.-B.; Resources, F.J.R.-A., O.R.L.-B. and E.I.-G.; Software, F.J.R.-A.; Supervision, E.E.G.-G. and E.I.-G.; Validation, J.M.C.-V., E.G.-C. and G.M.G.-A.; Visualization, J.M.C.-V. and O.R.L.-B.; Writing—original draft, F.J.R.-A.; Writing—review and editing, E.E.G.-G., E.T.-C., E.G.-C. and E.I.-G. All authors have read and agreed to the published version of the manuscript.

Funding: This research study's funds were provided by Universidad Autónoma de Baja California (UABC) through grant number 679.

Institutional Review Board Statement: Not applicable.

Informed Consent Statement: Not applicable.

Data Availability Statement: We share the dataset as supplementary material.

Acknowledgments: The authors are very grateful to PRODEP for supporting the academic groups to increase their degree of consolidation. We also want to thank the Faculty of Engineering and Technology Sciences (FCITEC) for all the facilities provided for the development of this project. F.J.R.-A. would like to thank SPSU-UABC (Union of Teachers University Improvement of the Autonomous University of Baja California) for the scholarship supporting their postgraduate studies.

Conflicts of Interest: The authors declare no conflict of interest. The funders had no role in the design of the study; in the collection, analyses, or interpretation of data; in the writing of the manuscript, or in the decision to publish the results.

Appendix A. Fragment of the Dataset Created for This Study

C3_Amplitude	C3_Frequency	C3_Phase	C1_Peak to Peak	C1_Negative Peak	C1_Positive Peak	Cz_Median	Cz_Mode	Cz_Mean	C2_RMS	C2_SD	C2_Summation	C4_Variance	C4_Kurtosis	C4_Skewness	Classes
108.86617	−29.32351	79.54266	0.26955	2.97236	1.51095	17.27037	17.22153	979.79113	295.73598	7.49178	1.60906	8.42745	0.01242	−57.88825	0
145.2712	−52.97385	92.29735	−1.91859	−11.14636	−0.38611	18.46683	18.46945	−543.63732	366.13252	5.52692	0.902	9.11907	0.00636	7.69257	0
100.73007	−38.61309	62.11698	−1.21691	0.21219	−0.23162	16.3806	16.38057	−951.47539	257.03075	3.80701	0.61077	8.92251	0.00794	24.9861	0
119.87798	−57.16337	62.71462	−1.69857	0.04493	−0.00493	19.4224	19.42449	18.93081	368.71271	3.19239	0.32758	8.68137	0.0046	−97.59919	0
125.2563	−57.19535	68.06094	−1.91499	−5.52267	−0.49701	19.50148	19.49685	−2956.4001	377.8044	3.11134	0.27458	8.8037	0.00691	−109.72379	0
125.67379	−57.61492	68.05887	−1.47821	−4.39237	0.02402	19.54563	19.54696	173.02863	379.61249	3.12059	0.25601	7.58664	0.00088	−37.67507	0
125.68931	−57.61504	68.07427	−1.74668	−4.33893	−0.02686	17.26488	17.26585	−284.69096	298.30936	3.5752	0.41033	6.55339	0.00225	150.20889	0
125.31995	−57.61353	67.70643	−1.70869	−4.66525	0.03645	16.42054	16.42135	213.89227	270.07081	3.84392	0.46781	5.07003	0.00898	−129.32524	0
60.11105	−26.70885	33.4022	−0.69864	−0.25313	−0.5557	13.09898	13.09417	−414.86556	162.42172	2.4878	0.25535	10.79111	0.01569	2.81889	0
190.91658	−95.98303	94.93355	−1.77143	−5.3341	−0.20248	20.87957	20.88366	−434.88338	437.30133	8.8172	0.0715	16.00605	0.00621	115.23688	0
81.42884	−35.41586	46.01297	−1.01522	−0.04502	−0.15462	13.69927	13.69939	−766.25766	193.38404	4.47048	0.48256	5.48752	0.01028	−51.56381	0
149.91035	−80.69967	69.21068	−1.06843	−9.09432	−0.07855	20.50509	20.50722	−237.70391	422.09271	4.16793	0.10647	8.36332	0.00551	140.77191	0
168.19124	−80.77541	87.41583	−1.01319	2.02131	−0.17725	22.09124	22.09226	−1142.61373	488.73057	4.06269	0.24355	9.24432	0.00596	10.96383	0
178.18538	−80.95155	97.23383	−1.91043	−1.48237	−0.38365	23.20017	23.19817	−2985.49963	542.32634	4.07673	0.37249	6.94932	0.007	124.37271	0
165.99797	−68.76385	97.23412	−1.80815	3.15668	−0.28765	24.48947	24.48931	−2400.00755	603.6636	4.07886	0.54094	8.45258	0.00216	21.81982	0
164.91541	−67.67608	97.23933	−1.7938	2.43564	0.06713	24.04032	24.04147	510.43837	585.46381	4.08828	0.56694	10.81424	0.00215	−45.78993	0
208.80963	−113.29443	95.5152	0.15291	−17.76571	3.27554	40.61923	40.51458	2222.06571	1621.20376	3.29414	−0.15438	39.40843	0.0081	−115.65555	0
86.64714	−32.25059	54.39655	−1.54822	−1.8633	−0.34174	12.82435	12.8229	−690.91512	159.61514	4.44967	0.70982	7.62719	0.00757	200.835	0
85.72714	−36.61694	49.1102	−1.34702	0.03854	−0.03464	14.51275	14.51493	37.44572	206.20682	3.25947	0.48182	7.04131	0.00579	64.3528	0
96.2026	−48.3727	47.8299	−1.55971	−0.02817	−0.13503	14.57802	14.57878	−715.35163	207.82778	3.31905	0.48414	7.16151	0.00606	−119.52097	0
96.61122	−48.34653	48.26469	−2.07812	−3.30589	−0.16501	15.34706	15.3469	−1257.20859	232.60851	3.21864	0.47966	7.57887	0.00601	−81.37703	0
115.44138	−48.12757	67.31381	−1.68488	−4.53478	−0.09991	15.84409	15.84484	−758.99683	250.80891	3.39552	0.49671	8.80287	0.00606	−125.87036	0
107.7065	−40.38883	67.31767	−1.99927	−3.99952	−0.03327	16.32575	16.32665	−337.47932	264.43673	3.53177	0.60548	9.58213	0.00605	−106.34569	0
107.3927	−40.07353	67.31917	−1.95398	−5.44053	−0.08078	15.80524	15.8058	−857.09752	246.15637	3.57552	0.57687	7.56182	0.00606	−111.23568	0
129.15821	−78.83456	50.32365	2.54778	−1.75504	1.9225	25.30673	25.26614	1164.45591	642.98665	4.72533	−0.92255	10.21739	0.00855	−179.19851	0
105.01136	−43.05677	61.95459	−1.74764	−2.98789	−0.22264	18.15878	18.1624	−366.03301	313.94184	3.64993	0.43932	11.21388	0.00608	−49.82756	0
103.4699	−39.51072	63.95918	−0.33653	3.66172	−0.0269	16.65289	16.65536	−105.03978	262.71902	3.74805	0.49868	9.27736	0.00584	−9.21414	0
170.39639	−95.58504	74.81135	−1.27524	−2.79184	−0.1146	18.94103	18.94256	−642.32131	344.03532	4.73865	0.20353	8.07856	0.00629	43.58595	0
170.18116	−95.78412	74.39704	−1.66529	−3.58502	−0.05165	18.01691	18.01833	−322.69376	315.32922	5.12737	0.08568	8.99191	0.00726	−17.34837	0
132.33885	−66.59612	65.74274	−1.70012	−4.83305	−0.01402	16.44777	16.44886	−260.7329	272.77158	3.90503	0.30343	6.0148	0.00643	233.40587	0
140.09181	−66.24984	73.84196	−1.52839	−5.4838	0.03415	16.85112	16.85209	129.02571	289.59865	3.76946	0.35227	5.85945	0.00643	76.20641	0
129.81297	−55.97095	73.84203	−1.68358	−2.65769	−0.06647	16.31524	16.31593	−642.59732	267.36145	3.97121	0.52604	6.43784	0.0083	−185.97406	0
74.47058	−33.17593	41.29465	−1.06796	−5.44768	−0.00179	15.44494	15.45545	−117.89723	219.18465	3.20773	0.39781	9.42659	0.01214	206.93585	0
101.9468	−44.0935	57.8533	−1.57687	−3.03288	−0.19162	16.61435	16.61664	−497.86008	271.47804	4.08116	0.52031	7.4292	0.00666	−29.1921	0
161.79353	−97.85084	63.94269	−1.63274	−3.40556	−0.08934	20.86456	20.86752	−279.22265	416.11531	4.82303	−0.00011	14.74093	0.00469	156.06326	0
168.28902	−98.98957	69.29944	−2.26611	−4.3866	−0.0091	18.43858	18.44054	−121.97746	333.03526	5.78385	0.23606	8.07245	0.00565	34.34179	0
118.41162	−49.11881	69.29281	−1.80372	1.16639	−0.16257	15.4754	15.47586	−957.41073	238.87176	4.28581	0.82451	7.86934	0.00609	−169.0244	0

Figure A1. Fragment of the dataset created for this study.

Appendix B. Front Panel of Software (App) Developed for EEG Signal Analysis

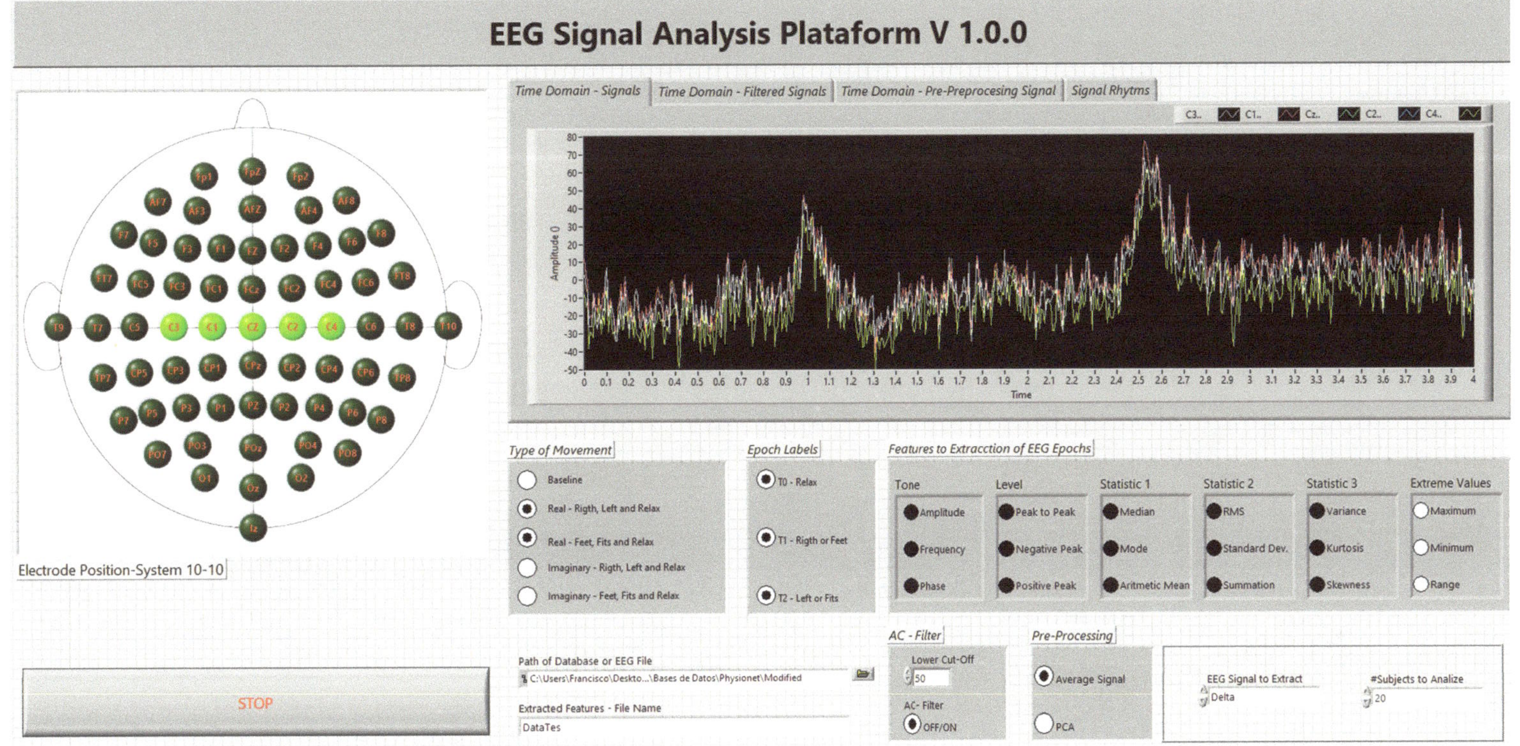

Figure A2. Front panel of software (App) developed for EEG signal analysis.

References

1. Savadkoohi, M.; Oladunni, T.; Thompson, L. A machine learning approach to epileptic seizure prediction using Electroencephalogram (EEG) Signal. *Biocybern. Biomed. Eng.* **2020**, *40*, 1328–1341. [CrossRef]
2. Lent, R.; Azevedo, F.C.; Andrade-Moraes, C.; Pinto, A. How many neurons do you have? Some dogmas of quantitative neuroscience under revision. *Eur. J. Neurosci.* **2012**, *35*, 1–9. [CrossRef] [PubMed]
3. Carter, R. *The Human Brain Book: An Illustrated Guide to Its Structure, Function, and Disorders*; Penguin: London, UK, 2019; Google-Books-ID: S8bhDwAAQBAJ.
4. Gross, J. Magnetoencephalography in Cognitive Neuroscience: A Primer. *Neuron* **2019**, *104*, 189–204. [CrossRef] [PubMed]
5. Ince, N.F.; Goksu, F.; Tewfik, A.H. ECoG Based Brain Computer Interface with Subset Selection. In *Biomedical Engineering Systems and Technologies*; Fred, A., Filipe, J., Gamboa, H., Eds.; Springer: Berlin/Heidelberg, Germany, 2009; pp. 357–374. [CrossRef]
6. Son, J.; Ai, L.; Lim, R.; Xu, T.; Colcombe, S.; Franco, A.R.; Cloud, J.; LaConte, S.; Lisinski, J.; Klein, A.; et al. Evaluating fMRI-Based Estimation of Eye Gaze During Naturalistic Viewing. *Cereb. Cortex* **2020**, *30*, 1171–1184. [CrossRef] [PubMed]
7. Coyle, S.; Ward, T.; Markham, C.; Mcdarby, G. On the suitability of near-infrared (NIR) systems for next-generation brain-computer interfaces. *Physiol. Meas.* **2004**, *25*, 815–822. [CrossRef]
8. Ramadan, R.A.; Vasilakos, A.V. Brain computer interface: Control signals review. *Neurocomputing* **2017**, *223*, 26–44. [CrossRef]
9. Fontanillo Lopez, C.A.; Li, G.; Zhang, D. Beyond Technologies of Electroencephalography-Based Brain-Computer Interfaces: A Systematic Review From Commercial and Ethical Aspects. *Front. Neurosci.* **2020**, *14*, 611130. [CrossRef]
10. Esqueda-Elizondo, J.J.; Juárez-Ramírez, R.; López-Bonilla, O.R.; García-Guerrero, E.E.; Galindo-Aldana, G.M.; Jiménez-Beristáin, L.; Serrano-Trujillo, A.; Tlelo-Cuautle, E.; Inzunza-González, E. Attention Measurement of an Autism Spectrum Disorder User Using EEG Signals: A Case Study. *Math. Comput. Appl.* **2022**, *27*, 21. [CrossRef]
11. Teplan, M. Fundamentals of Eeg Measurement. *Meas. Sci. Rev.* **2002**, *2*, 11.
12. Tiwari, N.; Edla, D.R.; Dodia, S.; Bablani, A. Brain computer interface: A comprehensive survey. *Biol. Inspired Cogn. Archit.* **2018**, *26*, 118–129. [CrossRef]
13. Luján, M.Á.; Jimeno, M.V.; Sotos, J.M.; Ricarte, J.J.; Borja, A.L. A Survey on EEG Signal Processing Techniques and Machine Learning: Applications to the Neurofeedback of Autobiographical Memory Deficits in Schizophrenia. *Electronics* **2021**, *10*, 3037. [CrossRef]
14. Shoeibi, A.; Sadeghi, D.; Moridian, P.; Ghassemi, N.; Heras, J.; Alizadehsani, R.; Khadem, A.; Kong, Y.; Nahavandi, S.; Zhang, Y.D.; et al. Automatic Diagnosis of Schizophrenia in EEG Signals Using CNN-LSTM Models. *Front. Neuroinform.* **2021**, *15*, 777977. [CrossRef] [PubMed]
15. Vrbancic, G.; Podgorelec, V. Automatic Classification of Motor Impairment Neural Disorders from EEG Signals Using Deep Convolutional Neural Networks. *Electron. Electr. Eng.* **2018**, *24*, 3–7. [CrossRef]
16. Abdulkader, S.N.; Atia, A.; Mostafa, M.S.M. Brain computer interfacing: Applications and challenges. *Egypt. Inform. J.* **2015**, *16*, 213–230. [CrossRef]
17. Neuper, C.; Müller, G.; Kübler, A.; Birbaumer, N.; Pfurtscheller, G. Clinical application of an EEG-based brain–computer interface: A case study in a patient with severe motor impairment. *Clin. Neurophysiol.* **2003**, *114*, 399. [CrossRef]
18. Bartur, G.; Pratt, H.; Soroker, N. Changes in mu and beta amplitude of the EEG during upper limb movement correlate with motor impairment and structural damage in subacute stroke. *Clin. Neurophysiol.* **2019**, *130*, 16440–16451. [CrossRef]
19. Sergeev, K.; Runnova, A.; Zhuravlev, M.; Kolokolov, O.; Akimova, N.; Kiselev, A.; Titova, A.; Slepnev, A.; Semenova, N.; Penzel, T. Wavelet skeletons in sleep EEG-monitoring as biomarkers of early diagnostics of mild cognitive impairment. *Chaos* **2021**, *31*, 073110. [CrossRef]
20. Samuel, O.W.; Xiangxin, L.; Yanjuan, G.; Pang, F.; Shixiong, C.; Guanglin, L. Motor imagery classification of upper limb movements based on spectral domain features of EEG patterns. In Proceedings of the 2017 39th Annual International Conference of the IEEE Engineering in Medicine and Biology Society (EMBC), Jeju, Korea, 11–15 July 2017.
21. Wang, H.; Song, Q.; Ma, T.; Cao, H.; Sun, Y. Study on Brain-Computer Interface Based on Mental Tasks. In Proceedings of the The 5th Annual IEEE International Conference on Cyber Technology in Automation, Control and Intelligent Systems, Shenyang, China, 8–12 June 2015; pp. 841–845. [CrossRef]
22. Zander, T.O.; Kothe, C. Towards passive brain–computer interfaces: Applying brain–computer interface technology to human–machine systems in general. *J. Neural Eng.* **2011**, *8*, 025005. [CrossRef]
23. Aggarwal, S.; Chugh, N. Signal processing techniques for motor imagery brain computer interface: A review. *Array* **2019**, *1–2*, 100003. [CrossRef]
24. Mudgal, S.K.; Sharma, S.K.; Chaturvedi, J.; Sharma, A. Brain computer interface advancement in neurosciences: Applications and issues. *Interdiscip. Neurosurg.* **2020**, *20*, 100694. [CrossRef]
25. Brunner, C.; Birbaumer, N.; Blankertz, B.; Guger, C.; Kübler, A.; Mattia, D.; Millán, J.d.R.; Miralles, F.; Nijholt, A.; Opisso, E.; et al. BNCI Horizon 2020: Towards a roadmap for the BCI community. *Brain-Comput. Interfaces* **2015**, *2*, 1–10. [CrossRef]
26. Jurcak, V.; Tsuzuki, D.; Dan, I. 10/20, 10/10, and 10/5 systems revisited: Their validity as relative head-surface-based positioning systems. *NeuroImage* **2007**, *34*, 1600–1611. [CrossRef]
27. Peng, H.; Hu, B.; Qi, Y.; Zhao, Q.; Ratcliffe, M. An improved EEG de-noising approach in electroencephalogram (EEG) for home care. In Proceedings of the 2011 5th International Conference on Pervasive Computing Technologies for Healthcare (PervasiveHealth) and Workshops, Dublin, Ireland, 23–26 May 2011; pp. 469–474. [CrossRef]

28. Khalid, S.; Khalil, T.; Nasreen, S. A survey of feature selection and feature extraction techniques in machine learning. In Proceedings of the 2014 Science and Information Conference, Karlovy Vary, Czech Republic, 27–28 December 2014; pp. 372–378. [CrossRef]
29. Saeidi, M.; Karwowski, W.; Farahani, F.V.; Fiok, K.; Taiar, R.; Hancock, P.A.; Al-Juaid, A. Neural Decoding of EEG Signals with Machine Learning: A Systematic Review. *Brain Sci.* **2021**, *11*, 1525. [CrossRef] [PubMed]
30. Stancin, I.; Cifrek, M.; Jovic, A. A Review of EEG Signal Features and Their Application in Driver Drowsiness Detection Systems. *Sensors* **2021**, *21*, 3786. [CrossRef] [PubMed]
31. Subasi, A.; Ismail Gursoy, M. EEG signal classification using PCA, ICA, LDA and support vector machines. *Expert Syst. Appl.* **2010**, *37*, 8659–8666. [CrossRef]
32. Yazdani, A.; Ebrahimi, T.; Hoffmann, U. Classification of EEG signals using Dempster Shafer theory and a k-nearest neighbor classifier. In Proceedings of the 2009 4th International IEEE/EMBS Conference on Neural Engineering, Antalya, Turkey, 29 April–2 May 2009; pp. 327–330. [CrossRef]
33. Edla, D.R.; Mangalorekar, K.; Dhavalikar, G.; Dodia, S. Classification of EEG data for human mental state analysis using Random Forest Classifier. *Procedia Comput. Sci.* **2018**, *132*, 1523–1532. [CrossRef]
34. Saragih, A.S.; Pamungkas, A.; Zain, B.Y.; Ahmed, W. Electroencephalogram (EEG) Signal Classification Using Artificial Neural Network to Control Electric Artificial Hand Movement. *IOP Conf. Ser. Mater. Sci. Eng.* **2020**, *938*, 012005. [CrossRef]
35. Han, Y.; Ma, Y.; Zhu, L.; Zhang, Y.; Li, L.; Zheng, W.; Guo, J.; Che, Y. *Study on Mind Controlled Robotic Arms by Collecting and Analyzing Brain Alpha Waves*; Atlantis Press: Amsterdam, The Netherlands, 2018; pp. 145–148. [CrossRef]
36. Casey, A.; Azhar, H.; Grzes, M.; Sakel, M. BCI controlled robotic arm as assistance to the rehabilitation of neurologically disabled patients. *Disabil. Rehabil. Assist. Technol.* **2021**, *16*, 525–537. [CrossRef]
37. Van Erp, J.; Lotte, F.; Tangermann, M. Brain-Computer Interfaces: Beyond Medical Applications. *Computer* **2012**, *45*, 26–34. [CrossRef]
38. Navarro-Espinoza, A.; López-Bonilla, O.R.; García-Guerrero, E.E.; Tlelo-Cuautle, E.; López-Mancilla, D.; Hernández-Mejía, C.; Inzunza-González, E. Traffic Flow Prediction for Smart Traffic Lights Using Machine Learning Algorithms. *Technologies* **2022**, *10*, 5. [CrossRef]
39. Cerrada, M.; Trujillo, L.; Hernández, D.E.; Correa Zevallos, H.A.; Macancela, J.C.; Cabrera, D.; Vinicio Sánchez, R. AutoML for Feature Selection and Model Tuning Applied to Fault Severity Diagnosis in Spur Gearboxes. *Math. Comput. Appl.* **2022**, *27*, 6. [CrossRef]
40. Enríquez Zárate, J.; Gómez López, M.d.l.A.; Carmona Troyo, J.A.; Trujillo, L. Analysis and Detection of Erosion in Wind Turbine Blades. *Math. Comput. Appl.* **2022**, *27*, 5. [CrossRef]
41. Janiesch, C.; Zschech, P.; Heinrich, K. Machine learning and deep learning. *Electron. Mark.* **2021**, *31*, 685–695. [CrossRef]
42. Fong-Mata, M.B.; García-Guerrero, E.E.; Mejía-Medina, D.A.; López-Bonilla, O.R.; Villarreal-Gómez, L.J.; Zamora-Arellano, F.; López-Mancilla, D.; Inzunza-González, E. An Artificial Neural Network Approach and a Data Augmentation Algorithm to Systematize the Diagnosis of Deep-Vein Thrombosis by Using Wells' Criteria. *Electronics* **2020**, *9*, 1810. [CrossRef]
43. Cho, J.H.; Jeong, J.H.; Shim, K.H.; Kim, D.J.; Lee, S.W. Classification of Hand Motions within EEG Signals for Non-Invasive BCI-Based Robot Hand Control. In Proceedings of the 2018 IEEE International Conference on Systems, Man, and Cybernetics (SMC), Miyazaki, Japan, 7–10 October 2018; pp. 515–518. [CrossRef]
44. Roy, G.; Bhoi, A.K.; Bhaumik, S. A Comparative Approach for MI-Based EEG Signals Classification Using Energy, Power and Entropy. *IRBM* **2021**. [CrossRef]
45. You, Y.; Chen, W.; Zhang, T. Motor imagery EEG classification based on flexible analytic wavelet transform. *Biomed. Signal Process. Control* **2020**, *62*, 102069. [CrossRef]
46. Faiz, M.Z.A.; Al-Hamadani, A.A. Online Brain Computer Interface Based Five Classes EEG To Control Humanoid Robotic Hand. In Proceedings of the 2019 42nd International Conference on Telecommunications and Signal Processing (TSP), Budapest, Hungary, 1–3 July 2019; pp. 406–410. [CrossRef]
47. LeCun, Y.; Bengio, Y.; Hinton, G. Deep learning. *Nature* **2015**, *521*, 436–444. [CrossRef] [PubMed]
48. Zhang, Y.; Ling, C. A strategy to apply machine learning to small datasets in materials science. *NPJ Comput. Mater.* **2018**, *4*, 25. [CrossRef]
49. Rudin, C. Stop explaining black box machine learning models for high stakes decisions and use interpretable models instead. *Nat. Mach. Intell.* **2019**, *1*, 206–215. [CrossRef]
50. Alomari, M.H.; Samaha, A.; AlKamha, K. Automated Classification of L/R Hand Movement EEG Signals using Advanced Feature Extraction and Machine Learning. *arXiv* **2013**, arXiv: 1312.2877. https://doi.org/10.14569/IJACSA.2013.040628.
51. Pinheiro, O.; Alves, L.; Souza, J. EEG Signals Classification: Motor Imagery for Driving an Intelligent Wheelchair. *IEEE Lat. Am. Trans.* **2018**, *16*, 254–259. [CrossRef]
52. Bousseta, R.; El Ouakouak, I.; Gharbi, M.; Regragui, F. EEG Based Brain Computer Interface for Controlling a Robot Arm Movement Through Thought. *IRBM* **2018**, *39*, 129–135. [CrossRef]
53. Tang, Z.; Sun, S.; Zhang, S.; Chen, Y.; Li, C.; Chen, S. A Brain-Machine Interface Based on ERD/ERS for an Upper-Limb Exoskeleton Control. *Sensors* **2016**, *16*, 2050. [CrossRef] [PubMed]
54. Kant, P.; Laskar, S.H.; Hazarika, J.; Mahamune, R. CWT Based Transfer Learning for Motor Imagery Classification for Brain computer Interfaces. *J. Neurosci. Methods* **2020**, *345*, 108886. [CrossRef] [PubMed]

55. Kaur, D.; Uslu, S.; Rittichier, K.J.; Durresi, A. Trustworthy Artificial Intelligence: A Review. *ACM Comput. Surv.* **2022**, *55*, 1–38. [CrossRef]

56. Theissler, A.; Thomas, M.; Burch, M.; Gerschner, F. ConfusionVis: Comparative evaluation and selection of multi-class classifiers based on confusion matrices. *Knowl.-Based Syst.* **2022**, *247*, 108651. [CrossRef]

57. Sabharwal, R.; Miah, S.J. An intelligent literature review: Adopting inductive approach to define machine learning applications in the clinical domain. *J. Big Data* **2022**, *9*, 53. [CrossRef]

58. Haque, R.; Islam, N.; Islam, M.; Ahsan, M.M. A Comparative Analysis on Suicidal Ideation Detection Using NLP, Machine, and Deep Learning. *Technologies* **2022**, *10*, 57. [CrossRef]

59. Contreras-Luján, E.E.; García-Guerrero, E.E.; López-Bonilla, O.R.; Tlelo-Cuautle, E.; López-Mancilla, D.; Inzunza-González, E. Evaluation of Machine Learning Algorithms for Early Diagnosis of Deep Venous Thrombosis. *Math. Comput. Appl.* **2022**, *27*, 24. [CrossRef]

60. Aboneh, T.; Rorissa, A.; Srinivasagan, R. Stacking-Based Ensemble Learning Method for Multi-Spectral Image Classification. *Technologies* **2022**, *10*, 17. [CrossRef]

61. Bi, L.; Fan, X.; Liu, Y. EEG-Based Brain-Controlled Mobile Robots: A Survey. *IEEE Trans. Hum.-Mach. Syst.* **2013**, *43*, 161–176. [CrossRef]

62. Abbasi, B.; Goldenholz, D.M. Machine learning applications in epilepsy. *Epilepsia* **2019**, *60*, 2037–2047. [CrossRef] [PubMed]

63. Majidov, I.; Whangbo, T. Efficient Classification of Motor Imagery Electroencephalography Signals Using Deep Learning Methods. *Sensors* **2019**, *19*, 1736. [CrossRef] [PubMed]

64. Craik, A.; He, Y.; Contreras-Vidal, J.L. Deep learning for electroencephalogram (EEG) classification tasks: A review. *J. Neural Eng.* **2019**, *16*, 031001. [CrossRef] [PubMed]

65. Padfield, N.; Zabalza, J.; Zhao, H.; Masero, V.; Ren, J. EEG-based brain-computer interfaces using motor-imagery: Techniques and challenges. *Sensors* **2019**, *19*, 1423. [CrossRef] [PubMed]

66. Lawhern, V.J.; Solon, A.J.; Waytowich, N.R.; Gordon, S.M.; Hung, C.P.; Lance, B.J. EEGNet: A compact convolutional neural network for EEG-based brain-computer interfaces. *J. Neural Eng.* **2018**, *15*, 056013. [CrossRef]

67. Lotte, F.; Bougrain, L.; Cichocki, A.; Clerc, M.; Congedo, M.; Rakotomamonjy, A.; Yger, F. A review of classification algorithms for EEG-based brain-computer interfaces: A 10 year update. *J. Neural Eng.* **2018**, *15*, 031005. [CrossRef]

68. Mridha, M.F.; Das, S.C.; Kabir, M.M.; Lima, A.A.; Islam, M.R.; Watanobe, Y. Brain-Computer Interface: Advancement and Challenges. *Sensors* **2021**, *21*, 5746. [CrossRef]

69. Samuel, O.W.; Geng, Y.; Li, X.; Li, G. Towards Efficient Decoding of Multiple Classes of Motor Imagery Limb Movements Based on EEG Spectral and Time Domain Descriptors. *J. Med. Syst.* **2017**, *41*, 194. [CrossRef]

70. Goldberger, A.L.; Amaral, L.A.N.; Glass, L.; Hausdorff, J.M.; Ivanov, P.C.; Mark, R.G.; Mietus, J.E.; Moody, G.B.; Peng, C.K.; Stanley, H.E. PhysioBank, PhysioToolkit, and PhysioNet. *Circulation* **2000**, *101*, e215–e220. [CrossRef]

71. Hosseini, M.P.; Hosseini, A.; Ahi, K. A Review on Machine Learning for EEG Signal Processing in Bioengineering. *IEEE Rev. Biomed. Eng.* **2021**, *14*, 204–218. [CrossRef] [PubMed]

72. Schalk, G.; McFarland, D.J.; Hinterberger, T.; Birbaumer, N.; Wolpaw, J.R. BCI2000: A general-purpose brain-computer interface (BCI) system. *IEEE Trans. Biomed. Eng.* **2004**, *51*, 1034–1043. [CrossRef] [PubMed]

73. Kemp, B.; Olivan, J. European data format 'plus' (EDF+), an EDF alike standard format for the exchange of physiological data. *Clin. Neurophysiol.* **2003**, *114*, 1755–1761. [CrossRef]

74. Neuper, C.; Pfurtscheller, G. Evidence for distinct beta resonance frequencies in human EEG related to specific sensorimotor cortical areas. *Clin. Neurophysiol.* **2001**, *112*, 2084–2097. [CrossRef]

75. Deecke, L.; Weinberg, H.; Brickett, P. Magnetic fields of the human brain accompanying voluntary movement: Bereitschaftsmagnetfeld. *Exp. Brain Res.* **1982**, *48*, 144–148. [CrossRef]

76. Hashimoto, Y.; Ushiba, J. EEG-based classification of imaginary left and right foot movements using beta rebound. *Clin. Neurophysiol.* **2013**, *124*, 2153–2160. [CrossRef]

77. Sleight, J.; Pillai, P.J.; Mohan, S. Classification of Executed and Imagined Motor Movement EEG Signals. *Comput. Sci.* **2009**.

78. Lee, K.; Liu, D.; Perroud, L.; Chavarriaga, R.; Millán, J.d.R. A brain-controlled exoskeleton with cascaded event-related desynchronization classifiers. *Spec. Issue New Res. Front. Intell. Auton. Syst.* **2017**, *90*, 15–23. [CrossRef]

79. Phinyomark, A.; Thongpanja, S.; Hu, H.; Phukpattaranont, P.; Limsakul, C. The Usefulness of Mean and Median Frequencies in Electromyography Analysis. In *Computational Intelligence in Electromyography Analysis: A Perspective on Current Applications and Future Challenges*; InTech: London, UK, 2012. [CrossRef]

80. Zwillinger, D.; Kokoska, S. *CRC Standard Probability and Statistics Tables and Formulae*; CRC Press: Boca Raton, FL, USA, 1999; Google-Books-ID: TB3RVEZ0UIMC.

81. Jatupaiboon, N.; Pan-ngum, S.; Israsena, P. Emotion classification using minimal EEG channels and frequency bands. In Proceedings of the 2013 10th International Joint Conference on Computer Science and Software Engineering (JCSSE), Lampang, Thailand, 2–30 June 2013; pp. 21–24. [CrossRef]

82. Al-Ani, A.; Al-Sukker, A. Effect of Feature and Channel Selection on EEG Classification. In Proceedings of the 2006 International Conference of the IEEE Engineering in Medicine and Biology Society, New York, NY, USA, 30 August–3 September 2006; pp. 2171–2174. [CrossRef]

83. Sadrawi, M.; Sun, W.Z.; Ma, M.M.; Yeh, Y.T.; Abbod, M.; Shieh, J.S. Ensemble Genetic Fuzzy Neuro Model Applied for the Emergency Medical Service via Unbalanced Data Evaluation. *Symmetry* **2018**, *10*, 71. [CrossRef]
84. Guang-Hui, F.; Feng, X.; Bing-Yang, Z.; Lun-Zhao, Y. Stable variable selection of class-imbalanced data with precision-recall criterion. *Chemom. Intell. Lab. Syst.* **2017**, *171*, 241–250. [CrossRef]
85. Castro, C.; Vargas-Viveros, E.; Sánchez, A.; Gutiérrez-López, E.; Flores, D. Parkinson's Disease Classification Using Artificial Neural Networks. In Proceedings of the VIII Latin American Conference on Biomedical Engineering and XLII National Conference on Biomedical Engineering, CLAIB 2019, Cancun, Mexico, 2–5 October 2019; Volume 75. [CrossRef]
86. Metrics for Multi-Class Classification: An Overview. *arXiv* **2020**, arXiv:2008.05756.
87. Vieira, S.M.; Kaymak, U.; Sousa, J.M.C. Cohen's kappa coefficient as a performance measure for feature selection. In Proceedings of the International Conference on Fuzzy Systems, Yantai, China, 10–12 August 2010; pp. 1–8. [CrossRef]
88. Matthews, B.W. Comparison of the predicted and observed secondary structure of T4 phage lysozyme. *Biochim. Biophys. Acta (BBA)-Protein Struct.* **1975**, *405*, 442–451. [CrossRef]
89. Aguirre-Castro, O.; García-Guerrero, E.; López-Bonilla, O.; Tlelo-Cuautle, E.; López-Mancilla, D.; Cárdenas-Valdez, J.; Olguín-Tiznado, J.; Inzunza-González, E. Evaluation of underwater image enhancement algorithms based on Retinex and its implementation on embedded systems. *Neurocomputing* **2022**, *494*, 148–159. [CrossRef]

MDPI AG
Grosspeteranlage 5
4052 Basel
Switzerland
Tel.: +41 61 683 77 34

Technologies Editorial Office
E-mail: technologies@mdpi.com
www.mdpi.com/journal/technologies